Machine Tool Vibrations and Cutting Dynamics

Brandon C. Gegg • C. Steve Suh
Albert C.J. Luo

Machine Tool Vibrations and Cutting Dynamics

Brandon C. Gegg
Dynacon Inc.
Winches and Handling Systems
831 Industrial Blvd
Bryan, TX 77803, USA
bgegg@dynacon.com

C. Steve Suh
Department of Mechanical Engineering
Texas A&M University
College Station, TX, USA
ssuh@tamu.edu

Albert C.J. Luo
Department of Mechanical
and Industrial Engineering
Southern Illinois University
Edwardsville, IL, USA
aluo@siue.ed

ISBN 978-1-4899-9753-1 ISBN 978-1-4419-9801-9 (eBook)
DOI 10.1007/978-1-4419-9801-9
Springer New York Dordrecht Heidelberg London

Softcover re-print of the Hardcover 1st edition 2011

Printed on acid-free paper

Springer is part of Springer Science+Business Media (www.springer.com)

Preface

Machining dynamics is a very old topic on manufacturing processes with consideration of cutting interruption, intermittency, and the coupled interaction between the tool- and work-piece for a better understanding of the underlying physics dictating material removal. In this book, cutting is treated as a nonsmooth system composed of three continuous dynamical subsystems. The corresponding boundaries are (1) the tool- and work-piece contact/impact boundary, (2) the onset/disappearance of cutting boundary, (3) the chip/tool friction boundary, and (4) the chip vanishing boundary. The complex motions in cutting dynamics are mainly caused by discontinuities, including *chip and tool-piece seizure* and complex *stick–slip motion*. Through the application of discontinuous system theory, a comprehensive understanding of the grazing phenomena induced by the frictional-velocity boundary and the loss of contact between the tool- and work-piece are discussed. Significant insights are to control machine-tool vibration and to develop chatter-free machine-tool concept.

The coupling, interaction, and evolution of different cutting states to mitigate machining instability and to enable better machine-tool design are addressed in this book. The monograph presents a sound foundation upon which engineering professionals, practicing and in-training alike, could explore with rigor to make advance in manufacturing, machine-tool design, and machining chatter control. Research professionals in the general areas of nonlinear dynamics and nonlinear control would also find the volume informative in qualitative and quantitative terms as to how semistable interrupted periodic motions lead to unstable motions.

The research and endeavor needed for the creation of *Machine Tool Vibrations and Cutting Dynamics* necessarily put a burden on family life. Without the unwavering support of our families, the completion of the book would not have been possible. Our sincere gratitude also goes to our editor at Springer Science, Steven Elliot, for his professionalism and encouragement, and to the institutions we are associated with for the collegiate support.

Brandon C. Gegg
C. Steve Suh
Albert C.J. Luo

Contents

Chapter 1
Introduction

Manufacturing processes involving material removal have been heavily studied over the past century. This book presents a foundation for the future of manufacturing research. As the industry is aimed to become proficient at the micro- and nanoscale levels of high-speed manufacturing, the dynamics of machining systems needs to be completely established throughout the entire process. Availability of the book is henceforth important to bridge engineering education to the need for high precision, high-yield micro- and nanomachining.

This book covers the fundamentals of cutting dynamics from the perspective of the discontinuous system theory developed by Dr. Albert C.J. Luo. It establishes the coupling, interaction, and evolution of different cutting states to mitigate machining instability and to enable better machine-tool design.

The main features that would contribute to the in-depth understanding of machining dynamics are

- Complete comprehension of the underlying dynamics of cutting and interruptions in cutting motions.
- Operation of the machine-tool system over a broad range of operating conditions with minimal vibration; such as high-speed operation to achieve high-quality finish of the machined surface.
- Increased rate of production to maximize profit and minimize operating and maintenance costs.
- Concentrating on the apparent discontinuities allows the nature of the complex machine-tool system motions to be fully understood. The application of nonsmooth system theory to establish and interpret complex intermittent cutting is both novel and unique. The impact on the area of discontinuous systems is in the understanding of intermittent cutting due to the grazing phenomena produced by the frictional-velocity boundary and by the loss of contact between the tool and work-piece.
- Development of concepts for cutting instability control and machine-tool design applicable to high-speed cutting process.

B.C. Gegg et al., *Machine Tool Vibrations and Cutting Dynamics*,
DOI 10.1007/978-1-4419-9801-9_1,

In engineering, discontinuous dynamical systems exist everywhere. One usually uses continuous models to describe discontinuous dynamical systems. Such continuous models cannot provide suitable predictions of discontinuous dynamical systems. To better understand discontinuous systems, one should realize that discontinuous models will provide an adequate and real predication of engineering systems. Thus, one considers a global discontinuous system consisting of many continuous subsystems in different domains. For each continuous subsystem, it possesses dynamical properties different from the adjacent continuous subsystems. Because of such a difference between two adjacent subsystems, the switchability and/or transport laws on their boundaries should be addressed. To investigate such discontinuous systems, one needs to focus on the time-independent boundary between two dynamical systems. In fact, the boundary relative to time is more popular. In addition to introducing the discontinuity in manufacturing systems, discontinuous dynamical systems will be presented in order to understand the machine tool vibration and cutting dynamics. A brief survey on discontinuous dynamical systems will be given through two practical problems. Finally, the book layout is presented, and the summarization of all chapters of the main body of the book is also given.

1.1 Machining Problems

The manufacturing process has been scrutinized in nearly all aspects in the past several decades. Most of the studies are focused on various approaches to the manufacturing process, such as *mechanics of materials*, *energy*, and *dynamics* techniques. The earliest studies of manufacturing systems were presented in Merchant (1945a, b). Merchant developed a theory predicting the shear angle solution from the principle of *minimum work*. From a materials point of view, Oxley (1961) studied the *mechanics* of metal cutting by examining the shear plane solutions with ideal slip-line theory (Childs 2007).

The *stick–slip* motion has been termed in machining studies in Rubenstien and Storie (1969). The *stick–slip* phenomenon is predominantly observed in dynamical systems and contact material flow problems. The *plasticity* theory was applied to metal cutting in Shouchry (1979). A rounded tool edge was studied for the stagnation depths of material flow in Basuray (1977). The stagnation point (or neutral angle, defining the source of work-piece material flow) was derived by equating the *power* of the plowing tool and cutting forces.

The boundary conditions for a comparison between *seizure* and *sliding* motions of a chip based on the ratio of seized area to real contact area was studied in Wright et al. (1979). The study shows that the *sticking* region of the chip on the tool rake surface is characterized by a constant shear stress at the chip/tool interface. The fundamental classification of chip structures in the past research and a current model was developed to predict and discuss the stability of a certain type of chip formation. The stability of a chip structure is mainly determined by the seizure of the work-piece chip material to the tool-piece rake surface in Astakhov et al. (1997). Two structures contributed to this phenomenon are the continuous and fragmentary hump-backed chips.

Son et al. (2005) extended the work of Basuray (1977) on a rounded edge tool to determine the minimum cutting depth. Experimental results showed that the surface quality was best when cutting was conducted near the minimum cutting depth, and continuous chip formation was observed. Liu and Melkote (2006) derived surface roughness due to stress fields, feed rates, and tool edge radius. Son et al. (2006) considered vibration cutting applied to determine the response with respect to minimum cutting thickness. Recent studies including the stick–slip phenomena, which specifically point out the stick–slip in the cutting process, are typically modeled by the finite element methods, validated via approximate methods and high-speed photography by Simoneau et al. (2006), Woon et al. (2008), Wahi and Chatterjee (2008), and Vela-Martinez et al. (2008).

The breakdown of the manufacturing process has been traditionally completed by considering continuous processes. However, recently researchers have begun to recognize the importance in comprehensive modeling of the process. For instance, following the dynamics approach to model a machine tool will lead to a web of interacting continuous systems. Through this complex network, the prediction of realistic phenomena (such as frictional chatter, regenerative chatter, cutting to plowing transition transient effects, etc.) is possible. To model vibration of the machine-tool problem, many mechanical models have been developed.

Popular models are those given by a one-degree-of-freedom oscillator (e.g., Moon and Kalmar-Nagy 2001) and two-degree-of-freedom oscillator (e.g., Wu and Liu 1984a, b). In many cases, the single degree of freedom model is not adequate to describe such vibration of the machine tool in cutting process (e.g., Wiercigroch and Budak 2001). Thus, in Moon and Kalmar-Nagy (2001), the two-degree-of-freedom oscillators were developed with the practical combinations of nonlinearities through mode coupling and the loss of contact with the work-piece, etc. Analytical investigations of machine tools in cutting process were studied through the two-degree-of-freedom oscillator. The chaotic dynamics of the machine-tool system were also investigated (e.g., Grabec 1988).

Moon and Kalmar-Nagy (2001) reviewed various models of complex dynamics in machine-tool systems. Wiercigroch and Budak (2001) reviewed fundamental cutting forces and discussed sources of nonlinearities in metal cutting similar to those discussed by Moon and Kalmar-Nagy (2001). Fang and Jawahir (2002) gave a survey on restricted contact machining operations including: delay models, nonlinear stiffness, hysteretic cutting forces, viscoelasticity, and nonlinear cutting forces. Through these studies, a clear description of the nature of nonlinearities in the machine-tool systems has *not* been provided. However, Luo and Gegg (2007a, b, c) applied a general theory of discontinuous systems on connectable domains to a forced dry-friction oscillator. The stick and nonstick motions and grazing phenomenon with respect to a friction (velocity) boundary were presented through the vector fields of the oscillator. The necessary and sufficient conditions for the passability of the motion from one continuous system to another were presented. Such an idea is extended, and the model for the dynamics of a machine tool in the cutting, noncutting, and chip seizure processes is discussed in this book.

In previous studies, Wu and Liu (1984a, b) presented an analytical model of cutting dynamics to explain chatter vibration, friction, and mode coupling effects. The two-degree-of-freedom system includes friction on the rake face of the machine tool, modeled by a traveling belt. Wiercigroch (1994) studied the stick–slip phenomenon of a machine tool, associated with a traveling belt analogy. The intermittent loss of cutting in Chandiramani and Pothala (2006) and chip stick–slip motion in Gegg et al. (2008) have been initially studied. Wiercigroch (1997a) modeled cutting forces for a two-degree-of-freedom machine-tool model through multiple discontinuities; which considered loss of contact with the chip. Wiercigroch and Cheng (1997b) developed an orthogonal cutting model with stochastic dynamics, and grazing and stick motion were observed. Wiercigroch and de Kraker (2000) reviewed the traditional approaches to nonsmooth systems with applications ranging from one- and two-degree-of-freedom systems to multiple degrees of freedom in orthogonal machine-tool systems.

Warminski et al. (2003) studied a nonlinear cutting force model with multiple discontinuities based on Grabec's model in Grabec (1986, 1988). The system was analyzed through a perturbation scheme. The various types of machine-tool orientations in the manufacturing environment consist of the fixed work-piece with a rotating and/or traversing tool-piece and the fixed tool-piece with a rotating and/or traversing work-piece, as shown in Fig. 1.1a. Limits on the machining systems include but not limited to operation over a broad range of conditions, quality of the work-piece finish, rate of production, and maintenance intervals.

For a better understanding of the underlying dynamics of machining systems, it is necessary for such limits to be altered. For instance, the ever-present drive toward higher operating speeds in manufacturing requires an advanced understanding and analysis. To study the cutting dynamics, consider a specific problem first. A basic representation of the machine-tool network can be described by three general situations: (a) no contact between the tool-piece and work-piece, (b) the tool contacting the work-piece but no cutting, and (c) tool in contact with the work-piece with cutting. For case (c), the force at the contact point is sufficient to produce cutting of material and friction on the tool rake face of the tool is present, as shown in Fig. 1.1b, c. The friction forces generated on the surface are dependent on relative velocity and discontinuous. This boundary is for stick–slip motions. A multitude of discontinuities including the cutting and thrust forces, elastic deformation, and stagnation effects are a natural occurrence in machining systems. The discontinuities for machine-tool systems are displacement boundaries (loss of contact with the work-piece), force boundaries (onset of cutting), and velocity boundaries (chip/tool rake and work-piece/tool flank stick–slip).

The machine-tool system presented in this book is inspired by the work of Wu and Liu (1984a, b) and Grabec (1986, 1988). However, Wu and Liu (1984a, b) analyzed a two-degree-of-freedom model of the tool-piece with applied external forces

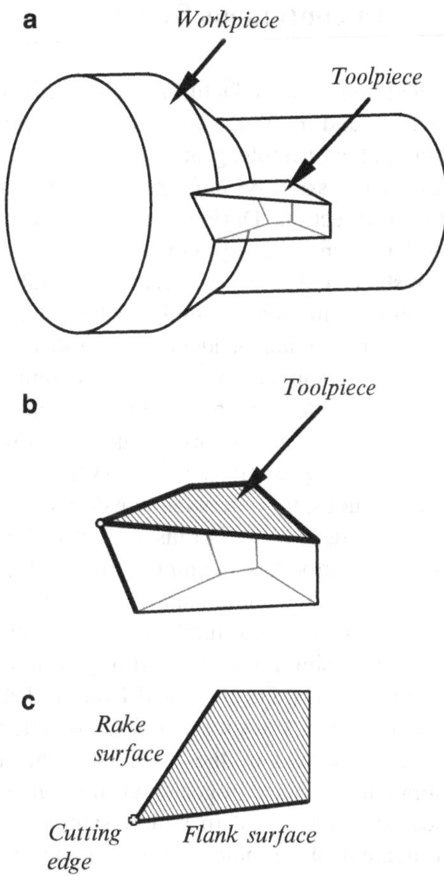

Fig. 1.1 Physical setup: (**a**) tool- and work-piece sample configuration, (**b**) tool-piece with rake and flank surface bolded, (**c**) two-dimensional tool-piece surface definition

acting on the rake and flank tool-piece faces. The system of discontinuous forces has been investigated by Berger et al. (1992), Wiercigroch and Cheng (1997), and Warminski et al. (2003). The boundaries are paramount to describe the underlying physics and interpretation of bifurcations observed in machine-tool systems. The chatter vibration in dynamical cutting process was investigated (e.g., Kim and Lee 1991, Tarng et al. 1994). Such machine-tool systems should be further described for the understanding of cutting dynamics in manufacturing. The loss of cutting and contact can be easily described through discontinuities in machine-tool dynamical systems.

The cutting dynamics in this book is focused on the (a) underlying dynamics of interrupted cutting motions in a machining system and (b) qualitative and quantitative descriptions of interrupted cutting periodic motions. The force model contained several discontinuities is defined by no contact, contact without cutting, and contact with cutting for the tool- and work-pieces.

1.2 Discontinuous Systems

In addition to manufacturing systems, in mechanical engineering, there are two common and important contacts between two dynamical systems, i.e., impact and friction. For example, gear transmission systems possess impact and frictions. Such gear transmission systems are used to transmit power between parallel shafts or to change direction. During the power transmission, a pair of two gears forms a resultant dynamical system. Each gear has its own dynamical system connected with shafts and bearings. Because two subsystems are without any connection, the power transmission is completed through the impact and frictions. Because both the subsystems are independent to each other except for impacting and sliding together, such two dynamical systems have a common time-varying boundary for impacts, which cause domains for the two dynamical systems to be time varying.

In the early investigation, one used a piecewise stiffness model to investigate dynamics of gear transmission systems. Although such a dynamical system is discontinuous, the corresponding domains for vector fields of the dynamical system are time independent. For instance, den Hartog and Mikina (1932) used a piecewise linear system without damping to model gear transmission systems, and the symmetric periodic motion in such a system was investigated. For low-speed gear systems, such a linear model gave a reasonable prediction of gear-tooth vibrations. With increasing rotation speed in gear transmission systems, vibrations and noise become serious. Ozguven and Houser (1988) gave a survey on the mathematical models of gear transmission systems. The piecewise linear model and the impact model were two of the main mechanical models to investigate the origin of vibration and noise in gear transmission systems. Natsiavas (1998) investigated a piecewise linear system with a symmetric trilinear spring, and the stability and bifurcation of periodic motions in such a system were investigated through the variation of initial conditions. Based on a piecewise linear model, the dynamics of gear transmission systems were investigated in Comparin and Singh (1989) and Theodossiades and Natsiavas (2000). Pfeiffer (1984) presented an impact model of gear transmissions, and the theoretical and experimental investigations on regular and chaotic motions in the gear box were later carried out in Karagiannis and Pfeiffer (1991).

To model vibrations in gear transmission systems, Luo and Chen (2005) gave an analytical prediction of the simplest, periodic motion through a piecewise linear, impacting system. In addition, the corresponding grazing of periodic motions was observed, and chaotic motions were simulated numerically through such a piecewise linear system. From the local singularity theory in Luo (2005a, b), the grazing mechanism of the strange fragmentation of such a piecewise linear system was discussed by Luo and Chen (2006). Luo and Chen (2007) used the mapping structure technique to analytically predict arbitrary periodic motions of such a piecewise linear system. In this piecewise linear model, it was assumed that impact locations were fixed, and the perfectly plastic impact was considered. Separation of the two gears is occurred at the same location as the gear impact. Compared with the existing models, this model can give a better prediction of periodic motions in gear transmission systems, but the related assumptions may not be realistic to practical transmission systems because all the aforementioned investigations are based on a time-independent boundary or a given motion boundary. To consider the dynamical systems with the time-varying boundary, Luo and O'Connor (2007a, b) proposed a mechanical model to determine the mechanism of impacting chatter and stick in gear transmission systems. The corresponding analytical conditions for such impacting chatter and stick were developed.

In mechanical engineering, the friction contact between two surfaces of two bodies is an important connection in motion transmissions (e.g., clutch systems and brake systems), again, because two systems are independent except for friction contact. Such a problem will have time-varying boundary and domains. For such a friction problem, it should return back to the 30s of last century. den Hartog (1931) investigated the periodic motion of the forced, damped, linear oscillator contacting a surface with friction. Levitan (1960) investigated the existence of periodic motions in a friction oscillator with a periodically driven base. Filippov (1964) investigated the motion existence of a Coulomb friction oscillator and presented a differential equation theory with discontinuous right-hand sides. The differential inclusion was introduced via the set-valued analysis for the sliding motion along the discontinuous boundary. The investigations of discontinuous differential equations were summarized in Filippov (1988). However, the Filippov's theory mainly focused on the existence and uniqueness of solutions for nonsmooth dynamical systems. Such a differential equation theory with discontinuity is difficult to apply to practical problems. Luo (2005a) developed a general theory to handle the local singularity of discontinuous dynamical systems. To determine the sliding and source motions in discontinuous dynamical systems, the imaginary, sink, and source flows were introduced in Luo (2005b). For detailed discussions, one can refer to Luo (2006, 2009).

On the other hand, Hundal (1979) used a periodic, continuous function to investigate the frequency-amplitude response of such a friction oscillator. Shaw (1986) investigated nonstick, periodic motions of a friction oscillator through Poincaré mapping. Feeny (1992) analytically investigated the nonsmoothness of the Coulomb friction oscillator. To verify the analytic results, Feeny and Moon (1994) investigated chaotic dynamics of a dry-friction oscillator experimentally and numerically. Feeny (1994) gave a systematical discussion of the nonlinear dynamical mechanism of stick–slip motion of friction oscillators. Hinrichs et al. (1997) investigated the nonlinear phenomena in an impact and friction oscillator under external excitations (see also Hinrichs et al. 1998). Natsiavas (1998) developed an algorithm to numerically determine the periodic motion and the corresponding stability of piecewise linear oscillators with viscous and dry-friction damping (see also Natsiavas and Verros 1999). Ko et al. (2001) investigated the friction-induced vibrations with and without external excitations. Andreaus and Casini (2002) gave a closed form solution of a Coulomb friction–impact model without external excitations. Thomsen and Fidlin (2003) gave an approximate estimate of response amplitude for stick–slip motion in a nonlinear friction oscillator.

Kim and Perkins (2003) investigated stick–slip motions in a friction oscillator via the harmonic balance/Galerkin method. Li and Feng (2004) investigated the bifurcation and chaos in a friction-induced oscillator with a nonlinear friction model. Pilipchuk and Tan (2004) investigated the dynamical behaviors of a two-degree-of-freedom mass–damper–spring system contacting on a decelerating rigid strip with friction. Awrejcewicz and Pyryev (2004) gave an investigation on frictional periodic processes by acceleration or braking of a shaft–pad system. In 2007, Hetzler et al. (2007) considered a nonlinear friction model to analytically investigate the Hopf-bifurcation in a sliding friction oscillator with application to the low-frequency disk brake noise.

In the aforementioned investigations, the conditions for motion switchability to the discontinuous boundary were not considered enough. Luo and Gegg (2006a) used the local singularity theory of Luo (2005a, 2006) to develop the force criteria for motion switchability on the velocity boundary in a harmonically driven linear oscillator with dry friction (see also Luo and Gegg 2006b). Through such an investigation, the traditional eigenvalue analysis may not be useful for motion switching at the discontinuous boundary. Lu (2007) mathematically proved the existence of such a periodic motion in such a friction oscillator. Luo and Gegg (2007a, b) investigated the dynamics of a friction-induced oscillator contacting on time-varying belts with friction. Recently, to model the disk brake system, many researchers still considered the mechanical model as in Hetzler et al. (2007). Luo and Thapa (2007) proposed a new model to model the brake system consists of two oscillators, and the two oscillators are connected through a contacting surface with friction. Based on this model, the nonlinear dynamical behaviors of a brake system under a periodical excitation were investigated.

In recent decades, the other developments in nonsmooth dynamical system will be addressed as well. Feigin (1970) investigated the C-bifurcation in piecewise-continuous systems via the Floquet theory of mappings, and the motion complexity was classified by the eigenvalues of mappings, which can be referred to recent publications (e.g., Feigin 1995, di Bernardo et al. 1999). The C-bifurcation is also termed as grazing bifurcation by many researchers. Nordmark (1991) used "grazing" terminology to describe the grazing phenomena in a simple impact oscillator. No strict mathematical description was given, but the grazing condition (i.e., the velocity $\mathrm{d}x/\mathrm{d}t = 0$ for displacement x) in such an impact oscillator was obtained. From Luo (2005a, 2006), it is noted that such a grazing condition is a necessary condition only. The grazing is the tangency between an n-dimensional flow curve of dynamical systems and the discontinuous boundary surface. From differential geometry point of view, Luo (2005a) gave the strict mathematic definition of the "grazing," and the necessary and sufficient conditions of the general discontinuous boundary were presented (see also Luo 2006). Nordmark's result is a special case. Nusse and Yorke (1992) used the simple discrete mapping from Nordmark's impact oscillator and showed the bifurcation phenomena numerically. Based on the numerical observation, the sudden change bifurcation in the numerical simulation is called the *border-collision bifurcation*. So, the similar discrete mappings in discontinuous dynamical system were further developed. Especially,

Dankowicz and Nordmark (2000) developed a discontinuous mapping from a general way to investigate the grazing bifurcation, and the discontinuous mapping is based on the Taylor series expansion in the neighborhood of the discontinuous boundary. Following the same idea, di Bernardo et al. (2001a, b; 2002) developed a normal form to describe the grazing bifurcation. In addition, di Bernardo et al. (2001c) used the normal form to obtain the discontinuous mapping and numerically observed such a border-collision bifurcation through such a discontinuous mapping. From such a discontinuous mapping and the normal form, the aforementioned bifurcation theory structure was developed for the so-called co-dimension one dynamical system.

The discontinuous mapping and normal forms on the discontinuous boundary were developed from the Taylor series expansion in the neighborhood of the discontinuous boundary. However, the normal form requires the vector field with the C^r-continuity and the corresponding convergence, where the order r is the highest order of the total power numbers in each term of normal form. For piecewise linear and nonlinear systems, the C^1-continuity of the vector field cannot provide enough mathematical base to develop the normal form. The normal form also cannot be used to investigate global periodic motions in such a discontinuous system. Leine et al. (2000) used the Filippov theory to investigate bifurcations in nonlinear discontinuous systems. However, the discontinuous mapping techniques were employed to determine the bifurcation via the Floquet multiplier. For more discussion about the traditional analysis of bifurcation in nonsmooth dynamical systems, the reader can refer to Zhusubaliyev and Mosekilde (2003). Based on the recent research, the Floquet multiplier also may not be adequate for periodic motions involved with the grazing and sliding motions in nonsmooth dynamical systems. Therefore, Luo (2005a) proposed a general theory for the local singularity of nonsmooth dynamical systems on connectable domains (see also Luo 2006). To resolve the difficulty, Luo (2007b) developed a general theory for the switching possibilities of flow on the boundary from the passable to nonpassable one, and so on. From recent developments in Luo (2007a, 2008a, b), a generalized theory for discontinuous systems on time-varying domains was presented in Luo (2009). For recent development of discontinuous dynamical systems, one can refer to Luo (2011).

1.3 Book Layout

To help readers easily read this book, the main contents are summarized as follows.

In Chap. 2, as in Luo (2011), the passability of a flow to the separation boundary of two different dynamical systems is presented. The accessible and inaccessible subdomains are introduced first for a theory of discontinuous dynamic systems. On the accessible domains, the corresponding dynamic systems are introduced. The flow orientation and singular sets of the separation boundary are discussed. The passability and tangency (grazing) of a flow to the separation boundary between two adjacent accessible domains are presented, and the necessary and sufficient conditions for such passability and tangency of the flow to the boundary are presented. The product of the normal components of vector fields to the boundary is presented, and the corresponding conditions for the flow passability to the boundary are discussed.

In Chap. 3, the passability of a flow to the separation boundary for two friction-induced oscillators is presented for a better understanding of cutting mechanism in manufacturing. The friction-induced oscillator with a constant velocity belt is presented first. Through this practical example, the theory for flow singularity and passability in discontinuous dynamical systems is more understandable. To further build the basic concepts and knowledge of discontinuous dynamical systems for cutting dynamics, the friction-induced oscillator with a time-varying velocity belt is addressed. From the theory of discontinuous dynamical systems, the analytical conditions for stick and grazing motions to the velocity boundary are presented, and intuitive illustrations are used to help one understand the physical meaning of the mathematical conditions.

In Chap. 4, a mechanical model is developed for cutting dynamics caused by coupled interactions among the tool-piece, work-piece, and chip during the complex cutting operation. Four distinctive regions are used to describe four distinct cutting processes. The switching from one motion to another in the adjacent region is to characterize the specific tool–workpiece interaction at the boundary, and such a switching is a key to understand cutting mechanism. The domains and boundaries for cutting dynamical systems are discussed using the discontinuous system theory in Chap. 2. The passable motions and nonpassable flows to the boundaries are discussed, and the corresponding criteria are given. In addition, the switching bifurcations from passable motion (nonpasssable motion) to nonpassable motion (passable motion) are developed for cutting process.

In Chap. 5, complex motions for cutting process in manufacturing are discussed. Based on the boundaries and domain, the switching planes and generic mappings are introduced to analyze complex motions. From generic mappings, complex cutting motions varying with system parameters are presented. The motion complexity can be measured through mapping structures. Periodic cutting motions in manufacturing can be predicted analytically. The motion switching is determined by the force product criteria. Numerical illustrations of cutting motions near the chip seizure for the machine-tool system are given for a better understanding of complex cutting process in manufacturing.

References

Andreaus, U. and Casini, P., 2002, Friction oscillator excited by moving base and colliding with a rigid or deformable obstacle, *International Journal of Non-Linear Mechanics*, **37**, pp. 117–133.

Astakhov, V.P., Shvets, S.V. and Osman, M.O.M., 1997, Chip structure classification based on mechanics of its formation, *Journal of Materials Processing Technology*, **71**, pp. 247–257.

Awrejcewicz, J. and Pyryev, Y., 2004, Tribological periodical processes exhibited by acceleration or braking of a shaft-pad system, *Communications in Nonlinear Science and Numerical Simulation*, **9**, pp. 603–614.

Basuray, P.M., 1977, Transition from ploughing to cutting during machining with blunt tools, *Wear*, **43**, pp. 341–349.

Berger, B.S., Rokni, M. and Minis, I., 1992, The nonlinear dynamics of metal cutting, *International Journal of Engineering Science*, **30**, pp. 1433–1440.

Chandiramani, N.K. and Pothala, T., 2006, Dynamics of 2-DOF of regenerative chatter during turning, *Journal of Sound and Vibration*, **290**, pp. 488–464.

Childs, T.H.C., 2007, Numerical experiments on the influence of material and other variables on plane strain continuous chip formation in metal machining, *International Journal of Mechanical Sciences*, **48**, pp. 307–322.

Comparin, R.J. and Singh, R., 1989, Nonlinear frequency response characteristics of an impact pair, *Journal of Sound and Vibration*, **134**, pp. 259–290.

Dankowicz, H. and Nordmark A.B., 2000, On the origin and bifurcations of stick-slip oscillations, *Physica D*, **136**, pp. 280–302.

den Hartog, J.P., 1931, Forced vibrations with Coulomb and viscous damping, *Transactions of the American Society of Mechanical Engineers*, **53**, pp. 107–115.

den Hartog, J.P. and Mikina, S.J., 1932, Forced vibrations with non-linear spring constants, *ASME Journal of Applied Mechanics*, **58**, pp. 157–164.

di Bernardo, M., Budd, C.J. and Champneys, A.R., 2001a, Grazing and Border-collision in piecewise-smooth systems: a unified analytical framework, *Physical Review Letters*, **86**, pp. 2553–2556.

di Bernardo, M., Budd. C.J. and Champneys, A.R., 2001b, Normal form maps for grazing bifurcation in n-dimensional piecewise-smooth dynamical systems, *Physica D*, **160**, pp. 222–254.

di Bernardo, M., Budd. C.J. and Champneys, A.R., 2001c, Corner-collision implies border-collision bifurcation, *Physica D*, **154**, pp. 171–194.

di Bernardo, M., Feigin, M.I., Hogan, S.J. and Homer, M.E., 1999, Local analysis of C-bifurcations in n-dimensional piecewise-smooth dynamical systems, *Chaos, Solitons and Fractals*, **10**, pp. 1881–1908.

di Bernardo, M., Kowalczyk, P. and Nordmark, A.B., 2002, Bifurcation of dynamical systems with sliding: derivation of normal form mappings, *Physica D*, **170**, pp. 175–205.

Fang, N. and Jawahir, I.S., 2002, Analytical predictions and experimental validation of cutting force ratio, chip thickness, and chip back-flow angle in restricted contact machining using the universal slip-line model, *International Journal of Machine Tools & Manufacture*, **42**, pp. 681–694.

Feeny, B.F., 1992, A non-smooth Coulomb friction oscillator, *Physics D*, **59**, pp. 25–38.

Feeny, B.F., 1994, The nonlinear dynamics of oscillators with stick-slip friction, in: A. Guran, F. Pfeiffer and K. Popp (eds), *Dynamics with Friction*, World Scientific: River Edge, pp. 36–92.

Feeny, B.F. and Moon, F.C., 1994, Chaos in a forced dry-friction oscillator: experiments and numerical modeling, *Journal of Sound and Vibration*, **170**, pp. 303–323.

Feigin, M.I., 1970, Doubling of the oscillation period with C-bifurcation in piecewise-continuous systems, *PMM*, **34**, pp. 861–869.

Feigin, M.I., 1995, The increasingly complex structure of the bifurcation tree of a piecewise-smooth system, *Journal of Applied Mathematics and Mechanics*, **59**, pp. 853–863.

Filippov, A.F., 1964, Differential equations with discontinuous right-hand side, *American Mathematical Society Translations, Series 2*, **42**, pp. 199–231.

Filippov, A.F., 1988, *Differential Equations with Discontinuous Righthand Sides*, Kluwer Academic Publishers: Dordrecht.

Gegg, B.C., Suh, C.S. and Luo, A.C.J., 2008, Chip Stick and Slip Periodic Motions of a Machine Tool in the Cutting Process, *ASME Manufacturing Science and Engineering Conference Proceedings*, MSEC ICMP2008/DYN-72052, October 7th–10th.

Grabec, I., 1986, Chaos generated by the cutting process, *Physics Letters A*, **117**(8), pp. 384–386.

Grabec, I., 1988, Chaotic Dynamics of the cutting process, *International Journal of Machine Tools and Manufacturing*, **28**, pp. 19–32.

Hetzler, H., Schwarzer, D. and Seemann, W., 2007, Analytical investigation of steady-state stability and Hopf-bifurcation occurring in sliding friction oscillators with application to low-frequency disc brake noise, *Communications in Nonlinear Science and Numerical Simulation*, **12**, pp. 83–99.

Hinrichs, N., Oestreich, M. and Popp, K., 1997, Dynamics of oscillators with impact and friction, *Chaos, Solitons and Fractals*, **8**, pp. 535–558.

Hinrichs, N., Oestreich, M. and Popp, K., 1998, On the modeling of friction oscillators, *Journal of Sound and Vibration*, **216**, pp. 435–459.

Hundal, M.S., 1979, Response of a base excited system with Coulomb and viscous friction, *Journal of Sound and Vibration*, **64**, pp. 371–378

Karagiannis, K. and Pfeiffer, F., 1991, Theoretical and experimental investigations of gear box, *Nonlinear Dynamics*, **2**, pp. 367–387.

Kim, J.S. and Lee, B.H., 1991, An analytical model of dynamic cutting forces in chatter vibration, *International Journal of Machine Tools Manufacturing*, **31**, pp. 371–381.

Kim, W.J. and Perkins, N.C., 2003, Harmonic balance/Galerkin method for non-smooth dynamical system, *Journal of Sound and Vibration*, **261**, pp. 213–224.

Ko, P.L., Taponat, M.-C. and Pfaifer, R., 2001, Friction-induced vibration-with and without external disturbance, *Tribology International*, **34**, pp. 7–24.

Leine, R.I., van Campen, D.H. and van de Vrande, 2000, Bifurcations in nonlinear discontinuous systems, *Nonlinear Dynamics*, **23**, pp. 105–164.

Levitan, E.S., 1960, Forced oscillation of a spring-mass system having combined Coulomb and viscous damping, *Journal of the Acoustical Society of America*, **32**, pp. 1265–1269.

Li, Y. and Feng, Z.C., 2004, Bifurcation and chaos in friction-induced vibration, *Communications in Nonlinear Science and Numerical Simulation*, **9**, pp. 633–647.

Liu, K. and Melkote, S.N., 2006, Effect of plastic side flow on surface roughness in micro-turning process, *International Journal of Machine Tools and Manufacture*, **46**, pp. 1778–1785.

Lu, C., 2007, Existence of slip and stick periodic motions in a non-smooth dynamical system, *Chaos, Solitons and Fractals*, **35**, pp. 949–959.

Luo, A.C.J., 2005a, A theory for non-smooth dynamical systems on connectable domains, *Communication in Nonlinear Science and Numerical Simulation*, **10**, pp. 1–55.

Luo, A.C.J., 2005b, Imaginary, sink and source flows in the vicinity of the separatrix of non-smooth dynamic system, *Journal of Sound and Vibration*, **285**, pp. 443–456.

Luo, A.C.J., 2006, *Singularity and Dynamics on Discontinuous Vector Fields*, Elsevier: Amsterdam.

Luo, A.C.J., 2007a, Differential Geometry of Flows in Nonlinear Dynamical Systems, *Proceedings of IDECT'07*, ASME International Design Engineering Technical Conferences, September 4–7, 2007, Las Vegas, Nevada, USA. DETC2007-84754.

Luo, A.C.J., 2007b, On flow switching bifurcations in discontinuous dynamical system, *Communications in Nonlinear Science and Numerical Simulation*, **12**, pp. 100–116.

Luo, A.C.J., 2008a, A theory for flow switchability in discontinuous dynamical systems, *Nonlinear Analysis: Hybrid Systems*, **2**(4), 1030–1061.

Luo, A.C.J., 2008b, Global Transversality, Resonance and Chaotic Dynamics, World Scientific: Singapore.

Luo, A.C.J., 2009, Discontinuous Dynamical Systems on Time-varying Domains, Heidelberg: HEP-Springer.

Luo, A.C.J., 2011, *Discontinuous Dynamical Systems*, HEP- Springer: Heidelberg.

Luo, A.C.J. and Chen, L.D., 2005, Periodic motion and grazing in a harmonically forced, piecewise linear, oscillator with impacts, *Chaos, Solitons and Fractals*, **24**, pp. 567–578.

Luo, A.C.J. and Chen, L.D., 2006, The grazing mechanism of the strange attractor fragmentation of a harmonically forced, piecewise, linear oscillator with impacts, *IMechE Part K: Journal of Multi-body Dynamics*, **220**, pp. 35–51.

Luo, A.C.J. and Chen, L.D., 2007, Arbitrary periodic motions and grazing switching of a forced piecewise-linear, impacting oscillator, *ASME Journal of Vibration and Acoustics*, **129**, pp. 276–284.

Luo, A.C.J. and Gegg, B.C., 2006a, On the mechanism of stick and non-stick periodic motion in a forced oscillator including dry-friction, *ASME Journal of Vibration and Acoustics*, **128**, pp. 97–105.

Luo, A.C.J. and Gegg, B.C., 2006b, Stick and non-stick periodic motions in a periodically forced, linear oscillator with dry friction, *Journal of Sound and Vibration*, **291**, pp. 132–168.

Luo, A.C.J. and Gegg, B.C., 2007a, Grazing phenomena in a periodically forced, linear oscillator with dry friction, *Communications in Nonlinear Science and Numerical Simulation*, **11**(7), pp. 777–802.

Luo, A.C.J. and Gegg, B.C., 2007b, Periodic motions in a periodically forced oscillator moving on an oscillating belt with dry friction, *ASME Journal of Computational and Nonlinear Dynamics*, **1**, pp. 212–220.

Luo, A.C.J. and Gegg, B.C., 2007c, Dynamics of a periodically excited oscillator with dry friction on a sinusoidally time-varying, traveling surface, *International Journal of Bifurcation and Chaos*, **16**, pp. 3539–3566.

Luo, A.C.J. and O'Connor, D., 2007a, Nonlinear dynamics of a gear transmission system, Part I: mechanism of impacting chatter with stick, *Proceedings of IDETC'07*, 2007 ASME International Design Engineering Conferences and Exposition, September 4–7, 2007, Las Vegas, Nevada. IDETC2007–34881.

Luo, A.C.J. and O'Connor, D., 2007b, Nonlinear dynamics of a gear transmission system, Part II: periodic impacting chatter and stick, *Proceedings of IMECE'07*, 2007 ASME International Mechanical Engineering Congress and Exposition, November 10–16, 2007, Seattle, Washington. IMECE2007–43192.

Luo, A.C.J. and Thapa, S., 2007, On nonlinear dynamics of simplified brake dynamical systems, *Proceedings of IMECE2007*, 2007 ASME International Mechanical Engineering Congress and Exposition, November 5–10, 2007, Seattle, Washington, USA. IMECE2007–42349.

Merchant, M.E., 1945a, Mechanics of the metal cutting process. I. Orthogonal cutting and a type 2 chip, *Journal of Applied Physics*, **16**, pp. 267–275.

Merchant, M.E., 1945b, Mechanics of metal cutting process. II Plasticity conditions in orthogonal cutting, *Journal of Applied Physics*, **16**(5), pp. 318–324.

Moon, F.C. and Kalmar-Nagy, T., 2001, Nonlinear models for complex dynamics in cutting Materials, *Philosophical Transactions of the Royal Society of London A*, **359**, pp. 695–711.

Natsiavas, S. (1998), Stability of piecewise linear oscillators with viscous and dry friction damping, *Journal of Sound and Vibration*, **217**, pp. 507–522.

Natsiavas, S. and Verros, G., 1999, Dynamics of oscillators with strongly nonlinear asymmetric damping, *Nonlinear Dynamics*, **20**, pp. 221–246.

Nordmark, A.B., 1991, Non-periodic motion caused by grazing incidence in an impact oscillator, *Journal of Sound and Vibration*, **145**, pp. 279–297.

Nusse, H.E. and Yorke J.A., 1992, Border-collision bifurcations including 'period two to period three' for piecewise smooth systems, *Physica D*, 1992, **57**, pp. 39–57.

Oxley, P.L., 1961, Mechanics of metal cutting, *International Journal of Machine Tool Design Research*, **1**, pp. 89–97.

Ozguven, H.N. and Houser, D.R., 1988, Mathematical models used in gear dynamics-a Review, *Journal of Sound and Vibration*, **121**(3), pp. 383–411.

Pfeiffer, F., 1984, Mechanische systems mit unstetigen ubergangen, *Ingeniuer-Archiv*, **54**, pp. 232–240.

Pilipchuk, V.N. and Tan, C.A., 2004, Creep-slip capture as a possible source of squeal during decelerating sliding, *Nonlinear Dynamics* **35**, pp. 258–285.

Rubenstien, C. and Storie, R.M., 1969, The cutting of polymers, *International Journal of Machine Tool Design and Research*, **9**, pp. 117–130.

Shaw, S.W., 1986, On the dynamic response of a system with dry-friction, *Journal of Sound and Vibration*, **108**, pp. 305–325.

Shouchry, A., 1979, Metal cutting and plasticity theory, *Wear*, **55**, pp. 313–329.

Simoneau, A., Ng, E. and Elbestawi, M.A., 2006, Chip formation during microscale cutting of a medium carbon steel, *International Journal of Machine Tools and Manufacture*, **46**, pp. 467–481.

Son, S., Lim, H. and Ahn, J., 2005, Effects of the friction coefficient on the minimum cutting thickness in micro cutting, *International Journal of Machine Tools and Manufacture*, **45**, pp. 529–535.

Son, S., Lim, H. and Ahn, J., 2006, The effect of vibration cutting on minimum cutting thickness, *International Journal of Machine Tools and Manufacture*, **46**, pp. 2066–2072.

Tarng, Y.S., Young, H.T. and Lee, B.Y., 1994, An analytical model of chatter vibration in metal cutting, *International Journal of Machine Tools Manufacturing*, **34**, pp. 183–197.

Theodossiades, S. and Natsiavas, S., 2000, Non-linear dynamics of gear-pair systems with periodic stiffness and backlash, *Journal of Sound and Vibration*, **229**(2), pp. 287–310.

Thomsen, J.J. and Fidlin, A., 2003, Analytical approximations for stick-slip vibration amplitudes, *International Journal of Non-Linear Mechanics*, **38**, pp. 389–403.

Vela-Martinez, L., Jauregui-Correa, J.C., Rubio-Cerda, E., Herrera-Ruiz, G. and Lozano-Guzman, A., 2008, Analysis of compliance between the cutting tool and the workpiece on the stability of a turning process, *International Journal of Machine Tools and Manufacture*, **48**, pp. 1054–1062.

Wahi, P. and Chatterjee, A., 2008, Self-interrupted regenerative metal cutting in turning, *International Journal of Non-linear Mechanics*, **43**, pp. 111–123.

Warminski, J., Litak, G., Cartmell, M.P., Khanin, R. and Wiercigroch, M., 2003, Approximate analytical solution for primary chatter in the non-linear metal cutting model, *Journal of Sound and Vibration*, **259**(4), pp. 917–933.

Wiercigroch, M., 1994, A note on the switch function for the stick-slip phenomenon, *Journal of Sound and Vibration*, **175**(5), pp. 700–704.

Wiercigroch, M. and Cheng, A.H-D., 1997, Chaotic and stochastic dynamics of orthogonal metal cutting, *Chaos, Solitons and Fractals*, **8**, pp. 715–726.

Wiercigroch, M., 1997, Chaotic vibration of a simple model of the machine tool-cutting process system, *Transactions of the ASME: Journal of Vibration and Acoustics*, **119**, pp. 468–475.

Wiercigroch, M. and de Kraker, B., 2000, Applied nonlinear dynamics and chaos of mechanical systems with discontinuities, *World Scientific Series A*, **28**.

Wiercigroch, M. and Budak, E., 2001, Sources of nonlinearities, chatter generation and suppression in metal cutting, *Philosophical Transactions of the Royal Society of London A*, **359**, pp. 663–693.

Woon, K.S., Rahman, M., Neo, K.S. and Liu, K., 2008, The effect of tool edge radius on the contact phenomenon of tool-based micromachining, *International Journal of Machine Tools and Manufacture*, **48**(12–13), pp. 1395–1407.

Wright, P.K., Horne, J.G. and Tabor, D., 1979, Boundary conditions at the chip-tool interface in machining: comparison between seizure and sliding friction, *Wear*, **54**, pp. 371–390.

Wu, D.W. and Liu, C.R., 1984a, An analytical model of cutting dynamics. Part 1: Model building, *The American Society of Mechanical Engineers*, 84-WA/Prod-**20**, pp. 107–111.

Wu, D.W. and Liu, C.R., 1984b, An analytical model of cutting dynamics. Part 2: Verification, *The American Society of Mechanical Engineers*, 84-WA/Prod-**21**, pp. 112–118.

Zhusubaliyev, Z. and Mosekilde, E., 2003, *Bifurcations and Chaos in Piecewise-smooth Dynamical Systems*, World Scientific: Singapore.

Chapter 2
Discontinuous System Theory

As in Luo (2011), the passability of a flow to the separation boundary of two different dynamical systems is presented. The accessible and inaccessible subdomains are introduced first for a theory of discontinuous dynamic systems. On the accessible domains, the corresponding dynamic systems are introduced. The flow orientation and singular sets of the separation boundary are discussed. The passability and tangency (grazing) of a flow to the separation boundary between two adjacent accessible domains are presented, and the necessary and sufficient conditions for such passability and tangency of the flow to the boundary are presented. The product of the normal components of vector fields to the boundary is presented, and the corresponding conditions for the flow passability to the boundary are discussed.

2.1 Domain Accessibility

For any discontinuous dynamical system, there are many vector fields defined on different domains in phase space, and such distinct vector fields between two vector fields in two adjacent domains cause flows at the boundary of the domains to be nonsmooth or discontinuous. To investigate the dynamics of discontinuous dynamical systems, consider a discontinuous dynamical system on a universal domain $\mho \subset R^n$, and the passability of a flow from one domain to its adjacent domains is discussed first. Thus, subdomains Ω_α $(\alpha \in I,\ I = \{1, 2, \cdots, N\})$ of the universal domain $\mho$ are introduced and the vector fields on the subdomains may be defined differently. If there is a vector field on a subdomain, then this subdomain is said to be an accessible domain. Otherwise, such a domain is said to be an inaccessible domain. Thus, the domain accessibility can provide a design possibility for discontinuous dynamical systems. The corresponding definitions of the domain accessibility are given as follows.

Definition 2.1. A subdomain in the universal domain $\mho$ in a discontinuous dynamical system is termed an *accessible* subdomain, if at least a specific, continuous vector field can be defined on such a subdomain.

B.C. Gegg et al., *Machine Tool Vibrations and Cutting Dynamics*,
DOI 10.1007/978-1-4419-9801-9_2,

Definition 2.2. A subdomain in a universal domain $\mho$ in discontinuous dynamical systems is termed an *inaccessible* subdomain, if no any vector fields can be defined on such a subdomain.

Since the accessible and inaccessible subdomains exist in discontinuous dynamical systems, the universal domain $\mho$ is classified into connectable and separable domains. The connectable domain is defined as follows.

Definition 2.3. A domain $\mho$ in phase space is termed a *connectable domain* if all the accessible subdomains of the universal domain can be connected without any inaccessible subdomain.

Similarly, a definition of the separable domain is given as follows.

Definition 2.4. A domain is termed a *separable domain*, if the accessible subdomains in the universal domain are separated by inaccessible domains.

Since any discontinuous dynamical system possesses different vector fields defined on each accessible subdomain, the corresponding dynamical behaviors in those accessible subdomains Ω_α are distinguishing. The different behaviors in distinct subdomains cause flow complexity in the domain $\mho$ of discontinuous dynamical systems. The boundary between two adjacent, accessible subdomains is a bridge of dynamical behaviors in two domains for flow continuity. Any connectable domain is bounded by the universal boundary $S \subseteq R^{n-1}$, and each subdomain is bounded by the subdomain boundary surface $S_{\alpha\beta} \subset R^{n-1}$ $(\alpha, \beta \in I)$ with or without the partial universal boundary. For instance, consider an n-dimensional connectable domain in phase space, as shown in Fig. 2.1a through an n_1-dimensional, subvector $\mathbf{x}_{n_1}$ and an n_2-dimensional, subvector $\mathbf{x}_{n_2}$ $(n_1 + n_2 = n)$. The shaded area Ω_α is a specific subdomain, and the other subdomains are white. The dark, solid curve represents the original boundary of the domain $\mho$. For the separable domain, there is at least an inaccessible subdomain to separate the accessible subdomains. The union of inaccessible subdomains is also called the "inaccessible sea." The inaccessible sea is the complement of the accessible subdomains to the universal (original) domain $\mho$. That is determined by $\Omega_0 = \mho \backslash \cup_{\alpha \in I} \Omega_\alpha$. The accessible subdomains in the domain $\mho$ are also called the "islands." For illustration of such a definition, a separable domain is shown in Fig. 2.1b. The thick curve is the boundary of the universal domain, and the gray area is the inaccessible sea. The white regions are the accessible domains (or islands). The hatched region represents a specific accessible subdomain (island).

From one accessible island to another, the transport laws are needed for motion continuity, which is not discussed in this book. For information about this topic, the reader can refer to Luo (2011).

2.2 Discontinuous Dynamical Systems

Consider a dynamic system consisting of N subdynamic systems in a universal domain $\mho \subset R^n$. The universal domain is divided into N accessible subdomains Ω_α $(\alpha \in I)$ and the union of inaccessible domain Ω_0. The union of all accessible

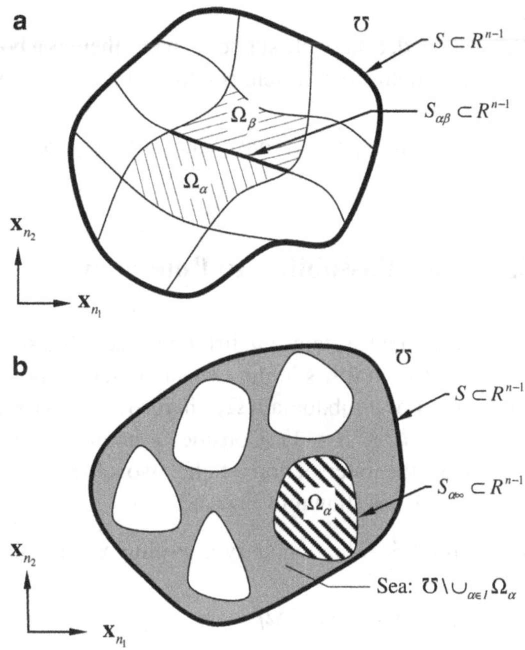

Fig. 2.1 Phase space: (**a**) connectable and (**b**) separable domains ($n_1 + n_2 = n$)

subdomains $\cup_{\alpha \in I} \Omega_\alpha$ and $℧ = \cup_{\alpha \in I} \Omega_\alpha \cup \Omega_0$ is the universal domain, as shown in Fig. 2.1 by an n_1-dimensional, subvector $\mathbf{x}_{n_1}$ and an n_2-dimensional, subvector $\mathbf{x}_{n_2}$ ($n_1 + n_2 = n$). For the connectable domain in Fig. 2.1a, $\Omega_0 = \varnothing$. In Fig. 2.1b, the union of the inaccessible subdomains is the sea, and $\Omega_0 = ℧ \backslash \cup_{\alpha \in I} \Omega_\alpha$ is the complement of the union of the accessible subdomain. On the αth open subdomain Ω_α, there is a C^{r_α}-continuous system ($r_\alpha \geq 1$) in the form of

$$\dot{\mathbf{x}}^{(\alpha)} \equiv \mathbf{F}^{(\alpha)}(\mathbf{x}^{(\alpha)}, t, \mathbf{p}_\alpha) \in R^n, \quad \mathbf{x}^{(\alpha)} = (x_1^{(\alpha)}, x_2^{(\alpha)}, \ldots, x_n^{(\alpha)})^{\mathrm{T}} \in \Omega_\alpha. \tag{2.1}$$

The time is denoted by t and $\dot{\mathbf{x}} = d\mathbf{x}/dt$. In an accessible subdomain Ω_α, the vector field $\mathbf{F}^{(\alpha)}(\mathbf{x}, t, \mathbf{p}_\alpha)$ with parameter vector $\mathbf{p}_\alpha = (p_\alpha^{(1)}, p_\alpha^{(2)}, \ldots, p_\alpha^{(l)})^{\mathrm{T}} \in R^l$ is C^{r_α}-continuous ($r_\alpha \geq 1$) in $\mathbf{x} \in \Omega_\alpha$ and for all time t; and the continuous flow in (2.1) $\mathbf{x}^{(\alpha)}(t) = \Phi^{(\alpha)}(\mathbf{x}^{(\alpha)}(t_0), t, \mathbf{p}_\alpha)$ with $\mathbf{x}^{(\alpha)}(t_0) = \Phi^{(\alpha)}(\mathbf{x}^{(\alpha)}(t_0), t_0, \mathbf{p}_\alpha)$ is C^{r+1}-continuous for time t.

For discontinuous dynamical systems, the following assumptions will be adopted herein.

H2.1
The flow switching between two adjacent subsystems is time-continuous.

H2.2
For an unbounded, accessible subdomain Ω_α, there is a bounded domain $D_\alpha \subset \Omega_\alpha$ and the corresponding vector field and its flow are bounded, i.e.,

$$||\mathbf{F}^{(\alpha)}|| \leq K_1(\text{const}) \quad \text{and} \quad ||\Phi^{(\alpha)}|| \leq K_2(\text{const}) \quad \text{on} \quad D_\alpha \quad \text{for } t \in [0, \infty). \tag{2.2}$$

H2.3
For a bounded, accessible subdomain Ω_α, there is a bounded domain $D_\alpha \subset \Omega_\alpha$ and the corresponding vector field is bounded, but the flow may be unbounded, i.e.,

$$||\mathbf{F}^{(\alpha)}|| \leq K_1(\text{const}) \text{ and } ||\Phi^{(\alpha)}|| \leq \infty \text{ on } D_\alpha \text{ for } t \in [0,\infty). \tag{2.3}$$

2.3 Flow Passability to Boundary

Since dynamical systems on different accessible subdomains are distinguishing, the relation between flows in the two subdomains should be developed herein for flow continuity. For a subdomain Ω_α, there are k_α-adjacent subdomains with k_α-pieces of boundaries ($k_\alpha \leq N-1$). Consider a boundary of any two adjacent subdomains, formed by the intersection of the two closed subdomains (i.e., $\partial\Omega_{ij} = \bar{\Omega}_i \cap \bar{\Omega}_j$, $i,j \in I, j \neq i$), as shown in Fig. 2.2.

Definition 2.5. The boundary in n-dimensional phase space is defined as

$$\begin{aligned} S_{ij} &\equiv \partial\Omega_{ij} = \bar{\Omega}_i \cap \bar{\Omega}_j \\ &= \left\{ \mathbf{x} | \varphi_{ij}(\mathbf{x},t,\lambda) = 0, \varphi_{ij} \text{ is } C^r \text{ - continuous } (r \geq 1) \right\} \subset R^{n-1}. \end{aligned} \tag{2.4}$$

Definition 2.6. The two subdomains Ω_i and Ω_j are *disjoint* if the boundary $\partial\Omega_{ij}$ is an empty set (i.e., $\partial\Omega_{ij} = \varnothing$).

From the definition, $\partial\Omega_{ij} = \partial\Omega_{ji}$. The flow on the boundary $\partial\Omega_{ij}$ can be determined by

$$\dot{\mathbf{x}}^{(0)} = \mathbf{F}^{(0)}(\mathbf{x}^{(0)},t) \text{ with } \varphi_{ij}(\mathbf{x}^{(0)},t,\lambda) = 0, \tag{2.5}$$

where $\mathbf{x}^{(0)} = (x_1^{(0)}, x_2^{(0)}, \ldots, x_n^{(0)})^{\mathrm{T}}$. With specific initial conditions, one always obtains different flows on $\varphi_{ij}(\mathbf{x}^{(0)},t,\lambda) = \varphi_{ij}(\mathbf{x}_0^{(0)},t_0,\lambda) = 0$.

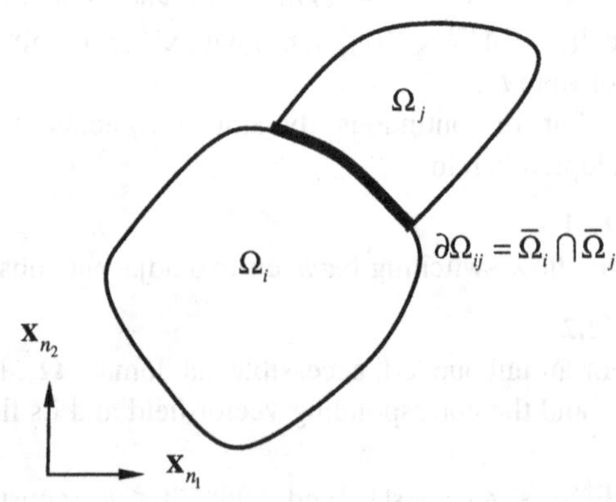

Fig. 2.2 Subdomains Ω_α and Ω_β, the corresponding boundary $\partial\Omega_{\alpha\beta}$

Definition 2.7. If the intersection of three or more subdomains,

$$\Gamma_{\alpha_1\alpha_2\ldots\alpha_k} \equiv \cap_{\alpha=\alpha_1}^{\alpha_k} \bar{\Omega}_\alpha \subset R^r \quad (r = 0, 1, \ldots, n-2), \tag{2.6}$$

where $\alpha_k \in I$ and $k \geq 3$ is nonempty, the subdomain intersection is termed the *singular* set.

For $r = 0$, the singular sets are singular points, which are also termed *the corner points or vertex*. In other words, any corner point is the intersection of n-linearly independent, $(n-1)$-dimensional boundary surfaces in an n-dimensional state space. For $r = 1$, the singular sets will be curves, which are termed the one-dimensional singular edges to the $(n-1)$-dimensional boundary. Similarly, any one-dimensional singular edge is the intersection of $(n-1)$-linearly independent, $(n-1)$-dimensional boundary surfaces in an n-dimensional state space. For $r \in \{2, 3, \ldots, n-2\}$, the singular sets are the r-dimensional singular surfaces to the $(n-1)$-dimensional discontinuous boundary. In Fig. 2.3, the singular set for three closed domains $\{\bar{\Omega}_i, \bar{\Omega}_j, \bar{\Omega}_k\}$ $(i, j, k \in I)$ is sketched. The circular symbols represent intersection sets. The largest solid circular symbol stands for the singular set Γ_{ijk}. The corresponding discontinuous boundaries relative to the singular set are labeled by $\partial\Omega_{ij}$, $\partial\Omega_{jk}$, and $\partial\Omega_{ik}$. The singular set possesses the hyperbolic or parabolic behavior depending on the properties of the separation boundary, which can be referred to Luo (2005, 2006, 2011). The flow on the singular sets can be similarly defined as in (2.5), by a dynamical system with the corresponding boundary constraints. The detailed discussion is given later.

Definition 2.8. For a discontinuous dynamical system in (2.1), there is a point $\mathbf{x}(t_m) \equiv \mathbf{x}_m \in \partial\Omega_{ij}$ at time t_m between two adjacent domains Ω_α $(\alpha = i, j)$. For an arbitrarily small $\varepsilon > 0$, there are two time intervals $[t_{m-\varepsilon}, t_m)$ and $(t_m, t_{m+\varepsilon}]$. Suppose $\mathbf{x}^{(i)}(t_{m-}) = \mathbf{x}_m = \mathbf{x}^{(j)}(t_{m+})$, then a resultant flow of two flows $\mathbf{x}^{(\alpha)}(t)$ $(\alpha = i, j)$ is called as a *semipassable flow* from domain Ω_i to Ω_j at point $(\mathbf{x}_m, t_m)$ to boundary

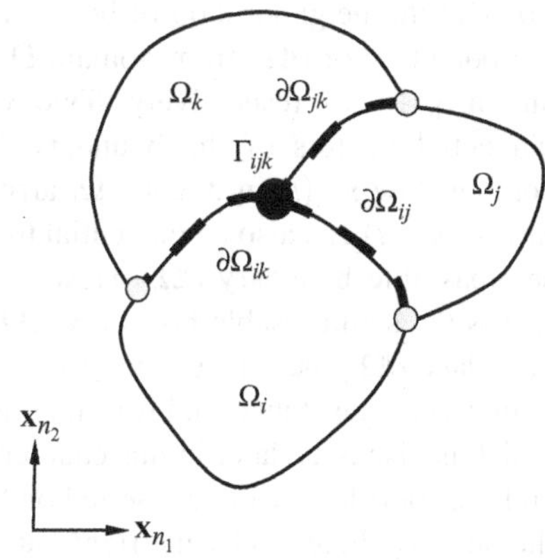

Fig. 2.3 A singular set for the intersection of three domains $\{\bar{\Omega}_i, \bar{\Omega}_j, \bar{\Omega}_k\}$ $(i, j, k \in I)$. The *circular symbols* represent intersection sets. The *large solid circular symbol* stands for the singular set Γ_{ijk}. The corresponding discontinuous boundaries are marked by $\partial\Omega_{ij}$, $\partial\Omega_{jk}$, and $\partial\Omega_{ik}$

$\partial\Omega_{ij}$ if the two flows $\mathbf{x}^{(\alpha)}(t)$ $(\alpha = i,j)$ in the neighborhood of $\partial\Omega_{ij}$ possess the following properties

$$\begin{aligned}
&\text{either} \left.\begin{aligned} \mathbf{n}^{\mathrm{T}}_{\partial\Omega_{ij}} \cdot [\mathbf{x}^{(i)}(t_{m-}) - \mathbf{x}^{(i)}(t_{m-\varepsilon})] > 0 \\ \mathbf{n}^{\mathrm{T}}_{\partial\Omega_{ij}} \cdot [\mathbf{x}^{(j)}(t_{m+\varepsilon}) - \mathbf{x}^{(j)}(t_{m+})] > 0 \end{aligned}\right\} \text{ for } \mathbf{n}_{\partial\Omega_{ij}} \to \Omega_j \\
&\text{or} \left.\begin{aligned} \mathbf{n}^{\mathrm{T}}_{\partial\Omega_{ij}} \cdot [\mathbf{x}^{(i)}(t_{m-}) - \mathbf{x}^{(i)}(t_{m-\varepsilon})] < 0 \\ \mathbf{n}^{\mathrm{T}}_{\partial\Omega_{ij}} \cdot [\mathbf{x}^{(j)}(t_{m+\varepsilon}) - \mathbf{x}^{(j)}(t_{m+})] < 0 \end{aligned}\right\} \text{ for } \mathbf{n}_{\partial\Omega_{ij}} \to \Omega_i,
\end{aligned} \tag{2.7}$$

where the normal vector of the boundary $\partial\Omega_{ij}$ is

$$\mathbf{n}_{\partial\Omega_{ij}} = \nabla\varphi_{ij}|_{\mathbf{x}=\mathbf{x}_m} = \left(\frac{\partial\varphi_{ij}}{\partial x_1}, \frac{\partial\varphi_{ij}}{\partial x_2}, \ldots, \frac{\partial\varphi_{ij}}{\partial x_n}\right)^{\mathrm{T}}|_{\mathbf{x}=\mathbf{x}_m}. \tag{2.8}$$

The notations $t_{m\pm\varepsilon} = t_m \pm \varepsilon$ and $t_{m\pm} = t_m \pm 0$ are used. $\mathbf{n}_{\partial\Omega_{ij}} \to \Omega_j$ represents that the normal vector of boundary at $(\mathbf{x}_m, t_m)$ points to domain Ω_j. In addition, a boundary $\partial\Omega_{ij}$ to semipassable flows $\mathbf{x}^{(\alpha)}(t)$ $(\alpha = i,j)$ from domain Ω_i to domain Ω_j is called the *semipassable* boundary (expressed by $\overrightarrow{\partial\Omega}_{ij}$). For a geometrical explanation of the semipassable flow to the boundary, consider a flow $\mathbf{x}^{(i)}(t)$ of discontinuous dynamical system in (2.1) passing through boundary $\partial\Omega_{ij}$ from domain Ω_i to domain Ω_j. At time t_m, the flow $\mathbf{x}^{(i)}(t)$ arrives to the boundary $\partial\Omega_{ij}$, and there is a small neighborhood $(t_{m-\varepsilon}, t_{m+\varepsilon})$ of time t_m, which is arbitrarily selected. Before the flow $\mathbf{x}^{(i)}(t)$ reaches to the boundary $\partial\Omega_{ij}$, a point $\mathbf{x}^{(i)}(t_{m-\varepsilon})$ lies in domain Ω_i. As $\varepsilon \to 0$, the time increments $\Delta t \equiv \varepsilon \to 0$. A point $\mathbf{x}_m$ on the boundary is the limit of $\mathbf{x}^{(i)}(t_{m-\varepsilon})$ as $\varepsilon \to 0$, and the point $\mathbf{x}_m$ must satisfy the boundary constraint of $\varphi_{ij}(\mathbf{x}, t) = 0$. After the flow $\mathbf{x}^{(i)}(t)$ passes through the boundary at point $\mathbf{x}_m$, the flow $\mathbf{x}^{(i)}(t)$ will switch to the flow $\mathbf{x}^{(j)}(t)$ on the side of domain Ω_j. $\mathbf{x}^{(j)}(t_{m+\varepsilon})$ is a point in the neighborhood of boundary, and a point $\mathbf{x}_m$ on the boundary is also the limit of $\mathbf{x}^{(j)}(t_{m+\varepsilon})$ as $\varepsilon \to 0$. The coming and leaving flow vectors are $\mathbf{x}^{(i)}(t_m) - \mathbf{x}^{(i)}(t_{m-\varepsilon})$ and $\mathbf{x}^{(j)}(t_{m+\varepsilon}) - \mathbf{x}^{(j)}(t_m)$, respectively. Whether the flow passes through the boundary or not is dependent on the properties of both coming and leaving flows in the neighborhood of boundary. The processes of a flow passing through the boundary of $\partial\Omega_{ij}$ from domain Ω_i to Ω_j are shown in Fig. 2.4 for $\mathbf{n}_{\partial\Omega_{ij}} \to \Omega_j$ and $\mathbf{n}_{\partial\Omega_{ij}} \to \Omega_i$, respectively. Two vectors $\mathbf{n}_{\partial\Omega_{ij}}$ and $\mathbf{t}_{\partial\Omega_{ij}}$ are the normal and tangential vectors of the boundary $\partial\Omega_{ij}$, determined by $\varphi_{ij}(\mathbf{x}, t) = 0$. When a coming flow $\mathbf{x}^{(i)}(t)$ in domain Ω_i arrives to the semipassable boundary $\overrightarrow{\partial\Omega}_{ij}$, the flow of $\mathbf{x}^{(i)}(t)$ can also be tangential to or bouncing on (or switching back from) the semipassable boundary $\partial\Omega_{ij}$. However, once a leaving flow $\mathbf{x}^{(j)}(t)$ in domain Ω_j leaves the semipassable boundary $\overrightarrow{\partial\Omega}_{ij}$, the leaving flow cannot pass through the boundary $\overrightarrow{\partial\Omega}_{ij}$, but the leaving flow $\mathbf{x}^{(j)}(t)$ can tangentially leave the semipassable boundary. Thus, tangential (or grazing) flows to the boundary are very important, which is discussed later in this chapter. In the following discussion, no any control and transport laws will be inserted on the boundary. The direction of $\mathbf{t}_{\partial\Omega_{ij}} \times \mathbf{n}_{\partial\Omega_{ij}}$ is the positive direction by the right-hand rule.

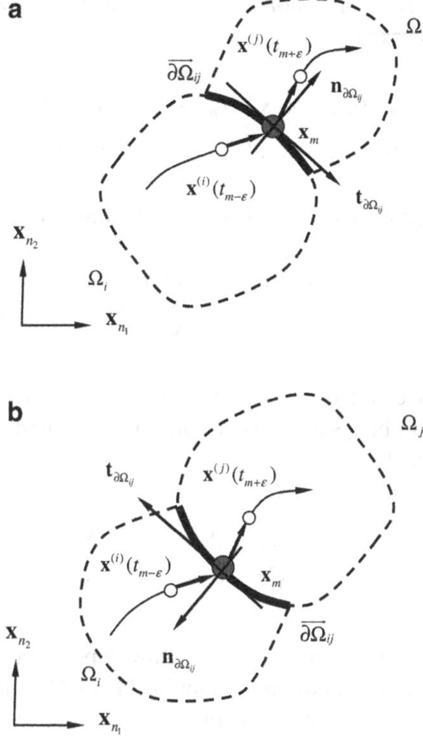

Fig. 2.4 A flow passing through the semipassable boundary $\overrightarrow{\partial\Omega}_{ij}$ from domain Ω_i to Ω_j: (**a**) $\mathbf{n}_{\partial\Omega_{ij}} \to \Omega_j$ and (**b**) $\mathbf{n}_{\partial\Omega_{ij}} \to \Omega_i$. $\mathbf{x}^{(i)}(t_{m-\varepsilon})$, $\mathbf{x}^{(j)}(t_{m+\varepsilon})$, and $\mathbf{x}_m$ are three points in Ω_i and Ω_j and on the boundary $\partial\Omega_{ij}$, respectively. Two vectors $\mathbf{n}_{\partial\Omega_{ij}}$ and $\mathbf{t}_{\partial\Omega_{ij}}$ are the normal and tangential vectors of $\partial\Omega_{ij}$

Theorem 2.1. *For a discontinuous dynamical system in (2.1), there is a point* $\mathbf{x}(t_m) \equiv \mathbf{x}_m \in \partial\Omega_{ij}$ *at time* t_m *between two adjacent domains* Ω_α $(\alpha = i, j)$. *For an arbitrarily small* $\varepsilon > 0$, *there are two time intervals* $[t_{m-\varepsilon}, t_m)$ *and* $(t_m, t_{m+\varepsilon}]$. *Suppose* $\mathbf{x}^{(i)}(t_{m-}) = \mathbf{x}_m = \mathbf{x}^{(j)}(t_{m+})$, *then two flows* $\mathbf{x}^{(i)}(t)$ *and* $\mathbf{x}^{(j)}(t)$ *are* $C^{r_i}_{[t_{m-\varepsilon},t_m)}$*- and* $C^{r_j}_{(t_m,t_{m+\varepsilon}]}$*-continuous* $(r_\alpha \geq 2, \alpha = i, j)$ *for time t, respectively.* $\|\mathrm{d}^{r_\alpha}\mathbf{x}^{(\alpha)}/\mathrm{d}t^{r_\alpha}\| < \infty$ $(\alpha = i, j)$. *The resultant flow of* $\mathbf{x}^{(i)}(t)$ *and* $\mathbf{x}^{(j)}(t)$ *at point* $(\mathbf{x}_m, t_m)$ *to the boundary* $\partial\Omega_{ij}$ *is semipassable from domain* Ω_i *to* Ω_j *if and only if*

$$\left.\begin{array}{ll}\text{either} & \left.\begin{array}{l}\mathbf{n}^{\mathrm{T}}_{\partial\Omega_{ij}} \cdot \dot{\mathbf{x}}^{(i)}(t_{m-}) > 0 \text{ and}\\ \mathbf{n}^{\mathrm{T}}_{\partial\Omega_{ij}} \cdot \dot{\mathbf{x}}^{(j)}(t_{m+}) > 0\end{array}\right\} \text{ for } \mathbf{n}_{\partial\Omega_{ij}} \to \Omega_j\\ \text{or} & \left.\begin{array}{l}\mathbf{n}^{\mathrm{T}}_{\partial\Omega_{ij}} \cdot \dot{\mathbf{x}}^{(i)}(t_{m-}) < 0 \text{ and}\\ \mathbf{n}^{\mathrm{T}}_{\partial\Omega_{ij}} \cdot \dot{\mathbf{x}}^{(j)}(t_{m+}) < 0\end{array}\right\} \text{ for } \mathbf{n}_{\partial\Omega_{ij}} \to \Omega_i.\end{array}\right\} \tag{2.9}$$

Proof. For a point $\mathbf{x}_m \in \partial\Omega_{ij}$ with $\mathbf{n}_{\partial\Omega_{ij}} \to \Omega_j$, suppose $\mathbf{x}^{(i)}(t_{m-}) = \mathbf{x}_m$ and $\mathbf{x}_m = \mathbf{x}^{(j)}(t_{m+})$, then the two flows $\mathbf{x}^{(i)}(t)$ and $\mathbf{x}^{(j)}(t)$ are $C^r_{[t_{m-\varepsilon},t_m)}$- and $C^r_{(t_m,t_{m+\varepsilon}]}$-continuous $(r \geq 2)$ for time t, respectively. $\|\ddot{\mathbf{x}}^{(\alpha)}(t)\| < \infty$ $(\alpha \in \{i, j\})$ for $0 < \varepsilon << 1$. Consider $a \in [t_{m-\varepsilon}, t_{m-})$ and $b \in (t_{m-}, t_{m+\varepsilon}]$. Application of the Taylor series expansion of $\mathbf{x}^{(\alpha)}(t_{m\pm\varepsilon})$ with $t_{m\pm\varepsilon} = t_m \pm \varepsilon$ ($\alpha \in \{i,j\}$) to $\mathbf{x}^{(\alpha)}(a)$ and $\mathbf{x}^{(\alpha)}(b)$ gives

$$\begin{aligned}\mathbf{x}^{(i)}(t_{m-\varepsilon}) &\equiv \mathbf{x}^{(i)}(t_{m-} - \varepsilon) = \mathbf{x}^{(i)}(a) + \dot{\mathbf{x}}^{(i)}(a)(t_{m-} - \varepsilon - a) + o(t_{m-} - \varepsilon - a),\\ \mathbf{x}^{(j)}(t_{m+\varepsilon}) &\equiv \mathbf{x}^{(j)}(t_{m+} + \varepsilon) = \mathbf{x}^{(j)}(b) + \dot{\mathbf{x}}^{(j)}(t_{m+} + \varepsilon - b) + o(t_{m+} + \varepsilon - b).\end{aligned}$$

Let $a \to t_{m-}$ and $b \to t_{m+}$, the limits of the foregoing equations lead to

$$\left.\begin{aligned}\mathbf{x}^{(i)}(t_{m-\varepsilon}) &\equiv \mathbf{x}^{(i)}(t_{m-} - \varepsilon) = \mathbf{x}^{(i)}(t_{m-}) - \dot{\mathbf{x}}^{(i)}(t_{m-})\varepsilon + o(\varepsilon),\\ \mathbf{x}^{(j)}(t_{m+\varepsilon}) &\equiv \mathbf{x}^{(j)}(t_{m+} + \varepsilon) = \mathbf{x}^{(j)}(t_{m+}) + \dot{\mathbf{x}}^{(j)}(t_{m+})\varepsilon + o(\varepsilon).\end{aligned}\right\}$$

Because of $0<\varepsilon<<1$, the ε^2 and higher-order terms of the foregoing equations can be ignored. Therefore, with the first equation of (2.9), the following relations exist,

$$\left.\begin{aligned}\mathbf{n}^{\mathrm{T}}_{\partial\Omega_{ij}} \cdot [\mathbf{x}^{(i)}(t_{m-}) - \mathbf{x}^{(i)}(t_{m-\varepsilon})] &= \mathbf{n}^{\mathrm{T}}_{\partial\Omega_{ij}} \cdot \dot{\mathbf{x}}^{(i)}(t_{m-})\varepsilon > 0,\\ \mathbf{n}^{\mathrm{T}}_{\partial\Omega_{ij}} \cdot [\mathbf{x}^{(j)}(t_{m+\varepsilon}) - \mathbf{x}^{(j)}(t_{m+})] &= \mathbf{n}^{\mathrm{T}}_{\partial\Omega_{ij}} \cdot \dot{\mathbf{x}}^{(j)}(t_{m+})\varepsilon > 0.\end{aligned}\right\}$$

From Definition 2.8, the flow at point $(\mathbf{x}_m, t_m)$ to boundary $\partial\Omega_{ij}$ with $\mathbf{n}_{\partial\Omega_{ij}} \to \Omega_j$ is semipassable from domain Ω_i to Ω_j under the condition in the first inequality equations of (2.9). In a similar manner, the flow at point $(\mathbf{x}_m, t_m)$ to boundary $\partial\Omega_{ij}$ with $\mathbf{n}_{\partial\Omega_{ij}} \to \Omega_i$ is semipassable under conditions in the second inequality equation in (2.9), and vice versa. ■

Theorem 2.2. *For a discontinuous dynamical system in (2.1), there is a point* $\mathbf{x}(t_m) \equiv \mathbf{x}_m \in \partial\Omega_{ij}$ *at time* t_m *between two adjacent domains* Ω_α ($\alpha = i,j$). *For an arbitrarily small* $\varepsilon > 0$, *there are two time intervals* $[t_{m-\varepsilon}, t_m)$ *and* $(t_m, t_{m+\varepsilon}]$. *Suppose* $\mathbf{x}^{(i)}(t_{m-}) = \mathbf{x}_m = \mathbf{x}^{(j)}(t_{m+})$, *then two vector fields of* $\mathbf{F}^{(i)}(\mathbf{x}, t, \mathbf{p}_i)$ *and* $\mathbf{F}^{(j)}(\mathbf{x}, t, \mathbf{p}_j)$ *are* $C^{r_i}_{[t_{m-\varepsilon}, t_m)}$*- and* $C^{r_j}_{(t_m, t_{m+\varepsilon}]}$*-continuous* ($r_\alpha \geq 1$, $\alpha = i,j$) *for time t, respectively.* $\|d^{r_\alpha+1}\mathbf{x}^{(\alpha)}/dt^{r_\alpha+1}\| < \infty$ ($\alpha = i,j$). *The resultant flow of two flows* $\mathbf{x}^{(i)}(t)$ *and* $\mathbf{x}^{(j)}(t)$ *at point* $(\mathbf{x}_m, t_m)$ *to boundary* $\partial\Omega_{ij}$ *is semipassable from domain* Ω_i *to* Ω_j *if and only if*

$$\begin{aligned}&\text{either}\quad \left.\begin{aligned}&\mathbf{n}^{\mathrm{T}}_{\partial\Omega_{ij}} \cdot \mathbf{F}^{(i)}(t_{m-}) > 0 \text{ and}\\ &\mathbf{n}^{\mathrm{T}}_{\partial\Omega_{ij}} \cdot \mathbf{F}^{(j)}(t_{m+}) > 0\end{aligned}\right\} \quad \text{for } \mathbf{n}_{\partial\Omega_{ij}} \to \Omega_j\\ &\text{or}\quad \left.\begin{aligned}&\mathbf{n}^{\mathrm{T}}_{\partial\Omega_{ij}} \cdot \mathbf{F}^{(i)}(t_{m-}) < 0 \text{ and}\\ &\mathbf{n}^{\mathrm{T}}_{\partial\Omega_{ij}} \cdot \mathbf{F}^{(j)}(t_{m+}) < 0\end{aligned}\right\} \quad \text{for } \mathbf{n}_{\partial\Omega_{ij}} \to \Omega_i,\end{aligned} \tag{2.10}$$

where $\mathbf{F}^{(i)}(t_{m-}) \equiv \mathbf{F}^{(i)}(\mathbf{x}_m, t_{m-}, \mathbf{p}_i)$ *and* $\mathbf{F}^{(j)}(t_{m+}) \equiv \mathbf{F}^{(j)}(\mathbf{x}_m, t_{m+}, \mathbf{p}_j)$.

Proof. For a point $\mathbf{x}_m \in \partial\Omega_{ij}$ with $\mathbf{n}_{\partial\Omega_{ij}} \to \Omega_j$, $\mathbf{x}^{(i)}(t_{m-}) = \mathbf{x}_m = \mathbf{x}^{(j)}(t_{m+})$. With (2.1), the first inequality equation of (2.10) gives

$$\left.\begin{aligned} \mathbf{n}^{\mathrm{T}}_{\partial\Omega_{ij}} \cdot \dot{\mathbf{x}}^{(i)}(t_{m-}) = \mathbf{n}^{\mathrm{T}}_{\partial\Omega_{ij}} \cdot \mathbf{F}^{(i)}(t_{m-}) > 0 \text{ and} \\ \mathbf{n}^{\mathrm{T}}_{\partial\Omega_{ij}} \cdot \dot{\mathbf{x}}^{(j)}(t_{m+}) = \mathbf{n}^{\mathrm{T}}_{\partial\Omega_{ij}} \cdot \mathbf{F}^{(j)}(t_{m+}) > 0. \end{aligned}\right\}$$

From Theorem 2.1 and Definition 2.8, the resultant flow at point $(\mathbf{x}_m, t_m)$ to boundary $\partial\Omega_{ij}$ with $\mathbf{n}_{\partial\Omega_{ij}} \to \Omega_j$ is semipassable. In a similar fashion, the resultant flow of two flows $\mathbf{x}^{(i)}(t)$ and $\mathbf{x}^{(j)}(t)$ at point $(\mathbf{x}_m, t_m)$ to boundary $\partial\Omega_{ij}$ with $\mathbf{n}_{\partial\Omega_{ij}} \to \Omega_i$ is semipassable under conditions in the second inequality equations of (2.10). ■

Definition 2.9. For a discontinuous dynamical system in (2.1), there is a point $\mathbf{x}(t_m) \equiv \mathbf{x}_m \in \partial\Omega_{ij}$ at time t_m between two adjacent domains Ω_α $(\alpha = i, j)$. For an arbitrarily small $\varepsilon > 0$, there is a time interval $[t_{m-\varepsilon}, t_m)$. Suppose $\mathbf{x}^{(\alpha)}(t_{m-}) = \mathbf{x}_m$, then two flows $\mathbf{x}^{(i)}(t)$ and $\mathbf{x}^{(j)}(t)$ are called *nonpassable flows of the first kind* at point $(\mathbf{x}_m, t_m)$ to boundary $\partial\Omega_{ij}$ (or termed *sink flows* at point $(\mathbf{x}_m, t_m)$ to boundary $\partial\Omega_{ij}$) if flows $\mathbf{x}^{(i)}(t)$ and $\mathbf{x}^{(j)}(t)$ in vicinity of $\partial\Omega_{ij}$ possess the following properties

$$\begin{aligned} &\text{either} \left.\begin{aligned} \mathbf{n}^{\mathrm{T}}_{\partial\Omega_{ij}} \cdot [\mathbf{x}^{(i)}(t_{m-}) - \mathbf{x}^{(i)}(t_{m-\varepsilon})] > 0 \text{ and} \\ \mathbf{n}^{\mathrm{T}}_{\partial\Omega_{ij}} \cdot [\mathbf{x}^{(j)}(t_{m-}) - \mathbf{x}^{(j)}(t_{m-\varepsilon})] < 0 \end{aligned}\right\} \text{for } \mathbf{n}_{\partial\Omega_{ij}} \to \Omega_j \\ &\text{or} \left.\begin{aligned} \mathbf{n}^{\mathrm{T}}_{\partial\Omega_{ij}} \cdot [\mathbf{x}^{(i)}(t_{m-}) - \mathbf{x}^{(i)}(t_{m-\varepsilon})] < 0 \text{ and} \\ \mathbf{n}^{\mathrm{T}}_{\partial\Omega_{ij}} \cdot [\mathbf{x}^{(j)}(t_{m-}) - \mathbf{x}^{(j)}(t_{m-\varepsilon})] > 0 \end{aligned}\right\} \text{for } \mathbf{n}_{\partial\Omega_{ij}} \to \Omega_i. \end{aligned} \tag{2.11}$$

Definition 2.10. For a discontinuous dynamical system in (2.1), there is a point $\mathbf{x}(t_m) \equiv \mathbf{x}_m \in \partial\Omega_{ij}$ at time t_m between two adjacent domains Ω_α $(\alpha = i, j)$. For an arbitrarily small $\varepsilon > 0$, there is a time interval $(t_m, t_{m+\varepsilon}]$. Suppose $\mathbf{x}^{(\alpha)}(t_{m+}) = \mathbf{x}_m$, then two flows $\mathbf{x}^{(i)}(t)$ and $\mathbf{x}^{(j)}(t)$ are called *nonpassable flows of the second kind* at point $(\mathbf{x}_m, t_m)$ to boundary $\partial\Omega_{ij}$ (or termed *source flows* at point $(\mathbf{x}_m, t_m)$ to boundary $\partial\Omega_{ij}$) if the flows $\mathbf{x}^{(i)}(t)$ and $\mathbf{x}^{(j)}(t)$ in neighborhood of $\partial\Omega_{ij}$ possess the following properties

$$\begin{aligned} &\text{either} \left.\begin{aligned} \mathbf{n}^{\mathrm{T}}_{\partial\Omega_{ij}} \cdot [\mathbf{x}^{(i)}(t_{m+\varepsilon}) - \mathbf{x}^{(i)}(t_{m+})] < 0 \text{ and} \\ \mathbf{n}^{\mathrm{T}}_{\partial\Omega_{ij}} \cdot [\mathbf{x}^{(j)}(t_{m+\varepsilon}) - \mathbf{x}^{(j)}(t_{m+})] > 0 \end{aligned}\right\} \text{for } \mathbf{n}_{\partial\Omega_{ij}} \to \Omega_j \\ &\text{or} \left.\begin{aligned} \mathbf{n}^{\mathrm{T}}_{\partial\Omega_{ij}} \cdot [\mathbf{x}^{(i)}(t_{m+\varepsilon}) - \mathbf{x}^{(i)}(t_{m+})] > 0 \text{ and} \\ \mathbf{n}^{\mathrm{T}}_{\partial\Omega_{ij}} \cdot [\mathbf{x}^{(j)}(t_{m+\varepsilon}) - \mathbf{x}^{(j)}(t_{m+})] < 0 \end{aligned}\right\} \text{for } \mathbf{n}_{\partial\Omega_{ij}} \to \Omega_i. \end{aligned} \tag{2.12}$$

The boundary $\partial\Omega_{ij}$ for two *sink* flows $\mathbf{x}^{(i)}(t)$ and $\mathbf{x}^{(j)}(t)$ at point $(\mathbf{x}_m, t_m)$ is called a *nonpassable boundary of the first kind*, donated by $\widetilde{\partial\Omega}_{ij}$ (or termed *a sink boundary* between Ω_i and Ω_j). The boundary $\partial\Omega_{ij}$ for two *source* flows $\mathbf{x}^{(i)}(t)$ and

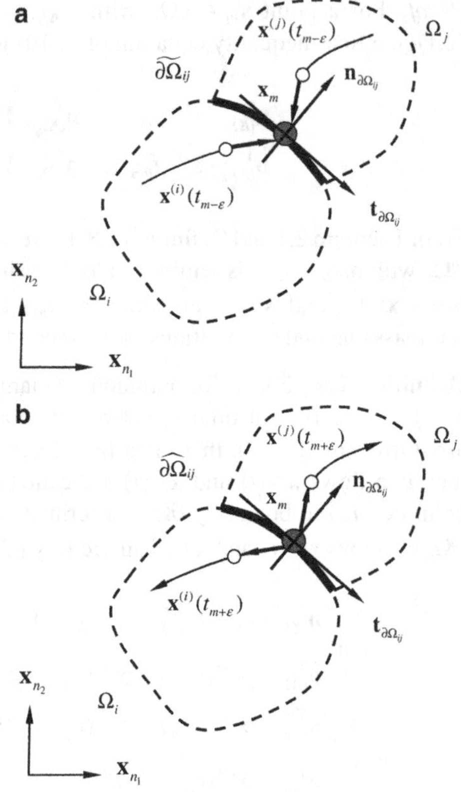

Fig. 2.5 Nonpassable flow to the boundary $\partial\Omega_{ij}$ with $\mathbf{n}_{\partial\Omega_{ij}} \to \Omega_j$: (**a**) sink flow to $\widetilde{\partial\Omega}_{ij}$ (or the nonpassable flow of the first kind), (**b**) source flow to $\widehat{\partial\Omega}_{ij}$ (or the nonpassable flow of the second kind). $\mathbf{x}_m \equiv (\mathbf{x}_{n_1}(t_m), \mathbf{x}_{n_2}(t_m))^{\mathrm{T}}$, $\mathbf{x}^{(\alpha)}(t_{m\pm\varepsilon}) \equiv (\mathbf{x}^{(\alpha)}_{n_1}(t_{m\pm\varepsilon}), \mathbf{x}^{(\alpha)}_{n_2}(t_{m\pm\varepsilon}))^{\mathrm{T}}$, and $\alpha = \{i,j\}$ where $t_{m\pm\varepsilon} = t_m \pm \varepsilon$ for an arbitrary small $\varepsilon > 0$

$\mathbf{x}^{(j)}(t)$ at point $(\mathbf{x}_m, t_m)$ is called a *nonpassable boundary of the second kind*, denoted by $\widehat{\partial\Omega}_{ij}$ (or termed *a source boundary* between Ω_i and Ω_j). The sink and source flows to the boundary $\partial\Omega_{ij}$ between Ω_i and Ω_j are illustrated in Fig. 2.5a, b. The flows in the neighborhood of boundary $\partial\Omega_{ij}$ are depicted. When a flow $\mathbf{x}^{(\alpha)}(t)$ $(\alpha = i,j)$ in domain Ω_α arrives to the nonpassable boundary of the first kind $\widetilde{\partial\Omega}_{ij}$, the flow can be either tangential to or sliding on the nonpassable boundary $\widetilde{\partial\Omega}_{ij}$. For the nonpassable boundary of the second kind $\widehat{\partial\Omega}_{ij}$, a flow $\mathbf{x}^{(\alpha)}(t)$ $(\alpha = i,j)$ in the domain Ω_α can be either tangential to or bouncing on the nonpassable boundary $\widehat{\partial\Omega}_{ij}$. In this chapter, only the flows tangential to the nonpassable boundary are discussed.

Theorem 2.3. *For a discontinuous dynamical system in (2.1), there is a point* $\mathbf{x}(t_m) \equiv \mathbf{x}_m \in \partial\Omega_{ij}$ *at time* t_m *between two adjacent domains* Ω_α $(\alpha = i,j)$. *For an arbitrarily small* $\varepsilon > 0$, *there is a time interval* $[t_{m-\varepsilon}, t_m)$. *Suppose* $\mathbf{x}^{(\alpha)}(t_{m-}) = \mathbf{x}_m$, *then the flow* $\mathbf{x}^{(\alpha)}(t)$ *is* $C^{r_\alpha}_{[t_{m-\varepsilon},t_m)}$*-continuous for time* t and $\|\mathrm{d}^{r_\alpha}\mathbf{x}^{(\alpha)}/\mathrm{d}t^{r_\alpha}\| < \infty$ $(r_\alpha \geq 2)$. *Two flows* $\mathbf{x}^{(i)}(t)$ *and* $\mathbf{x}^{(j)}(t)$ *at point* $(\mathbf{x}_m, t_m)$ *to boundary* $\partial\Omega_{ij}$ *are nonpassable flows of the first kind (or sink flows) if and only if*

$$\left.\begin{array}{ll}\text{either} & \left.\begin{array}{l}\mathbf{n}_{\partial\Omega_{ij}}^{\mathrm{T}}\cdot\dot{\mathbf{x}}^{(i)}(t_{m-})>0\\ \mathbf{n}_{\partial\Omega_{ij}}^{\mathrm{T}}\cdot\dot{\mathbf{x}}^{(j)}(t_{m-})<0\end{array}\right\}\text{ for } \mathbf{n}_{\partial\Omega_{ij}}\to\Omega_j\\ \text{or} & \left.\begin{array}{l}\mathbf{n}_{\partial\Omega_{ij}}^{\mathrm{T}}\cdot\dot{\mathbf{x}}^{(i)}(t_{m-})<0\\ \mathbf{n}_{\partial\Omega_{ij}}^{\mathrm{T}}\cdot\dot{\mathbf{x}}^{(j)}(t_{m-})>0\end{array}\right\}\text{ for } \mathbf{n}_{\partial\Omega_{ij}}\to\Omega_i.\end{array}\right. \tag{2.13}$$

Proof. Following the proof procedure of Theorem 2.1, Theorem 2.3 can be proved. ■

Theorem 2.4. *For a discontinuous dynamical system in (2.1), there is a point $\mathbf{x}(t_m)\equiv\mathbf{x}_m\in\partial\Omega_{ij}$ at t_m between two adjacent domains Ω_α $(\alpha=i,j)$. For an arbitrarily small $\varepsilon>0$, there is a time interval $[t_{m-\varepsilon},t_m)$. $\mathbf{x}^{(\alpha)}(t_{m-})=\mathbf{x}_m$. The vector field $\mathbf{F}^{(\alpha)}(\mathbf{x},t,\mathbf{p}_\alpha)$ is $C^{r_\alpha}_{[t_{m-\varepsilon},t_m)}$-continuous and $\|\mathrm{d}^{r_\alpha+1}\mathbf{x}^{(\alpha)}/\mathrm{d}t^{r_\alpha+1}\|<\infty$ $(r_\alpha\geq1)$. Two flows $\mathbf{x}^{(i)}(t)$ and $\mathbf{x}^{(j)}(t)$ at point $(\mathbf{x}_m,t_m)$ to boundary $\partial\Omega_{ij}$ are nonpassable flows of the first kind (or sink flows) if and only if*

$$\left.\begin{array}{ll}\text{either} & \left.\begin{array}{l}\mathbf{n}_{\partial\Omega_{ij}}^{\mathrm{T}}\cdot\mathbf{F}^{(i)}(t_{m-})>0\text{ and}\\ \mathbf{n}_{\partial\Omega_{ij}}^{\mathrm{T}}\cdot\mathbf{F}^{(j)}(t_{m-})<0\end{array}\right\}\text{ for } \mathbf{n}_{\partial\Omega_{ij}}\to\Omega_j\\ \text{or} & \left.\begin{array}{l}\mathbf{n}_{\partial\Omega_{ij}}^{\mathrm{T}}\cdot\mathbf{F}^{(i)}(t_{m-})<0\text{ and}\\ \mathbf{n}_{\partial\Omega_{ij}}^{\mathrm{T}}\cdot\mathbf{F}^{(j)}(t_{m-})>0\end{array}\right\}\text{ for } \mathbf{n}_{\partial\Omega_{ij}}\to\Omega_i,\end{array}\right. \tag{2.14}$$

where $\mathbf{F}^{(\alpha)}(t_{m-})\triangleq\mathbf{F}^{(\alpha)}(\mathbf{x},t_{m-},\mathbf{p}_\alpha)$ $(\alpha\in\{i,j\})$.

Proof. Following the proof procedure of Theorem 2.2, Theorem 2.4 can be easily proved. ■

Theorem 2.5. *For a discontinuous dynamical system in (2.1), there is a point $\mathbf{x}(t_m)\equiv\mathbf{x}_m\in\partial\Omega_{ij}$ at t_m between two adjacent domains Ω_α $(\alpha=i,j)$. For an arbitrarily small $\varepsilon>0$, there is a time interval $(t_m,t_{m+\varepsilon}]$. Suppose $\mathbf{x}^{(\alpha)}(t_{m+})=\mathbf{x}_m$, then $\mathbf{x}^{(\alpha)}(t)$ is $C^{r_\alpha}_{(t_m,t_{m+\varepsilon}]}$-continuous for time t with $\|\mathrm{d}^{r_\alpha}\mathbf{x}^{(\alpha)}/\mathrm{d}t^{r_\alpha}\|<\infty$ $(r_\alpha\geq2)$. Two flows $\mathbf{x}^{(i)}(t)$ and $\mathbf{x}^{(j)}(t)$ at point $(\mathbf{x}_m,t_m)$ to boundary $\partial\Omega_{ij}$ are nonpassable flows of the second kind (or source flows) if and only if*

$$\left.\begin{array}{ll}\text{either} & \left.\begin{array}{l}\mathbf{n}_{\partial\Omega_{ij}}^{\mathrm{T}}\cdot\dot{\mathbf{x}}^{(i)}(t_{m+})<0\\ \mathbf{n}_{\partial\Omega_{ij}}^{\mathrm{T}}\cdot\dot{\mathbf{x}}^{(j)}(t_{m+})>0\end{array}\right\}\text{ for } \mathbf{n}_{\partial\Omega_{ij}}\to\Omega_j\\ \text{or} & \left.\begin{array}{l}\mathbf{n}_{\partial\Omega_{ij}}^{\mathrm{T}}\cdot\dot{\mathbf{x}}^{(i)}(t_{m+})>0\\ \mathbf{n}_{\partial\Omega_{ij}}^{\mathrm{T}}\cdot\dot{\mathbf{x}}^{(j)}(t_{m+})<0\end{array}\right\}\text{ for } \mathbf{n}_{\partial\Omega_{ij}}\to\Omega_i.\end{array}\right. \tag{2.15}$$

Proof. Following the procedure of the proof of Theorem 2.1, Theorem 2.5 can be proved. ■

Theorem 2.6. *For a discontinuous dynamical system in (2.1), there is a point* $\mathbf{x}(t_m) \equiv \mathbf{x}_m \in \partial\Omega_{ij}$ *at time* t_m *between two adjacent domains* Ω_α $(\alpha = i,j)$*. For an arbitrarily small* $\varepsilon > 0$*, there is a time interval* $(t_m, t_{m+\varepsilon}]$*. Suppose* $\mathbf{x}^{(\alpha)}(t_{m+}) = \mathbf{x}_m$*, then the vector field* $\mathbf{F}^{(\alpha)}(\mathbf{x}, t, \mathbf{p}_\alpha)$ *is* $C^{r_\alpha}_{[t_{m-\varepsilon}, t_m)}$*-continuous and* $\|\mathrm{d}^{r_\alpha+1}\mathbf{x}^{(\alpha)}/\mathrm{d}t^{r_\alpha+1}\| < \infty$ $(r_\alpha \geq 1)$*. Two flows* $\mathbf{x}^{(i)}(t)$ *and* $\mathbf{x}^{(j)}(t)$ *at point* $(\mathbf{x}_m, t_m)$ *to boundary* $\partial\Omega_{ij}$ *are nonpassable flows of the second kind (or source flows) if and only if*

$$
\begin{aligned}
&\text{either} && \left.\begin{array}{l} \mathbf{n}^{\mathrm{T}}_{\partial\Omega_{ij}} \cdot \mathbf{F}^{(i)}(t_{m+}) < 0 \\ \mathbf{n}^{\mathrm{T}}_{\partial\Omega_{ij}} \cdot \mathbf{F}^{(j)}(t_{m+}) > 0 \end{array}\right\} && \text{for } \mathbf{n}_{\partial\Omega_{ij}} \to \Omega_j \\
&\text{or} && \left.\begin{array}{l} \mathbf{n}^{\mathrm{T}}_{\partial\Omega_{ij}} \cdot \mathbf{F}^{(i)}(t_{m-}) > 0 \\ \mathbf{n}^{\mathrm{T}}_{\partial\Omega_{ij}} \cdot \mathbf{F}^{(j)}(t_{m+}) < 0 \end{array}\right\} && \text{for } \mathbf{n}_{\partial\Omega_{ij}} \to \Omega_i,
\end{aligned}
\tag{2.16}
$$

where $\mathbf{F}^{(\alpha)}(t_{m+}) \triangleq \mathbf{F}^{(\alpha)}(\mathbf{x}, t_{m+}, \mathbf{p}_\alpha)$ $(\alpha = i,j)$*.*

Proof. Following the proof procedure of Theorem 2.2, Theorem 2.6 can be easily proved. ■

2.4 Tangential Flows to Boundary

In this section, the flow local singularity and tangential flow are discussed. The corresponding necessary and sufficient conditions are presented.

Definition 2.11. For a discontinuous dynamical system in (2.1), there is a point $\mathbf{x}(t_m) \equiv \mathbf{x}_m \in \partial\Omega_{ij}$ at time t_m between two adjacent domains Ω_α $(\alpha = i,j)$. For an arbitrarily small $\varepsilon > 0$, there are two time intervals (i.e., $[t_{m-\varepsilon}, t_m)$ and $(t_m, t_{m+\varepsilon}]$). Suppose $\mathbf{x}^{(\alpha)}(t_{m\pm}) = \mathbf{x}_m$ $(\alpha \in \{i,j\})$, then a flow $\mathbf{x}^{(\alpha)}(t)$ is $C^{r_\alpha}_{[t_{m-\varepsilon}, t_m)}$- and/or $C^{r_\alpha}_{(t_m, t_{m+\varepsilon}]}$-continuous $(r_\alpha \geq 2)$. A point $(\mathbf{x}_m, t_m)$ on boundary $\partial\Omega_{ij}$ is critical to flow $\mathbf{x}^{(\alpha)}(t)$ if

$$
\mathbf{n}^{\mathrm{T}}_{\partial\Omega_{ij}} \cdot \dot{\mathbf{x}}^{(\alpha)}(t_{m-}) = 0 \text{ and /or } \mathbf{n}^{\mathrm{T}}_{\partial\Omega_{ij}} \cdot \dot{\mathbf{x}}^{(\alpha)}(t_{m+}) = 0. \tag{2.17}
$$

Theorem 2.7. *For a discontinuous dynamical system in (2.1), there is a point* $\mathbf{x}(t_m) \equiv \mathbf{x}_m \in \partial\Omega_{ij}$ *at* t_m *between two adjacent domains* Ω_α $(\alpha = i,j)$*. For an arbitrarily small* $\varepsilon > 0$*, there are two time intervals (i.e.,* $[t_{m-\varepsilon}, t_m)$ *and* $(t_m, t_{m+\varepsilon}]$*). Suppose* $\mathbf{x}^{(\alpha)}(t_{m\pm}) = \mathbf{x}_m$ $(\alpha \in \{i,j\})$*, then a flow* $\mathbf{x}^{(\alpha)}(t)$ *is* $C^{r_\alpha}_{[t_{m-\varepsilon}, t_m)}$*- and/or* $C^{r_\alpha}_{(t_m, t_{m+\varepsilon}]}$*-continuous* $(r_\alpha \geq 2)$*. The vector field* $\mathbf{F}^{(\alpha)}(\mathbf{x}, t, \mathbf{p}_\alpha)$ *is* $C^{r_\alpha-1}_{[t_{m-\varepsilon}, t_m)}$*- and* $C^{r_\alpha-1}_{(t_m, t_{m+\varepsilon}]}$*-continuous for time t, respectively.* $\|\mathrm{d}^{r_\alpha+1}\mathbf{x}^{(\alpha)}/\mathrm{d}t^{r_\alpha+1}\| < \infty$*. A point* $(\mathbf{x}_m, t_m)$ *on the boundary* $\partial\Omega_{ij}$ *is critical to flow* $\mathbf{x}^{(\alpha)}(t)$ *if and only if*

$$
\mathbf{n}^{\mathrm{T}}_{\partial\Omega_{ij}} \cdot \mathbf{F}^{(\alpha)}(t_{m-}) = 0 \text{ and/or } \mathbf{n}^{\mathrm{T}}_{\partial\Omega_{ij}} \cdot \mathbf{F}^{(\alpha)}(t_{m+}) = 0, \tag{2.18}
$$

where $\mathbf{F}^{(\alpha)}(t_{m\pm}) = \mathbf{F}^{(\alpha)}(\mathbf{x}, t_{m\pm}, \mathbf{p}_\alpha)$*.*

Proof. Using (2.1) and Definition 2.11, Theorem 2.7 can be proved. ∎

The tangential vector of the coming and leaving flows $\mathbf{x}^{(\alpha)}(t_{m\pm})$ to the boundary $\partial\Omega_{ij}$ in domain Ω_α ($\alpha \in \{i,j\}$) is normal to the normal vector of the boundary, so the coming flow is tangential to the boundary.

Definition 2.12. For a discontinuous dynamical system in (2.1), there is a point $\mathbf{x}(t_m) \equiv \mathbf{x}_m \in \partial\Omega_{ij}$ at time t_m between two adjacent domains Ω_α ($\alpha = i,j$). For an arbitrarily small $\varepsilon > 0$, there are two time intervals (i.e., $[t_{m-\varepsilon}, t_m)$ and $(t_m, t_{m+\varepsilon}]$). Suppose $\mathbf{x}^{(\alpha)}(t_{m\pm}) = \mathbf{x}_m$ ($\alpha \in \{i,j\}$), then a flow $\mathbf{x}^{(\alpha)}(t)$ is $C^{r_\alpha}_{[t_{m-\varepsilon},t_m)}$- and $C^{r_\alpha}_{(t_m,t_{m+\varepsilon}]}$-continuous ($r_\alpha \geq 1$) for time t. The flow $\mathbf{x}^{(\alpha)}(t)$ in Ω_α is tangential to boundary $\partial\Omega_{ij}$ at point $(\mathbf{x}_m, t_m)$ if the following conditions hold.

$$\mathbf{n}^{\mathrm{T}}_{\partial\Omega_{ij}} \cdot \dot{\mathbf{x}}^{(\alpha)}(t_{m\pm}) = 0. \tag{2.19}$$

$$\text{either} \quad \left.\begin{array}{l} \mathbf{n}^{\mathrm{T}}_{\partial\Omega_{ij}} \cdot [\mathbf{x}^{(\alpha)}(t_{m-}) - \mathbf{x}^{(\alpha)}(t_{m-\varepsilon})] > 0 \\ \mathbf{n}^{\mathrm{T}}_{\partial\Omega_{ij}} \cdot [\mathbf{x}^{(\alpha)}(t_{m+\varepsilon}) - \mathbf{x}^{(\alpha)}(t_{m+})] < 0 \end{array}\right\} \quad \text{for } \mathbf{n}_{\partial\Omega_{ij}} \to \Omega_\beta, \tag{2.20}$$

$$\text{or} \quad \left.\begin{array}{l} \mathbf{n}^{\mathrm{T}}_{\partial\Omega_{ij}} \cdot [\mathbf{x}^{(\alpha)}(t_{m-}) - \mathbf{x}^{(\alpha)}(t_{m-\varepsilon})] < 0 \\ \mathbf{n}^{\mathrm{T}}_{\partial\Omega_{ij}} \cdot [\mathbf{x}^{(\alpha)}(t_{m+\varepsilon}) - \mathbf{x}^{(\alpha)}(t_{m+})] > 0 \end{array}\right\} \quad \text{for } \mathbf{n}_{\partial\Omega_{ij}} \to \Omega_\alpha, \tag{2.21}$$

where $\alpha, \beta \in \{i,j\}$ *but* $\beta \neq \alpha$.

The normal vector $\mathbf{n}_{\partial\Omega_{ij}}$ is normal to the tangential plane. Without any switching laws, equation (2.19) gives

$$\dot{\mathbf{x}}^{(\alpha)}(t_{m-}) = \dot{\mathbf{x}}^{(\alpha)}(t_{m+}) \quad \text{but} \quad \dot{\mathbf{x}}^{(\alpha)}(t_{m\pm}) \neq \dot{\mathbf{x}}^{(0)}(t_m). \tag{2.22}$$

The above equation implies that the flow $\mathbf{x}^{(\alpha)}$ on the boundary $\partial\Omega_{ij}$ is at least C^1-continuous. To demonstrate the above definition, consider a flow in domain Ω_i tangential to the boundary $\partial\Omega_{ij}$ with $\mathbf{n}_{\partial\Omega_{ij}} \to \Omega_j$, as shown in Fig. 2.6. The gray-filled symbols represent two points ($\mathbf{x}^{(i)}_{m\pm\varepsilon} = \mathbf{x}^{(i)}(t_m \pm \varepsilon)$) on the flow before and after the tangency. The tangential point $\mathbf{x}_m$ on the boundary $\partial\Omega_{ij}$ is depicted by a large circular symbol. This tangential flow is also termed *a grazing flow*.

Theorem 2.8. *For a discontinuous dynamical system in (2.1), there is a point* $\mathbf{x}(t_m) \equiv \mathbf{x}_m \in \partial\Omega_{ij}$ *at time* t_m *between two adjacent domains* Ω_α ($\alpha = i,j$). *For an arbitrarily small* $\varepsilon > 0$, *there are two time intervals (i.e.,* $[t_{m-\varepsilon}, t_m)$ *and* $(t_m, t_{m+\varepsilon}]$*). Suppose* $\mathbf{x}^{(\alpha)}(t_{m\pm}) = \mathbf{x}_m$ ($\alpha \in \{i,j\}$), *then a flow* $\mathbf{x}^{(\alpha)}(t)$ *is* $C^{r_\alpha}_{[t_{m-\varepsilon},t_m)}$*- and* $C^{r_\alpha}_{(t_m,t_{m+\varepsilon}]}$*-continuous* ($r_\alpha \geq 2$) *for time t.* $\|d^{r_\alpha}\mathbf{x}^{(\alpha)}/dt^{r_\alpha}\| < \infty$. *The flow* $\mathbf{x}^{(\alpha)}(t)$ *in* Ω_α *is tangential to boundary* $\partial\Omega_{ij}$ *at point* $(\mathbf{x}_m, t_m)$ *if and only if*

$$\mathbf{n}^{\mathrm{T}}_{\partial\Omega_{ij}} \cdot \dot{\mathbf{x}}^{(\alpha)}(t_{m\pm}) = 0, \tag{2.23}$$

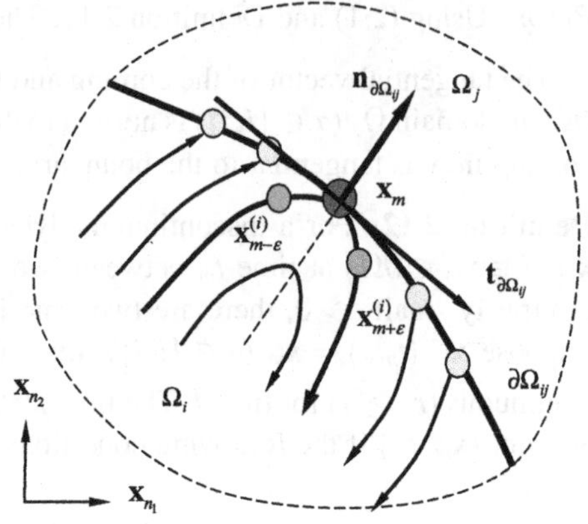

Fig. 2.6 A flow in domain Ω_i tangential to boundary $\partial\Omega_{ij}$ with $\mathbf{n}_{\partial\Omega_{ij}} \to \Omega_j$. The *gray-filled symbols* represent two points ($\mathbf{x}_{m-\varepsilon}^{(i)}$ and $\mathbf{x}_{m+\varepsilon}^{(i)}$) on the flow before and after the tangency. The tangential point $\mathbf{x}_m$ on the boundary $\partial\Omega_{ij}$ is depicted by a *large circular symbol*

$$\left.\begin{array}{l}\text{either} \left.\begin{array}{l}\mathbf{n}_{\partial\Omega_{ij}}^{\mathrm{T}} \cdot \dot{\mathbf{x}}^{(\alpha)}(t_{m-\varepsilon}) > 0 \\ \mathbf{n}_{\partial\Omega_{ij}}^{\mathrm{T}} \cdot \dot{\mathbf{x}}^{(\alpha)}(t_{m+\varepsilon}) < 0\end{array}\right\} \text{ for } \mathbf{n}_{\partial\Omega_{ij}} \to \Omega_\beta \\ \text{or} \left.\begin{array}{l}\mathbf{n}_{\partial\Omega_{ij}}^{\mathrm{T}} \cdot \dot{\mathbf{x}}^{(\alpha)}(t_{m-\varepsilon}) < 0 \\ \mathbf{n}_{\partial\Omega_{ij}}^{\mathrm{T}} \cdot \dot{\mathbf{x}}^{(\alpha)}(t_{m+\varepsilon}) > 0\end{array}\right\} \text{ for } \mathbf{n}_{\partial\Omega_{ij}} \to \Omega_\alpha,\end{array}\right. \tag{2.24}$$

where $\alpha, \beta \in \{i, j\}$ *but* $\beta \neq \alpha$.

Proof. Equation (2.23) is identical to the first condition in (2.18). Consider

$$\begin{aligned}\mathbf{x}^{(\alpha)}(t_{m\pm}) &\equiv \mathbf{x}^{(\alpha)}(t_{m\pm} \pm \varepsilon \mp \varepsilon) = \mathbf{x}^{(\alpha)}(t_{m\pm} \pm \varepsilon) \mp \varepsilon\dot{\mathbf{x}}^{(\alpha)}(t_{m\pm} \pm \varepsilon) + o(\varepsilon), \\ &= \mathbf{x}^{(\alpha)}(t_{m\pm\varepsilon}) \mp \varepsilon\dot{\mathbf{x}}^{(\alpha)}(t_{m\pm\varepsilon}) + o(\varepsilon).\end{aligned}$$

For $0 < \varepsilon << 1$, the higher-order terms in the above equation can be ignored. Therefore,

$$\left.\begin{array}{l}\mathbf{n}_{\partial\Omega_{ij}}^{\mathrm{T}} \cdot [\mathbf{x}^{(\alpha)}(t_{m-}) - \mathbf{x}^{(\alpha)}(t_{m-\varepsilon})] = \varepsilon\mathbf{n}_{\partial\Omega_{ij}}^{\mathrm{T}} \cdot \dot{\mathbf{x}}^{(\alpha)}(t_{m-\varepsilon}), \\ \mathbf{n}_{\partial\Omega_{ij}}^{\mathrm{T}} \cdot [\mathbf{x}^{(\alpha)}(t_{m+\varepsilon}) - \mathbf{x}^{(\alpha)}(t_{m+})] = \varepsilon\mathbf{n}_{\partial\Omega_{ij}}^{\mathrm{T}} \cdot \dot{\mathbf{x}}^{(\alpha)}(t_{m+\varepsilon}).\end{array}\right\}$$

From (2.24), the first case is

$$\mathbf{n}_{\partial\Omega_{ij}}^{\mathrm{T}} \cdot \dot{\mathbf{x}}^{(\alpha)}(t_{m-\varepsilon}) > 0 \text{ and } \mathbf{n}_{\partial\Omega_{ij}}^{\mathrm{T}} \cdot \dot{\mathbf{x}}^{(\alpha)}(t_{m+\varepsilon}) < 0$$

from which (2.20) holds for $\partial\Omega_{ij}$ with $\mathbf{n}_{\partial\Omega_{ij}} \to \Omega_\beta$ ($\beta \neq \alpha$). However, the second case is

$$\mathbf{n}_{\partial\Omega_{ij}}^{\mathrm{T}} \cdot \dot{\mathbf{x}}^{(\alpha)}(t_{m-\varepsilon}) < 0 \text{ and } \mathbf{n}_{\partial\Omega_{ij}}^{\mathrm{T}} \cdot \dot{\mathbf{x}}^{(\alpha)}(t_{m+\varepsilon}) > 0,$$

from which (2.21) holds for $\partial\Omega_{ij}$ with $\mathbf{n}_{\partial\Omega_{ij}} \to \Omega_\alpha$. Therefore, from Definition 2.12, the flow $\mathbf{x}^{(\alpha)}(t)$ for time $t \in [t_{m-\varepsilon}, t_{m+\varepsilon}]$ in Ω_α is tangential to the boundary $\partial\Omega_{ij}$. ■

Notice that the aforementioned theorem can be used for surface boundary.

Theorem 2.9. *For a discontinuous dynamical system in (2.1), there is a point* $\mathbf{x}(t_m) \equiv \mathbf{x}_m \in \partial\Omega_{ij}$ *at* t_m *between two adjacent domains* Ω_α $(\alpha = i, j)$. *For an arbitrarily small* $\varepsilon > 0$, *there are two time intervals (i.e.,* $[t_{m-\varepsilon}, t_m)$ *and* $(t_m, t_{m+\varepsilon}]$*). Suppose* $\mathbf{x}^{(\alpha)}(t_{m\pm}) = \mathbf{x}_m$ $(\alpha \in \{i, j\})$, *then a vector field* $\mathbf{F}^{(\alpha)}(\mathbf{x}, t, \mathbf{p}_\alpha)$ *is* $C^{r_\alpha}_{[t_{m-\varepsilon}, t_m)}$*- and* $C^{r_\alpha}_{(t_m, t_{m+\varepsilon}]}$*-continuous* $(r_\alpha \geq 1)$ *for time t.* $\|\mathrm{d}^{r_\alpha+1}\mathbf{x}^{(\alpha)}/\mathrm{d}t^{r_\alpha+1}\| < \infty$. *The flow* $\mathbf{x}^{(\alpha)}(t)$ *in* Ω_α *is tangential to boundary* $\partial\Omega_{ij}$ *at point* $(\mathbf{x}_m, t_m)$ *if and only if*

$$\mathbf{n}^{\mathrm{T}}_{\partial\Omega_{ij}} \cdot \mathbf{F}^{(\alpha)}(t_{m\pm}) = 0, \tag{2.25}$$

$$\begin{aligned} &\text{either} \left.\begin{aligned} \mathbf{n}^{\mathrm{T}}_{\partial\Omega_{ij}} \cdot \mathbf{F}^{(\alpha)}(t_{m-\varepsilon}) > 0 \\ \mathbf{n}^{\mathrm{T}}_{\partial\Omega_{ij}} \cdot \mathbf{F}^{(\alpha)}(t_{m+\varepsilon}) < 0 \end{aligned}\right\} \text{for } \mathbf{n}_{\partial\Omega_{ij}} \to \Omega_\beta \\ &\text{or} \left.\begin{aligned} \mathbf{n}^{\mathrm{T}}_{\partial\Omega_{ij}} \cdot \mathbf{F}^{(\alpha)}(t_{m-\varepsilon}) < 0 \\ \mathbf{n}^{\mathrm{T}}_{\partial\Omega_{ij}} \cdot \mathbf{F}^{(\alpha)}(t_{m+\varepsilon}) > 0 \end{aligned}\right\} \text{for } \mathbf{n}_{\partial\Omega_{ij}} \to \Omega_\alpha, \end{aligned} \tag{2.26}$$

where $\alpha, \beta \in \{i, j\}$ *but* $\beta \neq \alpha$.

Proof. Using (2.1) and Theorem 2.8, Theorem 2.9 can be proved. ■

For simplicity, consider $(n-1)$-dimensional *planes* in state space as the separation boundary in discontinuous dynamical systems, and the corresponding tangency to the $(n-1)$-dimensional boundary planes is discussed as follows. Because the normal vector $\mathbf{n}_{\partial\Omega_{ij}}$ for the $(n-1)$-dimensional plane boundaries does not change with location, the corresponding conditions for a flow to tangential to such plane boundaries can help one understand the concept of a flow tangential to the general separation boundary in discontinuous dynamical systems. The $(n-1)$-dimensional surfaces as general separation boundaries are discussed in the next chapter.

Theorem 2.10. *For a discontinuous dynamical system in (2.1), there is a point* $\mathbf{x}(t_m) \equiv \mathbf{x}_m \in \partial\Omega_{ij}$ *at* t_m *on the* $(n-1)$*-dimensional plane boundary* $\partial\Omega_{ij}$ *between two adjacent domains* Ω_α $(\alpha = i, j)$. *For an arbitrarily small* $\varepsilon > 0$, *there are two time intervals (i.e.,* $[t_{m-\varepsilon}, t_m)$ *and* $(t_m, t_{m+\varepsilon}]$*). Suppose* $\mathbf{x}^{(\alpha)}(t_{m\pm}) = \mathbf{x}_m$ $(\alpha \in \{i, j\})$, *then a flow* $\mathbf{x}^{(\alpha)}(t)$ *is* $C^{r_\alpha}_{[t_{m-\varepsilon}, t_m)}$*- and* $C^{r_\alpha}_{(t_m, t_{m+\varepsilon}]}$*-continuous* $(r_\alpha \geq 3)$ *for time t and* $\|\mathrm{d}^{r_\alpha}\mathbf{x}^{(\alpha)}/\mathrm{d}t^{r_\alpha}\| < \infty$. *The flow* $\mathbf{x}^{(\alpha)}(t)$ *in* Ω_α *is tangential to the* $(n-1)$*-dimensional plane boundary* $\partial\Omega_{ij}$ *at point* $(\mathbf{x}_m, t_m)$ *if and only if*

$$\mathbf{n}^{\mathrm{T}}_{\partial\Omega_{ij}} \cdot \dot{\mathbf{x}}^{(\alpha)}(t_{m\pm}) = 0, \tag{2.27}$$

$$\left.\begin{aligned} &\text{either } \mathbf{n}^{\mathrm{T}}_{\partial\Omega_{ij}} \cdot \ddot{\mathbf{x}}^{(\alpha)}(t_{m\pm}) < 0 \quad \text{for } \mathbf{n}_{\partial\Omega_{ij}} \to \Omega_\beta \\ &\text{or } \quad \mathbf{n}^{\mathrm{T}}_{\partial\Omega_{ij}} \cdot \ddot{\mathbf{x}}^{(\alpha)}(t_{m\pm}) > 0 \quad \text{for } \mathbf{n}_{\partial\Omega_{ij}} \to \Omega_\alpha, \end{aligned}\right\} \tag{2.28}$$

where $\alpha, \beta \in \{i, j\}$ *but* $\beta \neq \alpha$.

Proof. Equation (2.27) is identical to (2.19), thus the first condition in (2.19) is satisfied. From Definition 2.12, consider the boundary $\partial\Omega_{ij}$ with $\mathbf{n}_{\partial\Omega_{ij}} \to \Omega_\beta$ ($\beta \neq \alpha$) first. Suppose $\mathbf{x}^{(\alpha)}(t_{m\pm}) = \mathbf{x}_m$ ($\alpha \in \{i,j\}$) and a flow $\mathbf{x}^{(\alpha)}(t)$ is $C^{r_\alpha}_{[t_{m-\varepsilon},t_m)}$-and $C^{r_\alpha}_{(t_m,t_{m+\varepsilon}]}$-continuous ($r_\alpha \geq 3$) for time t, then for $a \in [t_{m-\varepsilon}, t_m)$ and $a \in (t_m, t_{m+\varepsilon}]$, the Taylor series expansion of $\mathbf{x}^{(\alpha)}(t_{m\pm\varepsilon})$ to $\mathbf{x}^{(\alpha)}(a)$ up to the third-order term is given as follows

$$\begin{aligned}\mathbf{x}^{(\alpha)}(t_{m\pm\varepsilon}) \equiv \mathbf{x}^{(\alpha)}(t_{m\pm} - \varepsilon) &= \mathbf{x}^{(\alpha)}(a) + \dot{\mathbf{x}}^{(\alpha)}(a)(t_{m\pm} \pm \varepsilon - a)\\ &+ \ddot{\mathbf{x}}^{(\alpha)}(a)(t_{m\pm} \pm \varepsilon - a)^2 + o((t_{m\pm} \pm \varepsilon - a)^2).\end{aligned}$$

As $a \to t_{m\pm}$, the limit of the foregoing equation leads to

$$\mathbf{x}^{(\alpha)}(t_{m\pm\varepsilon}) \equiv \mathbf{x}^{(\alpha)}(t_m \pm \varepsilon) = \mathbf{x}^{(\alpha)}(t_{m\pm}) \pm \dot{\mathbf{x}}^{(\alpha)}(t_{m\pm})\varepsilon + \ddot{\mathbf{x}}^{(\alpha)}(t_{m\pm})\varepsilon^2 + o(\varepsilon^2).$$

The ignorance of the ε^3 and higher-order terms, deformation of the above equation, and left multiplication of $\mathbf{n}_{\partial\Omega_{ij}}$ gives

$$\mathbf{n}^{\mathrm{T}}_{\partial\Omega_{ij}} \cdot [\mathbf{x}^{(\alpha)}(t_{m-}) - \mathbf{x}^{(\alpha)}(t_{m-\varepsilon})] = \mathbf{n}^{\mathrm{T}}_{\partial\Omega_{ij}} \cdot \dot{\mathbf{x}}^{(\alpha)}(t_{m-})\varepsilon - \mathbf{n}^{\mathrm{T}}_{\partial\Omega_{ij}} \cdot \ddot{\mathbf{x}}^{(\alpha)}(t_{m-})\varepsilon^2,$$

$$\mathbf{n}^{\mathrm{T}}_{\partial\Omega_{ij}} \cdot [\mathbf{x}^{(\alpha)}(t_{m+\varepsilon}) - \mathbf{x}^{(\alpha)}(t_{m+})] = \mathbf{n}^{\mathrm{T}}_{\partial\Omega_{ij}} \cdot \dot{\mathbf{x}}^{(\alpha)}(t_{m+})\varepsilon + \mathbf{n}^{\mathrm{T}}_{\partial\Omega_{ij}} \cdot \ddot{\mathbf{x}}^{(\alpha)}(t_{m+})\varepsilon^2.$$

With (2.27), one can obtain

$$\mathbf{n}^{\mathrm{T}}_{\partial\Omega_{ij}} \cdot [\mathbf{x}^{(\alpha)}(t_{m-}) - \mathbf{x}^{(\alpha)}(t_{m-\varepsilon})] = -\mathbf{n}^{\mathrm{T}}_{\partial\Omega_{ij}} \cdot \ddot{\mathbf{x}}^{(\alpha)}(t_{m-})\varepsilon^2,$$

$$\mathbf{n}^{\mathrm{T}}_{\partial\Omega_{ij}} \cdot [\mathbf{x}^{(\alpha)}(t_{m+\varepsilon}) - \mathbf{x}^{(\alpha)}(t_{m+})] = \mathbf{n}^{\mathrm{T}}_{\partial\Omega_{ij}} \cdot \ddot{\mathbf{x}}^{(\alpha)}(t_{m+})\varepsilon^2.$$

For the plane boundary $\partial\Omega_{ij}$ with $\mathbf{n}_{\partial\Omega_{ij}} \to \Omega_\beta$, using the first inequality equation of (2.28), the foregoing two equations lead to

$$\mathbf{n}^{\mathrm{T}}_{\partial\Omega_{ij}} \cdot [\mathbf{x}^{(\alpha)}(t_{m-}) - \mathbf{x}^{(\alpha)}(t_{m-\varepsilon})] = -\mathbf{n}^{\mathrm{T}}_{\partial\Omega_{ij}} \cdot \ddot{\mathbf{x}}^{(\alpha)}(t_{m-})\varepsilon^2 > 0,$$

$$\mathbf{n}^{\mathrm{T}}_{\partial\Omega_{ij}} \cdot [\mathbf{x}^{(\alpha)}(t_{m+\varepsilon}) - \mathbf{x}^{(\alpha)}(t_{m+})] = \mathbf{n}^{\mathrm{T}}_{\partial\Omega_{ij}} \cdot \ddot{\mathbf{x}}^{(\alpha)}(t_{m+})\varepsilon^2 < 0.$$

From Definition 2.12, the first inequality equation of (2.28) is obtained. Similarly, using the second inequality of (2.28), one can obtain

$$\mathbf{n}^{\mathrm{T}}_{\partial\Omega_{ij}} \cdot [\mathbf{x}^{(\alpha)}(t_{m-}) - \mathbf{x}^{(\alpha)}(t_{m-\varepsilon})] = -\mathbf{n}^{\mathrm{T}}_{\partial\Omega_{ij}} \cdot \ddot{\mathbf{x}}^{(\alpha)}(t_{m-})\varepsilon^2 < 0,$$

$$\mathbf{n}^{\mathrm{T}}_{\partial\Omega_{ij}} \cdot [\mathbf{x}^{(\alpha)}(t_{m+\varepsilon}) - \mathbf{x}^{(\alpha)}(t_{m+})] = \mathbf{n}^{\mathrm{T}}_{\partial\Omega_{ij}} \cdot \ddot{\mathbf{x}}^{(\alpha)}(t_{m+})\varepsilon^2 > 0.$$

for the boundary $\partial\Omega_{ij}$ with $\mathbf{n}_{\partial\Omega_{ij}} \to \Omega_\alpha$. Therefore, under (2.28), the flow $\mathbf{x}^{(\alpha)}(t)$ in domain Ω_α is tangential to the plane boundary $\partial\Omega_{ij}$, vice versa. ■

Theorem 2.11. *For a discontinuous dynamical system in (2.1), there is a point $\mathbf{x}(t_m) \equiv \mathbf{x}_m \in \partial\Omega_{ij}$ at time t_m on the $(n-1)$-dimensional plane boundary $\partial\Omega_{ij}$ between two adjacent domains Ω_α $(\alpha = i, j)$. For an arbitrarily small $\varepsilon > 0$, there are two time intervals (i.e., $[t_{m-\varepsilon}, t_m)$ and $(t_m, t_{m+\varepsilon}]$). Suppose $\mathbf{x}^{(\alpha)}(t_{m\pm}) = \mathbf{x}_m$ $(\alpha \in \{i, j\})$, then the vector field $\mathbf{F}^{(\alpha)}(\mathbf{x}, t, \mu_\alpha)$ is $C^{r_\alpha}_{[t_{m-\varepsilon}, t_m)}$- and $C^{r_\alpha}_{(t_m, t_{m+\varepsilon}]}$-continuous $(r_\alpha \geq 2)$ for time t and $\|d^{r_\alpha+1}\mathbf{x}^{(\alpha)}/dt^{r_\alpha+1}\| < \infty$. The flow $\mathbf{x}^{(\alpha)}(t)$ in Ω_α is tangential to the plane boundary $\partial\Omega_{ij}$ at point $(\mathbf{x}_m, t_m)$ if and only if*

$$\mathbf{n}^{\mathrm{T}}_{\partial\Omega_{ij}} \cdot \mathbf{F}^{(\alpha)}(t_{m\pm}) = 0, \tag{2.29}$$

$$\left.\begin{array}{ll} \text{either } \mathbf{n}^{\mathrm{T}}_{\partial\Omega_{ij}} \cdot D\mathbf{F}^{(\alpha)}(t_{m\pm}) < 0 & \text{for } \mathbf{n}_{\partial\Omega_{ij}} \to \Omega_\beta \\ \text{or } \quad \mathbf{n}^{\mathrm{T}}_{\partial\Omega_{ij}} \cdot D\mathbf{F}^{(\alpha)}(t_{m\pm}) > 0 & \text{for } \mathbf{n}_{\partial\Omega_{ij}} \to \Omega_\alpha, \end{array}\right\} \tag{2.30}$$

where $\alpha, \beta \in \{i, j\}$ but $\alpha \neq \beta$, and the total differentiation $(p, q \in \{1, 2, \ldots, n\})$

$$D\mathbf{F}^{(\alpha)}(t_{m\pm}) = \left\{ \left[\frac{\partial F^{(\alpha)}_p(\mathbf{x}, t, \mathbf{p}_\alpha)}{\partial x_q} \right]_{n\times n} \mathbf{F}^{(\alpha)}(t_{m\pm}) + \frac{\partial \mathbf{F}^{(\alpha)}(\mathbf{x}, t, \mathbf{p}_\alpha)}{\partial t} \right\} \Big|_{(\mathbf{x}_m, t_{m\pm})}. \tag{2.31}$$

Proof. Using (2.1) and (2.29), the first condition in (2.19) is satisfied. The derivative of (2.1) with respect to time t gives

$$\ddot{\mathbf{x}} \equiv D\mathbf{F}^{(\alpha)}(\mathbf{x}, t, \mathbf{p}_\alpha) = \left[\frac{\partial F^{(\alpha)}_p(\mathbf{x}, t, \mathbf{p}_\alpha)}{\partial x_q} \right]_{n\times n} \dot{\mathbf{x}} + \frac{\partial}{\partial t}\mathbf{F}^{(\alpha)}(\mathbf{x}, t, \mathbf{p}_\alpha).$$

For $t = t_{m\pm}$ and $\mathbf{x} = \mathbf{x}_m$, the left multiplication of $\mathbf{n}_{\partial\Omega_{ij}}$ to the above equation gives

$$\mathbf{n}^{\mathrm{T}}_{\partial\Omega_{ij}} \cdot \ddot{\mathbf{x}}(t_{m\pm}) = \mathbf{n}^{\mathrm{T}}_{\partial\Omega_{ij}} \cdot \left\{ \left[\frac{\partial F^{(\alpha)}_p(\mathbf{x}, t, \mathbf{p}_\alpha)}{\partial x_q} \right]_{n\times n} \mathbf{F}^{(\alpha)}(t_{m\pm}) + \frac{\partial \mathbf{F}^{(\alpha)}(\mathbf{x}, t, \mathbf{p}_\alpha)}{\partial t} \right\} \Big|_{(\mathbf{x}_m, t_{m\pm})},$$

where $\mathbf{F}^{(\alpha)}(\mathbf{x}_m, t_{m\pm}, \mathbf{p}_\alpha) \triangleq \mathbf{F}^{(\alpha)}(t_{m\pm})$. Using (2.30), the above equation leads to (2.28). From Theorem 2.10, the flow $\mathbf{x}^{(\alpha)}(t)$ in Ω_α is tangential to the plane boundary $\partial\Omega_{ij}$ at point $(\mathbf{x}_m, t_m)$, vice versa. ■

Definition 2.13. For a discontinuous dynamical system in (2.1), there is a point $\mathbf{x}(t_m) \equiv \mathbf{x}_m \in \partial\Omega_{ij}$ at time t_m on the $(n-1)$-dimensional *plane* boundary $\partial\Omega_{ij}$ between two adjacent domains Ω_α $(\alpha = i, j)$. For an arbitrarily small $\varepsilon > 0$, there are two time intervals (i.e., $[t_{m-\varepsilon}, t_m)$ and $(t_m, t_{m+\varepsilon}]$). Suppose $\mathbf{x}^{(\alpha)}(t_{m\pm}) = \mathbf{x}_m$

($\alpha \in \{i,j\}$), then a flow $\mathbf{x}^{(\alpha)}(t)$ is $C^{r_\alpha}_{[t_{m-\varepsilon},t_m)}$- and $C^{r_\alpha}_{(t_m,t_{m+\varepsilon}]}$-continuous ($r_\alpha \geq 2l_\alpha$) for time t. The flow $\mathbf{x}^{(\alpha)}(t)$ in Ω_α is tangential to the plane boundary $\partial\Omega_{ij}$ at point $(\mathbf{x}_m, t_m)$ with the $(2l_\alpha - 1)$th-order if

$$\mathbf{n}^{\mathrm{T}}_{\partial\Omega_{ij}} \cdot \frac{\mathrm{d}^{k_\alpha}\mathbf{x}^{(\alpha)}(t)}{\mathrm{d}t^{k_\alpha}}\Big|_{t=t_{m\pm}} = 0 \text{ for } k_\alpha = 1, 2, \ldots, 2l_\alpha - 1, \tag{2.32}$$

$$\mathbf{n}^{\mathrm{T}}_{\partial\Omega_{ij}} \cdot \frac{\mathrm{d}^{2l_\alpha}\mathbf{x}^{(\alpha)}(t)}{\mathrm{d}t^{2l_\alpha}}\Big|_{t=t_{m\pm}} \neq 0, \tag{2.33}$$

$$\text{either } \left.\begin{aligned} &\mathbf{n}^{\mathrm{T}}_{\partial\Omega_{ij}} \cdot [\mathbf{x}^{(\alpha)}(t_{m-}) - \mathbf{x}^{(\alpha)}(t_{m-\varepsilon})] > 0 \\ &\mathbf{n}^{\mathrm{T}}_{\partial\Omega_{ij}} \cdot [\mathbf{x}^{(\alpha)}(t_{m+\varepsilon}) - \mathbf{x}^{(\alpha)}(t_{m+})] < 0 \end{aligned}\right\} \text{ for } \mathbf{n}_{\partial\Omega_{ij}} \to \Omega_\beta \tag{2.34}$$

$$\text{or } \left.\begin{aligned} &\mathbf{n}^{\mathrm{T}}_{\partial\Omega_{ij}} \cdot [\mathbf{x}^{(\alpha)}(t_{m-}) - \mathbf{x}^{(\alpha)}(t_{m-\varepsilon})] < 0 \\ &\mathbf{n}^{\mathrm{T}}_{\partial\Omega_{ij}} \cdot [\mathbf{x}^{(\alpha)}(t_{m+\varepsilon}) - \mathbf{x}^{(\alpha)}(t_{m+})] > 0 \end{aligned}\right\} \text{ for } \mathbf{n}_{\partial\Omega_{ij}} \to \Omega_\alpha, \tag{2.35}$$

where $\alpha, \beta \in \{i,j\}$ but $\beta \neq \alpha$.

Theorem 2.12. *For a discontinuous dynamical system in (2.1), there is a point $\mathbf{x}(t_m) \equiv \mathbf{x}_m \in \partial\Omega_{ij}$ at time t_m on the $(n-1)$-dimensional plane boundary $\partial\Omega_{ij}$ between two adjacent domains Ω_α ($\alpha = i,j$). For an arbitrarily small $\varepsilon > 0$, there are two time intervals (i.e., $[t_{m-\varepsilon}, t_m)$ and $(t_m, t_{m+\varepsilon}]$). Suppose $\mathbf{x}^{(\alpha)}(t_{m\pm}) = \mathbf{x}_m$ ($\alpha \in \{i,j\}$), then a flow $\mathbf{x}^{(\alpha)}(t)$ is $C^{r_\alpha}_{[t_{m-\varepsilon},t_m)}$- and $C^{r_\alpha}_{(t_m,t_{m+\varepsilon}]}$-continuous ($r_\alpha \geq 2l_\alpha + 1$) for time t. $\|\mathrm{d}^{r_\alpha}\mathbf{x}^{(\alpha)}/\mathrm{d}t^{r_\alpha}\| < \infty$. The flow $\mathbf{x}^{(\alpha)}(t)$ in Ω_α is tangential to the plane boundary $\partial\Omega_{ij}$ at point $(\mathbf{x}_m, t_m)$ with the $(2l_\alpha - 1)$th-order if and only if*

$$\mathbf{n}^{\mathrm{T}}_{\partial\Omega_{ij}} \cdot \frac{\mathrm{d}^{k_\alpha}\mathbf{x}^{(\alpha)}(t)}{\mathrm{d}t^{k_\alpha}}\Big|_{t=t_{m\pm}} = 0 \quad \text{for } (k_\alpha = 1, 2, \ldots 2l_\alpha - 1), \tag{2.36}$$

$$\mathbf{n}^{\mathrm{T}}_{\partial\Omega_{ij}} \cdot \frac{\mathrm{d}^{2l_\alpha}\mathbf{x}^{(\alpha)}(t)}{\mathrm{d}t^{2l_\alpha}}\Big|_{t=t_{m\pm}} \neq 0, \tag{2.37}$$

$$\left.\begin{aligned} &\text{either } \mathbf{n}^{\mathrm{T}}_{\partial\Omega_{ij}} \cdot \frac{\mathrm{d}^{2l_\alpha}\mathbf{x}^{(\alpha)}(t)}{\mathrm{d}t^{2l_\alpha}}\Big|_{t=t_{m\pm}} < 0 \quad \text{for } \mathbf{n}_{\partial\Omega_{ij}} \to \Omega_\beta \\ &\text{or } \mathbf{n}^{\mathrm{T}}_{\partial\Omega_{ij}} \cdot \frac{\mathrm{d}^{2l_\alpha}\mathbf{x}^{(\alpha)}(t)}{\mathrm{d}t^{2l_\alpha}}\Big|_{t=t_{m\pm}} > 0 \quad \text{for } \mathbf{n}_{\partial\Omega_{ij}} \to \Omega_\alpha, \end{aligned}\right\} \tag{2.38}$$

where $\beta \in \{i,j\}$ but $\beta \neq \alpha$.

Proof. For (2.36) and (2.37), the first two conditions in Definition 2.13 are satisfied. Consider the boundary $\partial\Omega_{ij}$ with $\mathbf{n}_{\partial\Omega_{ij}} \to \Omega_\beta$ ($\beta \neq \alpha$) first. Choose

$a \in [t_{m-\varepsilon}, t_m)$ or $a \in (t_m, t_{m-\varepsilon}]$, and application of the Taylor series expansion of $\mathbf{x}^{(\alpha)}(t_{m\pm\varepsilon})$ to $\mathbf{x}^{(\alpha)}(a)$ and up to the $(2l_\alpha)$th-order term gives

$$\mathbf{x}^{(\alpha)}(t_{m\pm\varepsilon}) \equiv \mathbf{x}^{(\alpha)}(t_{m\pm} \pm \varepsilon) = \mathbf{x}^{(\alpha)}(a) + \sum\nolimits_{k_\alpha=1}^{2l_\alpha-1} \frac{\mathrm{d}^{k_\alpha}\mathbf{x}^{(\alpha)}(t)}{\mathrm{d}t^{k_\alpha}}\Big|_{t=a}(t_{m\pm} \pm \varepsilon - a)^{k_\alpha}$$
$$+ \frac{\mathrm{d}^{2l_\alpha}\mathbf{x}^{(\alpha)}(t)}{\mathrm{d}t^{2l_\alpha}}\Big|_{t=a}(t_{m\pm} \pm \varepsilon - a)^{2l_\alpha} + o((t_{m\pm} \pm \varepsilon - a)^{2l_\alpha}).$$

As $a \to t_{m\pm}$, the foregoing equation becomes

$$\mathbf{x}^{(\alpha)}(t_{m\pm\varepsilon}) \equiv \mathbf{x}^{(\alpha)}(t_{m\pm} \pm \varepsilon) = \mathbf{x}^{(\alpha)}(t_{m\pm}) + \sum\nolimits_{k_\alpha=1}^{2l_\alpha-1} \frac{\mathrm{d}^{k_\alpha}\mathbf{x}^{(\alpha)}(t)}{\mathrm{d}t^{k_\alpha}}\Big|_{t=t_{m\pm}}(\pm\varepsilon)^{k_\alpha}$$
$$+ \frac{\mathrm{d}^{2l_\alpha}\mathbf{x}^{(\alpha)}(t)}{\mathrm{d}t^{2l_\alpha}}\Big|_{t=t_{m\pm}}\varepsilon^{2l_\alpha} + o(\pm\varepsilon^{2l_\alpha}).$$

With (2.36) and (2.37), the deformation of the above equation and left multiplication of $\mathbf{n}_{\partial\Omega_{ij}}$ produces

$$\mathbf{n}_{\partial\Omega_{ij}}^{\mathrm{T}} \cdot [\mathbf{x}^{(\alpha)}(t_{m-}) - \mathbf{x}^{(\alpha)}(t_{m-\varepsilon})] = -\mathbf{n}_{\partial\Omega_{ij}}^{\mathrm{T}} \cdot \frac{\mathrm{d}^{2l_\alpha}\mathbf{x}^{(\alpha)}(t)}{\mathrm{d}t^{2l_\alpha}}\Big|_{t=t_{m-}}\varepsilon^{2l_\alpha},$$

$$\mathbf{n}_{\partial\Omega_{ij}}^{\mathrm{T}} \cdot [\mathbf{x}^{(\alpha)}(t_{m+\varepsilon}) - \mathbf{x}^{(\alpha)}(t_{m+})] = \mathbf{n}_{\partial\Omega_{ij}}^{\mathrm{T}} \cdot \frac{\mathrm{d}^{2l_\alpha}\mathbf{x}^{(\alpha)}(t)}{\mathrm{d}t^{2l_\alpha}}\Big|_{t=t_{m+}}\varepsilon^{2l_\alpha}.$$

Under (2.38), the condition in (2.34) is satisfied, and vice versa. Therefore, the flow $\mathbf{x}^{(\alpha)}(t)$ in domain Ω_α is tangential to $\partial\Omega_{ij}$ with the $(2l_\alpha - 1)$th-order for $\mathbf{n}_{\partial\Omega_{ij}} \to \Omega_\beta$. Similarly, under the condition in (2.38), the flow $\mathbf{x}^{(\alpha)}(t)$ in domain Ω_α is tangential to boundary $\partial\Omega_{ij}$ at point $(\mathbf{x}_m, t_m)$ with the $(2l_\alpha - 1)$th-order for $\mathbf{n}_{\partial\Omega_{ij}} \to \Omega_\alpha$. Hence, the theorem is proved. ■

Theorem 2.13. *For a discontinuous dynamical system in (2.1), there is a point $\mathbf{x}(t_m) \equiv \mathbf{x}_m \in \partial\Omega_{ij}$ at time t_m on the $(n-1)$-dimensional plane boundary $\partial\Omega_{ij}$ between two adjacent domains Ω_α $(\alpha = i,j)$. For an arbitrarily small $\varepsilon > 0$, there are two time intervals (i.e., $[t_{m-\varepsilon}, t_m)$ and $(t_m, t_{m+\varepsilon}]$). Suppose $\mathbf{x}^{(\alpha)}(t_{m\pm}) = \mathbf{x}_m$ $(\alpha \in \{i,j\})$, then the vector field $\mathbf{F}^{(\alpha)}(\mathbf{x}, t, \mathbf{p}_\alpha)$ is $C^{r_\alpha}_{[t_{m-\varepsilon}, t_m)}$- and $C^{r_\alpha}_{(t_m, t_{m+\varepsilon}]}$-continuous $(r_\alpha \geq 2l_\alpha)$ for time t. $\|\mathrm{d}^{r_\alpha+1}\mathbf{x}^{(\alpha)}/\mathrm{d}t^{r_\alpha+1}\| < \infty$. The flow $\mathbf{x}^{(\alpha)}(t)$ in Ω_α is tangential to the plane boundary $\partial\Omega_{ij}$ at point $(\mathbf{x}_m, t_m)$ with the $(2l_\alpha - 1)$th-order if and only if*

$$\mathbf{n}_{\partial\Omega_{ij}}^{\mathrm{T}} \cdot D^{k_\alpha-1}\mathbf{F}^{(\alpha)}(t_{m\pm}) = 0 \quad \text{for } k_\alpha = 1, 2, \ldots, 2l_\alpha - 1, \tag{2.39}$$

$$\mathbf{n}_{\partial\Omega_{ij}}^{\mathrm{T}} \cdot D^{2l_\alpha-1}\mathbf{F}^{(\alpha)}(t_{m\pm}) \neq 0, \tag{2.40}$$

$$\begin{aligned}&\text{either } \mathbf{n}^{\mathrm{T}}_{\partial\Omega_{ij}} \cdot D^{2l_\alpha-1}\mathbf{F}^{(\alpha)}(t_{m\pm}) < 0 \quad \text{for } \partial\Omega_{ij} \to \Omega_\beta\\ &\text{or} \quad\ \mathbf{n}^{\mathrm{T}}_{\partial\Omega_{ij}} \cdot D^{2l_\alpha-1}\mathbf{F}^{(\alpha)}(t_{m\pm}) > 0 \quad \text{for } \partial\Omega_{ij} \to \Omega_\alpha,\end{aligned} \tag{2.41}$$

where the total differentiation

$$D^{k_\alpha-1}\mathbf{F}^{(\alpha)}(t_m) = D^{k-2}\left\{\left[\frac{\partial F_p^{(\alpha)}(\mathbf{x},t,\mathbf{p}_\alpha)}{\partial x_q}\right]_{n\times n}\dot{\mathbf{x}} + \frac{\partial \mathbf{F}^{(\alpha)}(\mathbf{x},t,\mathbf{p}_\alpha)}{\partial t}\right\}\Big|_{(\mathbf{x}_m,t_m)}, \tag{2.42}$$

with $p,q \in \{1,2,\ldots,n\}$, $k_\alpha \in \{2,3,\ldots 2l_\alpha\}$, *and* $\beta \in \{i,j\}$ *but* $\alpha \neq \beta$.

Proof. The k_α-order derivative of (2.1) with respect to time gives

$$\begin{aligned}\frac{\mathrm{d}^{k_\alpha}\mathbf{x}^{(\alpha)}(t)}{\mathrm{d}t^{k_\alpha}}\Big|_{(\mathbf{x}_m,t_m)} &= \frac{\mathrm{d}^{k_\alpha-1}\dot{\mathbf{x}}^{(\alpha)}(t)}{\mathrm{d}t^{k_\alpha-1}}\Big|_{(\mathbf{x}_m,t_m)} = \frac{\mathrm{d}^{k_\alpha-1}\mathbf{F}^{(\alpha)}(\mathbf{x},t,\mathbf{p}_\alpha)}{\mathrm{d}t^{k_\alpha-1}}\Big|_{(\mathbf{x}_m,t_m)} \equiv D^{k_\alpha-1}\mathbf{F}^{(\alpha)}(t_m)\\ &= D^{k-2}\left\{\left[\frac{\partial F_p^{(\alpha)}(\mathbf{x},t,\mathbf{p}_\alpha)}{\partial x_q}\right]_{n\times n}\dot{\mathbf{x}} + \frac{\partial \mathbf{F}^{(\alpha)}(\mathbf{x},t,\mathbf{p}_\alpha)}{\partial t}\right\}\Big|_{(\mathbf{x}_m,t_m)}.\end{aligned}$$

Using the foregoing equation to the conditions in (2.39)–(2.42), the flow $\mathbf{x}^{(\alpha)}(t)$ in Ω_α is tangential to the plane boundary $\partial\Omega_{ij}$ at point $(\mathbf{x}_m, t_m)$ with the $(2l_\alpha - 1)$th order from Theorem 2.12. Therefore, this theorem is proved. ■

2.5 Switching Bifurcations of Passable Flows

In this section, the switching bifurcation between the passable and nonpassable flows to the boundary is discussed. In addition, the switching bifurcation between the sink and source flows on the boundary is also discussed. The switching bifurcations are defined first, and then the sufficient and necessary conditions for such switching bifurcations are developed. The *L*-functions of flows are introduced to develop criteria for the switching bifurcations from sufficient and necessary conditions.

Definition 2.14. For a discontinuous dynamical system in (2.1), there is a point $\mathbf{x}(t_m) = \mathbf{x}_m \in [\mathbf{x}_{m_1}, \mathbf{x}_{m_2}] \subset \overrightarrow{\partial\Omega}_{ij}$ for time t_m between two adjacent domains Ω_α $(\alpha = i,j)$. For an arbitrarily small $\varepsilon > 0$, there are two time intervals (i.e., $[t_{m-\varepsilon}, t_m)$ and $(t_m, t_{m+\varepsilon}]$), and $\mathbf{x}^{(i)}(t_{m-}) = \mathbf{x}_m = \mathbf{x}^{(j)}(t_{m\pm})$. The flows $\mathbf{x}^{(i)}(t)$ and $\mathbf{x}^{(j)}(t)$ are $C^{r_i}_{[t_{m-\varepsilon},t_m)}$- and $C^{r_j}_{[t_{m-\varepsilon},t_{m+\varepsilon}]}$-continuous $(r_\alpha \geq 1,\ \alpha = i,j)$ for time t, respectively. The tangential bifurcation of the flow $\mathbf{x}^{(j)}(t)$ at point $(\mathbf{x}_m, t_m)$ on the boundary $\overrightarrow{\partial\Omega}_{ij}$ is termed *the switching bifurcation of a flow* from the semipassable flow to the nonpassable flow of the first kind (or called *the sliding bifurcation from* $\overrightarrow{\partial\Omega}_{ij}$ *to* $\widetilde{\partial\Omega}_{ij}$) if

$$\mathbf{n}_{\partial\Omega_{ij}}^{\mathrm{T}} \cdot \dot{\mathbf{x}}^{(j)}(t_{m\pm}) = 0 \quad \text{and} \quad \mathbf{n}_{\partial\Omega_{ij}}^{\mathrm{T}} \cdot \dot{\mathbf{x}}^{(i)}(t_{m-}) \neq 0, \tag{2.43}$$

$$\text{either} \quad \left.\begin{array}{l} \mathbf{n}_{\partial\Omega_{ij}}^{\mathrm{T}} \cdot [\mathbf{x}^{(i)}(t_{m-}) - \mathbf{x}^{(i)}(t_{m-\varepsilon})] > 0 \\ \mathbf{n}_{\partial\Omega_{ij}}^{\mathrm{T}} \cdot [\mathbf{x}^{(j)}(t_{m-}) - \mathbf{x}^{(j)}(t_{m-\varepsilon})] < 0 \\ \mathbf{n}_{\partial\Omega_{ij}}^{\mathrm{T}} \cdot [\mathbf{x}^{(j)}(t_{m+\varepsilon}) - \mathbf{x}^{(j)}(t_{m+})] > 0 \end{array}\right\} \text{ for } \mathbf{n}_{\Omega_{ij}} \to \Omega_j \tag{2.44}$$

$$\text{or} \quad \left.\begin{array}{l} \mathbf{n}_{\partial\Omega_{ij}}^{\mathrm{T}} \cdot [\mathbf{x}^{(i)}(t_{m-}) - \mathbf{x}^{(i)}(t_{m-\varepsilon})] < 0 \\ \mathbf{n}_{\partial\Omega_{ij}}^{\mathrm{T}} \cdot [\mathbf{x}^{(j)}(t_{m-}) - \mathbf{x}^{(j)}(t_{m-\varepsilon})] > 0 \\ \mathbf{n}_{\partial\Omega_{ij}}^{\mathrm{T}} \cdot [\mathbf{x}^{(j)}(t_{m+\varepsilon}) - \mathbf{x}^{(j)}(t_{m+})] \end{array}\right\} \text{ for } \mathbf{n}_{\Omega_{ij}} \to \Omega_i. \tag{2.45}$$

Definition 2.15. For a discontinuous dynamical system in (2.1), there is a point $\mathbf{x}(t_m) = \mathbf{x}_m \in [\mathbf{x}_{m_1}, \mathbf{x}_{m_2}] \subset \overrightarrow{\partial\Omega}_{ij}$ for time t_m between two adjacent domains Ω_α ($\alpha = i, j$). For an arbitrarily small $\varepsilon > 0$, there are two time intervals (i.e., $[t_{m-\varepsilon}, t_m)$ and $(t_m, t_{m+\varepsilon}]$), and $\mathbf{x}^{(i)}(t_{m\pm}) = \mathbf{x}_m = \mathbf{x}^{(j)}(t_{m+})$. The flows $\mathbf{x}^{(i)}(t)$ and $\mathbf{x}^{(j)}(t)$ are $C^{r_i}_{[t_{m-\varepsilon}, t_{m+\varepsilon}]}$- and $C^{r_j}_{[t_{m+\varepsilon}, t_m)}$-continuous ($r_\alpha \geq 1$, $\alpha = i, j$) for time t, respectively. The tangential bifurcation of the flow $\mathbf{x}^{(i)}(t)$ at point $(\mathbf{x}_m, t_m)$ on the boundary $\overrightarrow{\partial\Omega}_{ij}$ is termed *the switching bifurcation of a flow from the passable flow to the nonpassable flow of the second kind* (or called *the source bifurcation from* $\overrightarrow{\partial\Omega}_{ij}$ *to* $\widehat{\partial\Omega}_{ij}$) if

$$\mathbf{n}_{\partial\Omega_{ij}}^{\mathrm{T}} \cdot \dot{\mathbf{x}}^{(i)}(t_{m\pm}) = 0 \quad \text{and} \quad \mathbf{n}_{\partial\Omega_{ij}}^{\mathrm{T}} \cdot \dot{\mathbf{x}}^{(j)}(t_{m\pm}) \neq 0, \tag{2.46}$$

$$\text{either} \quad \left.\begin{array}{l} \mathbf{n}_{\partial\Omega_{ij}}^{\mathrm{T}} \cdot [\mathbf{x}^{(i)}(t_{m-}) - \mathbf{x}^{(i)}(t_{m-\varepsilon})] > 0 \\ \mathbf{n}_{\partial\Omega_{ij}}^{\mathrm{T}} \cdot [\mathbf{x}^{(i)}(t_{m+\varepsilon}) - \mathbf{x}^{(i)}(t_{m+})] < 0 \\ \mathbf{n}_{\partial\Omega_{ij}}^{\mathrm{T}} \cdot [\mathbf{x}^{(j)}(t_{m+\varepsilon}) - \mathbf{x}^{(j)}(t_{m+})] > 0 \end{array}\right\} \text{ for } \mathbf{n}_{\Omega_{ij}} \to \Omega_j \tag{2.47}$$

$$\text{or} \quad \left.\begin{array}{l} \mathbf{n}_{\partial\Omega_{ij}}^{\mathrm{T}} \cdot [\mathbf{x}^{(i)}(t_{m-}) - \mathbf{x}^{(i)}(t_{m-\varepsilon})] < 0 \\ \mathbf{n}_{\partial\Omega_{ij}}^{\mathrm{T}} \cdot [\mathbf{x}^{(i)}(t_{m+\varepsilon}) - \mathbf{x}^{(i)}(t_{m+})] > 0 \\ \mathbf{n}_{\partial\Omega_{ij}}^{\mathrm{T}} \cdot [\mathbf{x}^{(j)}(t_{m+\varepsilon}) - \mathbf{x}^{(j)}(t_{m+})] < 0 \end{array}\right\} \text{ for } \mathbf{n}_{\Omega_{ij}} \to \Omega_i. \tag{2.48}$$

From the two definitions, the switching bifurcations of a flow from the semi-passable boundary to the nonpassable boundaries of the first and second kinds are

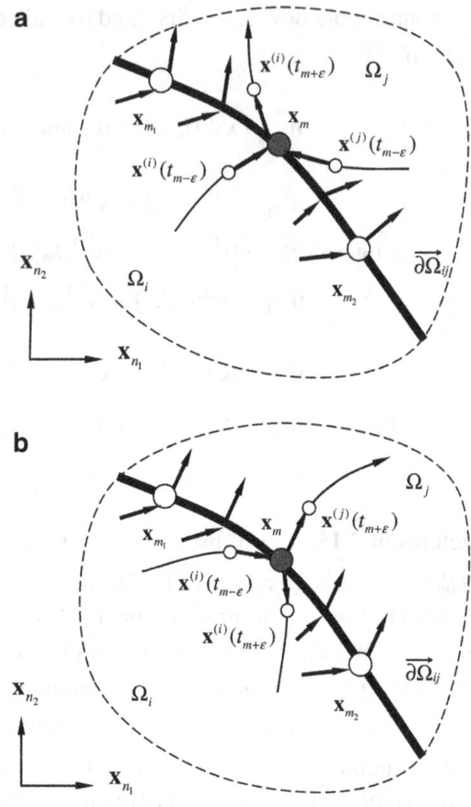

Fig. 2.7 (**a**) The sliding bifurcation and (**b**) the source bifurcation on the semipassable boundary $\overrightarrow{\partial\Omega}_{ij}$. Four points $\mathbf{x}^{(\alpha)}(t_{m\pm\varepsilon})$ $(\alpha \in \{i,j\})$ and $\mathbf{x}_m$ lie in the corresponding domains Ω_α and on the boundary $\partial\Omega_{ij}$, respectively

presented in Fig. 2.7. The source (or *sink*) bifurcation of a flow to the boundary requires the tangential bifurcation of the coming (or *leaving*) flow to the boundary. Similarly, the switching bifurcation of a passable flow from $\overrightarrow{\partial\Omega}_{ij}$ to $\overleftarrow{\partial\Omega}_{ij}$ is defined as follows.

Definition 2.16. For a discontinuous dynamical system in (2.1), there is a point $\mathbf{x}(t_m) = \mathbf{x}_m \in [\mathbf{x}_{m_1}, \mathbf{x}_{m_2}] \subset \overrightarrow{\partial\Omega}_{ij}$ for time t_m between two adjacent domains Ω_α $(\alpha = i,j)$. For an arbitrarily small $\varepsilon > 0$, there are two time intervals (i.e., $[t_{m-\varepsilon}, t_m)$ and $(t_m, t_{m+\varepsilon}]$) and $\mathbf{x}^{(i)}(t_{m\mp}) = \mathbf{x}_m = \mathbf{x}^{(j)}(t_{m\pm})$. The flows $\mathbf{x}^{(\alpha)}(t)$ $(\alpha = i,j)$ are $C^{r_\alpha}_{[t_{m-\varepsilon}, t_{m+\varepsilon}]}$-continuous $(r_\alpha \geq 1)$ for time t. The tangential bifurcation of the flow $\mathbf{x}^{(i)}(t)$ and $\mathbf{x}^{(j)}(t)$ at point $(\mathbf{x}_m, t_m)$ on the boundary $\overrightarrow{\partial\Omega}_{ij}$ is termed *the switching bifurcation* of a flow from $\overrightarrow{\partial\Omega}_{ij}$ to $\overleftarrow{\partial\Omega}_{ij}$ if

$$\mathbf{n}^{\mathbf{T}}_{\partial\Omega_{ij}} \cdot \dot{\mathbf{x}}^{(\alpha)}(t_{m\pm}) = 0 \quad \text{for} \quad \alpha = i,j \tag{2.49}$$

and

$$\text{either} \left.\begin{array}{l} \mathbf{n}_{\partial\Omega_{ij}}^{\mathrm{T}} \cdot [\mathbf{x}^{(i)}(t_{m-}) - \mathbf{x}^{(i)}(t_{m-\varepsilon})] > 0 \text{ and} \\ \mathbf{n}_{\partial\Omega_{ij}}^{\mathrm{T}} \cdot [\mathbf{x}^{(i)}(t_{m+\varepsilon}) - \mathbf{x}^{(i)}(t_{m+})] < 0 \\ \mathbf{n}_{\partial\Omega_{ij}}^{\mathrm{T}} \cdot [\mathbf{x}^{(j)}(t_{m-}) - \mathbf{x}^{(j)}(t_{m-\varepsilon})] < 0 \text{ and} \\ \mathbf{n}_{\partial\Omega_{ij}}^{\mathrm{T}} \cdot [\mathbf{x}^{(j)}(t_{m+\varepsilon}) - \mathbf{x}^{(j)}(t_{m+})] > 0 \end{array}\right\} \text{for } \mathbf{n}_{\Omega_{ij}} \to \Omega_j \tag{2.50}$$

$$\text{or} \left.\begin{array}{l} \mathbf{n}_{\partial\Omega_{ij}}^{\mathrm{T}} \cdot [\mathbf{x}^{(i)}(t_{m-}) - \mathbf{x}^{(i)}(t_{m-\varepsilon})] < 0 \text{ and} \\ \mathbf{n}_{\partial\Omega_{ij}}^{\mathrm{T}} \cdot [\mathbf{x}^{(i)}(t_{m+\varepsilon}) - \mathbf{x}^{(i)}(t_{m+})] > 0 \\ \mathbf{n}_{\partial\Omega_{ij}}^{\mathrm{T}} \cdot [\mathbf{x}^{(j)}(t_{m-}) - \mathbf{x}^{(j)}(t_{m-\varepsilon})] > 0 \text{ and} \\ \mathbf{n}_{\partial\Omega_{ij}}^{\mathrm{T}} \cdot [\mathbf{x}^{(j)}(t_{m+\varepsilon}) - \mathbf{x}^{(j)}(t_{m+})] < 0 \end{array}\right\} \text{for } \mathbf{n}_{\Omega_{ij}} \to \Omega_i. \tag{2.51}$$

The above definitions give the three possible switching bifurcations of the semipassable flow to the boundary $\overrightarrow{\partial\Omega}_{ij}$. The corresponding theorems can be stated for necessary and sufficient conditions. The proofs can be completed as in Theorems 2.8–2.10.

Theorem 2.14. *For a discontinuous dynamical system in (2.1), there is a point* $\mathbf{x}(t_m) = \mathbf{x}_m \in [\mathbf{x}_{m_1}, \mathbf{x}_{m_2}] \subset \overrightarrow{\partial\Omega}_{ij}$ *for time* t_m *between two adjacent domains* Ω_α $(\alpha = i,j)$. *For an arbitrarily small* $\varepsilon > 0$, *there are two time intervals (i.e.,* $[t_{m-\varepsilon}, t_m)$ *and* $(t_m, t_{m+\varepsilon}]$*), and* $\mathbf{x}^{(i)}(t_{m-}) = \mathbf{x}_m = \mathbf{x}^{(j)}(t_{m\pm})$. *The flows* $\mathbf{x}^{(i)}(t)$ *and* $\mathbf{x}^{(j)}(t)$ *are* $C^{r_i}_{[t_{m-\varepsilon},t_m)}$*- and* $C^{r_j}_{[t_{m-\varepsilon},t_{m+\varepsilon}]}$*-continuous* $(r_\alpha \geq 1,\ \alpha = i,j)$ *for time t, respectively. The sliding bifurcation of the flow* $\mathbf{x}^{(i)}(t) \cup \mathbf{x}^{(j)}(t)$ *at point* $(\mathbf{x}_m, t_m)$ *from* $\overrightarrow{\partial\Omega}_{ij}$ *to* $\widetilde{\partial\Omega}_{ij}$ *occurs if and only if*

$$\mathbf{n}_{\partial\Omega_{ij}}^{\mathrm{T}} \cdot \mathbf{F}^{(j)}(t_{m\pm}) = 0 \quad \text{and} \quad \mathbf{n}_{\partial\Omega_{ij}}^{\mathrm{T}} \cdot \mathbf{F}^{(i)}(t_{m-}) \neq 0, \tag{2.52}$$

$$\left.\begin{array}{ll} \text{either } \mathbf{n}_{\partial\Omega_{ij}}^{\mathrm{T}} \cdot \mathbf{F}^{(i)}(t_{m-}) > 0 & \text{for } \mathbf{n}_{\partial\Omega_{ij}} \to \Omega_j \\ \text{or} \quad \mathbf{n}_{\partial\Omega_{ij}}^{\mathrm{T}} \cdot \mathbf{F}^{(i)}(t_{m-}) < 0 & \text{for } \mathbf{n}_{\partial\Omega_{ij}} \to \Omega_i, \end{array}\right\} \tag{2.53}$$

$$\left.\begin{array}{l} \text{either} \left.\begin{array}{l} \mathbf{n}_{\partial\Omega_{ij}}^{\mathrm{T}} \cdot \mathbf{F}^{(j)}(t_{m-\varepsilon}) < 0 \\ \mathbf{n}_{\partial\Omega_{ij}}^{\mathrm{T}} \cdot \mathbf{F}^{(j)}(t_{m+\varepsilon}) > 0 \end{array}\right\} \text{for } \mathbf{n}_{\partial\Omega_{ij}} \to \Omega_j \\ \text{or} \left.\begin{array}{l} \mathbf{n}_{\partial\Omega_{ij}}^{\mathrm{T}} \cdot \mathbf{F}^{(j)}(t_{m-\varepsilon}) > 0 \\ \mathbf{n}_{\partial\Omega_{ij}}^{\mathrm{T}} \cdot \mathbf{F}^{(j)}(t_{m+\varepsilon}) < 0 \end{array}\right\} \text{for } \mathbf{n}_{\partial\Omega_{ij}} \to \Omega_i. \end{array}\right\} \tag{2.54}$$

Proof. Following the proof procedures in Theorems 2.8 and 2.9, the above theorem can be easily proved. ■

Theorem 2.15. *For a discontinuous dynamical system in (2.1), there is a point* $\mathbf{x}(t_m) = \mathbf{x}_m \in [\mathbf{x}_{m_1}, \mathbf{x}_{m_2}] \subset \overrightarrow{\partial\Omega}_{ij}$ *at time* t_m *on the* $(n-1)$*-dimensional plane* *boundary $\partial\Omega_{ij}$ between two adjacent domains Ω_α ($\alpha = i,j$). For an arbitrarily small $\varepsilon > 0$, there are two time intervals (i.e., $[t_{m-\varepsilon}, t_m)$ and $(t_m, t_{m+\varepsilon}]$), and $\mathbf{x}^{(i)}(t_{m-}) = \mathbf{x}_m = \mathbf{x}^{(j)}(t_{m\pm})$. The flows $\mathbf{x}^{(i)}(t)$ and $\mathbf{x}^{(j)}(t)$ are $C^{r_i}_{[t_{m-\varepsilon}, t_m)}$- and $C^{r_j}_{[t_{m-\varepsilon}, t_{m+\varepsilon}]}$-continuous ($r_\alpha \geq 2$, $\alpha = i,j$) for time t, respectively. The sliding bifurcation of the flow $\mathbf{x}^{(i)}(t) \cup \mathbf{x}^{(j)}(t)$ at point $(\mathbf{x}_m, t_m)$ from $\overrightarrow{\partial\Omega}_{ij}$ to $\widetilde{\partial\Omega}_{ij}$ occurs if and only if*

$$\mathbf{n}^{\mathrm{T}}_{\partial\Omega_{ij}} \cdot \mathbf{F}^{(j)}(t_{m\pm}) = 0 \quad \text{and} \quad \mathbf{n}^{\mathrm{T}}_{\partial\Omega_{ij}} \cdot \mathbf{F}^{(i)}(t_{m-}) \neq 0, \tag{2.55}$$

$$\left.\begin{array}{ll}\text{either} & \left.\begin{array}{l}\mathbf{n}^{\mathrm{T}}_{\partial\Omega_{ij}} \cdot \mathbf{F}^{(i)}(t_{m-}) > 0\\ \mathbf{n}^{\mathrm{T}}_{\partial\Omega_{ij}} \cdot D\mathbf{F}^{(j)}(t_{m\pm}) > 0\end{array}\right\} \text{for } \mathbf{n}_{\partial\Omega_{ij}} \to \Omega_j\\ \text{or} & \left.\begin{array}{l}\mathbf{n}^{\mathrm{T}}_{\partial\Omega_{ij}} \cdot \mathbf{F}^{(i)}(t_{m-}) < 0\\ \mathbf{n}^{\mathrm{T}}_{\partial\Omega_{ij}} \cdot D\mathbf{F}^{(j)}(t_{m\pm}) < 0\end{array}\right\} \text{for } \mathbf{n}_{\partial\Omega_{ij}} \to \Omega_i.\end{array}\right\} \tag{2.56}$$

Proof. Following the proof procedures of Theorems 2.10 and 2.11, the above theorem can be easily proved. ■

Theorem 2.16. *For a discontinuous dynamical system in (2.1), there is a point $\mathbf{x}(t_m) = \mathbf{x}_m \in [\mathbf{x}_{m_1}, \mathbf{x}_{m_2}] \subset \overrightarrow{\partial\Omega}_{ij}$ for time t_m between two adjacent domains Ω_α ($\alpha = i,j$). For an arbitrarily small $\varepsilon > 0$, there are two time intervals (i.e., $[t_{m-\varepsilon}, t_m)$ and $(t_m, t_{m+\varepsilon}]$), and $\mathbf{x}^{(i)}(t_{m\pm}) = \mathbf{x}_m = \mathbf{x}^{(j)}(t_{m+})$. The flows $\mathbf{x}^{(i)}(t)$ and $\mathbf{x}^{(j)}(t)$ are $C^{r_i}_{[t_{m-\varepsilon}, t_{m+\varepsilon}]}$- and $C^{r_j}_{[t_{m+\varepsilon}, t_m)}$-continuous ($r_\alpha \geq 1$, $\alpha = i,j$) for time t, respectively. The source bifurcation of the flow $\mathbf{x}^{(i)}(t) \cup \mathbf{x}^{(j)}(t)$ at point $(\mathbf{x}_m, t_m)$ from $\overrightarrow{\partial\Omega}_{ij}$ to $\widehat{\partial\Omega}_{ij}$ occurs if and only if*

$$\mathbf{n}^{\mathrm{T}}_{\partial\Omega_{ij}} \cdot \mathbf{F}^{(i)}(t_{m\pm}) = 0 \quad \text{and} \quad \mathbf{n}^{\mathrm{T}}_{\partial\Omega_{ij}} \cdot \mathbf{F}^{(j)}(t_{m+}) \neq 0, \tag{2.57}$$

$$\left.\begin{array}{lll}\text{either} & \mathbf{n}^{\mathrm{T}}_{\partial\Omega_{ij}} \cdot \mathbf{F}^{(j)}(t_{m+}) > 0 & \text{for } \mathbf{n}_{\partial\Omega_{ij}} \to \Omega_j\\ \text{or} & \mathbf{n}^{\mathrm{T}}_{\partial\Omega_{ij}} \cdot \mathbf{F}^{(j)}(t_{m+}) < 0 & \text{for } \mathbf{n}_{\partial\Omega_{ij}} \to \Omega_i,\end{array}\right\} \tag{2.58}$$

$$\left.\begin{array}{ll}\text{either} & \left.\begin{array}{l}\mathbf{n}^{\mathrm{T}}_{\partial\Omega_{ij}} \cdot \mathbf{F}^{(i)}(t_{m-\varepsilon}) > 0\\ \mathbf{n}^{\mathrm{T}}_{\partial\Omega_{ij}} \cdot \mathbf{F}^{(i)}(t_{m+\varepsilon}) < 0\end{array}\right\} \text{for } \mathbf{n}_{\partial\Omega_{ij}} \to \Omega_j\\ \text{or} & \left.\begin{array}{l}\mathbf{n}^{\mathrm{T}}_{\partial\Omega_{ij}} \cdot \mathbf{F}^{(i)}(t_{m-\varepsilon}) < 0\\ \mathbf{n}^{\mathrm{T}}_{\partial\Omega_{ij}} \cdot \mathbf{F}^{(i)}(t_{m+\varepsilon}) > 0\end{array}\right\} \text{for } \mathbf{n}_{\partial\Omega_{ij}} \to \Omega_i.\end{array}\right\} \tag{2.59}$$

Proof. Following the proof procedures of Theorems 2.8 and 2.9, the above theorem can be easily proved. ■

Theorem 2.17. *For a discontinuous dynamical system in (2.1), there is a point $\mathbf{x}(t_m) = \mathbf{x}_m \in [\mathbf{x}_{m_1}, \mathbf{x}_{m_2}] \subset \overrightarrow{\partial\Omega}_{ij}$ at time t_m on the $(n-1)$-dimensional plane* *boundary* $\partial\Omega_{ij}$ *between two adjacent domains* Ω_α $(\alpha = i, j)$. *For an arbitrarily small* $\varepsilon > 0$, *there are two time intervals (i.e.,* $[t_{m-\varepsilon}, t_m)$ *and* $(t_m, t_{m+\varepsilon}]$), *and* $\mathbf{x}^{(i)}(t_{m\pm}) = \mathbf{x}_m = \mathbf{x}^{(j)}(t_{m+})$. *The flows* $\mathbf{x}^{(i)}(t)$ *and* $\mathbf{x}^{(j)}(t)$ *are* $C^{r_i}_{[t_{m-\varepsilon}, t_{m+\varepsilon}]}$- *and* $C^{r_j}_{[t_{m+\varepsilon}, t_m)}$-*continuous* $(r_\alpha \geq 2, \alpha = i, j)$ *for time t, respectively. The source bifurcation of the flow* $\mathbf{x}^{(i)}(t) \cup \mathbf{x}^{(j)}(t)$ *at point* $(\mathbf{x}_m, t_m)$ *from* $\overrightarrow{\partial\Omega}_{ij}$ *to* $\widehat{\partial\Omega}_{ij}$ *occurs if and only if*

$$\mathbf{n}^{\mathrm{T}}_{\partial\Omega_{ij}} \cdot \mathbf{F}^{(i)}(t_{m\pm}) = 0 \quad \text{and} \quad \mathbf{n}^{\mathrm{T}}_{\partial\Omega_{ij}} \cdot \mathbf{F}^{(j)}(t_{m+}) \neq 0, \tag{2.60}$$

$$\left.\begin{array}{l} \text{either} \left.\begin{array}{l} \mathbf{n}^{\mathrm{T}}_{\partial\Omega_{ij}} \cdot \mathbf{F}^{(j)}(t_{m+}) > 0 \\ \mathbf{n}^{T}_{\partial\Omega_{ij}} \cdot D\mathbf{F}^{(i)}(t_{m\pm}) < 0 \end{array}\right\} \text{for } \mathbf{n}_{\partial\Omega_{ij}} \to \Omega_j \\ \text{or} \left.\begin{array}{l} \mathbf{n}^{\mathrm{T}}_{\partial\Omega_{ij}} \cdot \mathbf{F}^{(j)}(t_{m+}) < 0 \\ \mathbf{n}^{T}_{\partial\Omega_{ij}} \cdot D\mathbf{F}^{(i)}(t_{m\pm}) > 0 \end{array}\right\} \text{for } \mathbf{n}_{\partial\Omega_{ij}} \to \Omega_i. \end{array}\right\} \tag{2.61}$$

Proof. Following the proof procedures of Theorems 2.10 and 2.11, the above theorem can be easily proved. ■

Theorem 2.18. *For a discontinuous dynamical system in (2.1), there is a point* $\mathbf{x}(t_m) = \mathbf{x}_m \in [\mathbf{x}_{m_1}, \mathbf{x}_{m_2}] \subset \overrightarrow{\partial\Omega}_{ij}$ *for time* t_m *between two adjacent domains* Ω_α $(\alpha = i, j)$. *For an arbitrarily small* $\varepsilon > 0$, *there are two time intervals (i.e.,* $[t_{m-\varepsilon}, t_m)$ *and* $(t_m, t_{m+\varepsilon}]$), *and* $\mathbf{x}^{(i)}(t_{m\mp}) = \mathbf{x}_m = \mathbf{x}^{(j)}(t_{m\pm})$. *The flows* $\mathbf{x}^{(\alpha)}(t)$ $(\alpha = i, j)$ *are* $C^{r_\alpha}_{[t_{m-\varepsilon}, t_{m+\varepsilon}]}$-*continuous* $(r_\alpha \geq 1)$ *for time t. The switching bifurcation of the flow* $\mathbf{x}^{(i)}(t) \cup \mathbf{x}^{(j)}(t)$ *at point* $(\mathbf{x}_m, t_m)$ *from* $\overrightarrow{\partial\Omega}_{ij}$ *to* $\overleftarrow{\partial\Omega}_{ij}$ *occurs if and only if*

$$\mathbf{n}^{\mathrm{T}}_{\partial\Omega_{ij}} \cdot \mathbf{F}^{(i)}(t_{m\mp}) = 0 \quad \text{and} \quad \mathbf{n}^{\mathrm{T}}_{\partial\Omega_{ij}} \cdot \mathbf{F}^{(j)}(t_{m\pm}) = 0, \tag{2.62}$$

$$\left.\begin{array}{l} \text{either} \left.\begin{array}{l} \mathbf{n}^{\mathrm{T}}_{\partial\Omega_{ij}} \cdot \mathbf{F}^{(\alpha)}(t_{m-\varepsilon}) > 0 \\ \mathbf{n}^{\mathrm{T}}_{\partial\Omega_{ij}} \cdot \mathbf{F}^{(\alpha)}(t_{m+\varepsilon}) < 0 \end{array}\right\} \text{for } \mathbf{n}_{\partial\Omega_{ij}} \to \Omega_\beta \\ \text{or} \left.\begin{array}{l} \mathbf{n}^{\mathrm{T}}_{\partial\Omega_{ij}} \cdot \mathbf{F}^{(\alpha)}(t_{m-\varepsilon}) < 0 \\ \mathbf{n}^{\mathrm{T}}_{\partial\Omega_{ij}} \cdot \mathbf{F}^{(\alpha)}(t_{m+\varepsilon}) > 0 \end{array}\right\} \text{for } \mathbf{n}_{\partial\Omega_{ij}} \to \Omega_\alpha, \end{array}\right\} \tag{2.63}$$

with $\alpha, \beta = i, j$ *but* $\beta \neq \alpha$.

Proof. Following the proof procedures of Theorems 2.8 and 2.9, the above theorem can be easily proved. ■

Theorem 2.19. *For a discontinuous dynamical system in (2.1), there is a point* $\mathbf{x}(t_m) = \mathbf{x}_m \in [\mathbf{x}_{m_1}, \mathbf{x}_{m_2}] \subset \overrightarrow{\partial\Omega}_{ij}$ *at time* t_m *on the* $(n-1)$-*dimensional plane boundary* $\partial\Omega_{ij}$ *between two adjacent domains* Ω_α $(\alpha = i, j)$. *For an arbitrarily small*

$\varepsilon > 0$, there are two time intervals (i.e., $[t_{m-\varepsilon}, t_m)$ and $(t_m, t_{m+\varepsilon}]$), and $\mathbf{x}^{(i)}(t_{m\mp}) = \mathbf{x}_m = \mathbf{x}^{(j)}(t_{m\pm})$. The flows $\mathbf{x}^{(\alpha)}(t)$ $(\alpha = i, j)$ are $C^{r_\alpha}_{[t_{m-\varepsilon}, t_{m+\varepsilon}]}$-continuous $(r_\alpha \geq 1)$ for time t. The switching bifurcation of the flow $\mathbf{x}^{(i)}(t) \cup \mathbf{x}^{(j)}(t)$ at point $(\mathbf{x}_m, t_m)$ from $\overrightarrow{\partial\Omega}_{ij}$ to $\overleftarrow{\partial\Omega}_{ij}$ occurs if and only if

$$\mathbf{n}^{\mathrm{T}}_{\partial\Omega_{ij}} \cdot \mathbf{F}^{(i)}(t_{m\mp}) = 0 \quad \text{and} \quad \mathbf{n}^{\mathrm{T}}_{\partial\Omega_{ij}} \cdot \mathbf{F}^{(j)}(t_{m\pm}) = 0, \tag{2.64}$$

$$\left.\begin{array}{ll} \text{either} & \left.\begin{array}{l} \mathbf{n}^{\mathrm{T}}_{\partial\Omega_{ij}} \cdot D\mathbf{F}^{(i)}(t_{m\mp}) < 0 \\ \mathbf{n}^{\mathrm{T}}_{\partial\Omega_{ij}} \cdot D\mathbf{F}^{(j)}(t_{m\pm}) > 0 \end{array}\right\} \text{ for } \mathbf{n}_{\partial\Omega_{ij}} \to \Omega_j \\ \text{or} & \left.\begin{array}{l} \mathbf{n}^{\mathrm{T}}_{\partial\Omega_{ij}} \cdot D\mathbf{F}^{(i)}(t_{m\mp}) > 0 \\ \mathbf{n}^{\mathrm{T}}_{\partial\Omega_{ij}} \cdot D\mathbf{F}^{(j)}(t_{m\pm}) < 0 \end{array}\right\} \text{ for } \mathbf{n}_{\partial\Omega_{ij}} \to \Omega_i. \end{array}\right\} \tag{2.65}$$

Proof. Following the proof procedures of Theorems 2.10 and 2.11, the above theorem can be easily proved. ■

Definition 2.17. For a discontinuous dynamical system in (2.1), there is a point $\mathbf{x}(t_m) = \mathbf{x}_m \in [\mathbf{x}_{m_1}, \mathbf{x}_{m_2}] \subset \partial\Omega_{ij}$ for time t_m and $\mathbf{x}^{(\alpha)}(t_{m\pm}) = \mathbf{x}_m$ $(\alpha \in \{i, j\})$. The $L_{\alpha\beta}$-functions of flows to the boundary $\partial\Omega_{ij}$ is defined as

$$L_{\alpha\beta}(\mathbf{x}_m, t_m, \mathbf{p}_\alpha, \mathbf{p}_\beta) = [\mathbf{n}^{\mathrm{T}}_{\partial\Omega_{\alpha\beta}} \cdot \mathbf{F}^{(\alpha)}(t_{m\mp})] \times [\mathbf{n}^{\mathrm{T}}_{\partial\Omega_{\alpha\beta}} \cdot \mathbf{F}^{(\beta)}(t_{m\pm})], \tag{2.66}$$

where $\beta \in \{i, j\}$ but $\beta \neq \alpha$.

From the foregoing definition, the passable flows and nonpassable flows (including sink and source flows) at the boundary $\partial\Omega_{\alpha\beta}$, respectively, require the L-function satisfies

$$\left.\begin{array}{l} L_{\alpha\beta}(\mathbf{x}_m, t_m, \mathbf{p}_\alpha, \mathbf{p}_\beta) > 0 \text{ on } \overrightarrow{\partial\Omega}_{\alpha\beta}, \\ L_{\alpha\beta}(\mathbf{x}_m, t_m, \mathbf{p}_\alpha, \mathbf{p}_\beta) < 0 \text{ on } \overline{\partial\Omega}_{\alpha\beta} = \widetilde{\partial\Omega}_{\alpha\beta} \cup \widehat{\partial\Omega}_{\alpha\beta}. \end{array}\right\} \tag{2.67}$$

The switching bifurcation of a flow at point $(\mathbf{x}_m, t_m)$ on the boundary $\partial\Omega_{\alpha\beta}$ requires

$$L_{\alpha\beta}(\mathbf{x}_m, t_m, \mathbf{p}_\alpha, \mathbf{p}_\beta) = 0. \tag{2.68}$$

If the $L_{\alpha\beta}$-function of a flow is defined on one side of the neighborhood of the boundary $\partial\Omega_{\alpha\beta}$, one can obtain

$$L_{\alpha\alpha}(\mathbf{x}_{m\pm\varepsilon}, t_{m\pm\varepsilon}, \mathbf{p}_\alpha) = [\mathbf{n}^{\mathrm{T}}_{\partial\Omega_{\alpha\beta}} \cdot \mathbf{F}^{(\alpha)}(t_{m-\varepsilon})] \times [\mathbf{n}^{\mathrm{T}}_{\partial\Omega_{\alpha\beta}} \cdot \mathbf{F}^{(\alpha)}(t_{m+\varepsilon})]. \tag{2.69}$$

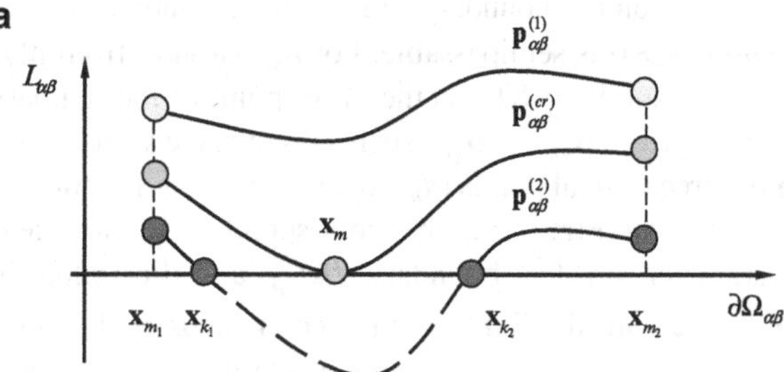

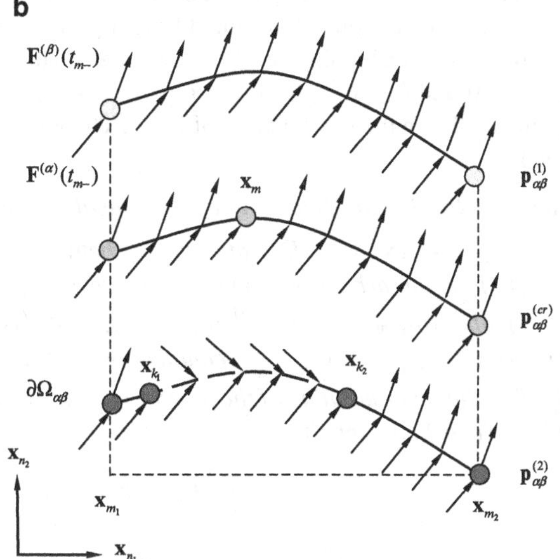

Fig. 2.8 (**a**) The $L_{\alpha\beta}$-functions of flows and (**b**) the vector fields between two points $\mathbf{x}_{m_1}$ and $\mathbf{x}_{m_2}$ on the boundary $\partial\Omega_{\alpha\beta}$. The point $\mathbf{x}_m$ for $\mathbf{p}_{\alpha\beta}^{(\mathrm{cr})}$ is the critical point for the switching bifurcation. Two points $\mathbf{x}_{k_1}$ and $\mathbf{x}_{k_2}$ are the onset and vanishing of the nonpassable flow for parameter on the boundary $\overrightarrow{\partial\Omega}_{\alpha\beta}$. The *dashed* and *solid curves* represent $L_{\alpha\beta}<0$ and $L_{\alpha\beta} \geq 0$, respectively

If $L_{\alpha\alpha}(\mathbf{x}_{m\pm\varepsilon}, t_{m\pm\varepsilon}, \mathbf{p}_\alpha) < 0$ and $\mathbf{n}^{\mathrm{T}}_{\partial\Omega_{\alpha\beta}} \cdot \mathbf{F}^{(\alpha)}(t_{m-}) = 0$, the flow $\mathbf{x}^{(\alpha)}(t)$ at $(\mathbf{x}_m, t_m)$ is tangential to the boundary $\partial\Omega_{\alpha\beta}$.

Consider the $L_{\alpha\beta}$-function varying with the parameter vector $\mathbf{p}_{ij} \in \{\mu_\alpha\}_{\alpha\in\{i,j\}}$ for the switching flow from $\overrightarrow{\partial\Omega}_{\alpha\beta}$ to $\widetilde{\partial\Omega}_{\alpha\beta}$. The $L_{\alpha\beta}$-functions of flows at different locations of the boundary are distinct. The $L_{\alpha\beta}$-functions of flows between two points $\mathbf{x}_{m_1}$ and $\mathbf{x}_{m_2}$ on the boundary $\partial\Omega_{\alpha\beta}$ are sketched in Fig. 2.8 for a parameter vector $\mathbf{p}_{\alpha\beta}$ between $\mathbf{p}_{\alpha\beta}^{(1)}$ and $\mathbf{p}_{\alpha\beta}^{(2)}$. For a specific value $\mathbf{p}_{\alpha\beta}^{(\mathrm{cr})}$ between $\mathbf{p}_{\alpha\beta}^{(1)}$ and $\mathbf{p}_{\alpha\beta}^{(2)}$, there is a point $\mathbf{x}_m$ on the boundary for the bifurcation of a flow switching from $\overrightarrow{\partial\Omega}_{ij}$ to $\widetilde{\partial\Omega}_{ij}$. Two points $\mathbf{x}_{k_1}$ and $\mathbf{x}_{k_2}$ are the onset and vanishing points of the sink flow for system parameter vector $\mathbf{p}_{\alpha\beta}$ on the boundary $\partial\Omega_{\alpha\beta}$. The dashed and solid curves represent $L_{\alpha\beta} < 0$ and $L_{\alpha\beta} \geq 0$, respectively. For parameter vector $\mathbf{p}_{\alpha\beta}$ varying from $\mathbf{p}_{\alpha\beta}^{(1)} \to \mathbf{p}_{\alpha\beta}^{(cr)}$, the $L_{\alpha\beta}$-function for a flow $\mathbf{x} \in (\mathbf{x}_{m_1}, \mathbf{x}_{m_2})$ on the boundary is positive (i.e., $L_{\alpha\beta} > 0$). Thus, the boundary $\partial\Omega_{\alpha\beta}$ is semipassable. For $\mathbf{p}_{\alpha\beta}$ varying from $\mathbf{p}_{\alpha\beta}^{(\mathrm{cr})} \to \mathbf{p}_{\alpha\beta}^{(2)}$, there are two ranges of $L_{\alpha\beta} > 0$ for $\mathbf{x} \in [\mathbf{x}_{m_1}, \mathbf{x}_{k_1}) \cup (\mathbf{x}_{k_2}, \mathbf{x}_{m_2}]$ and a range of $L_{\alpha\beta} < 0$ for $\mathbf{x} \in (\mathbf{x}_{k_1}, \mathbf{x}_{k_2})$. From (2.67), the flow at the portion of $\mathbf{x} \in$

$(\mathbf{x}_{k_1}, \mathbf{x}_{k_2})$ on the boundary $\partial\Omega_{\alpha\beta}$ is nonpassable. The flow at the portion of boundary with $L_{\alpha\beta} > 0$ is semipassable. For $\mathbf{p}_{\alpha\beta}$ varying from $\mathbf{p}_{\alpha\beta}^{(1)} \to \mathbf{p}_{\alpha\beta}^{(2)}$, the point $(\mathbf{x}_m, \mathbf{p}_{\alpha\beta}^{(cr)})$ on the boundary $\partial\Omega_{\alpha\beta}$ is the onset point of the nonpassable flow. However, for $\mathbf{p}_{\alpha\beta}$ varying from $\mathbf{p}_{\alpha\beta}^{(2)} \to \mathbf{p}_{\alpha\beta}^{(1)}$, such a point is the vanishing point of the nonpassable flow. For three critical points $(\mathbf{x}_m, \mathbf{x}_{k_1}, \mathbf{x}_{k_2})$, the $L_{\alpha\beta}$-function of flows is zero (i.e., $L_{\alpha\beta} = 0$). For $L_{\alpha\beta}$ in Fig. 2.8a, the corresponding vector fields varying with the system parameter on the boundary $\partial\Omega_{\alpha\beta}$ are illustrated in Fig. 2.8b. $\mathbf{F}^{(\alpha)}(t_{m-})$ and $\mathbf{F}^{(\beta)}(t_{m\pm})$ are the limits of the vector fields in domains Ω_α and Ω_β to the boundary $\partial\Omega_{\alpha\beta}$, respectively. This nonpassable flow on the boundary $\partial\Omega_{\alpha\beta}$ with $L_{\alpha\beta} < 0$ is a sink flow. The critical points $(\mathbf{x}_{k_1}, \mathbf{x}_{k_2})$ have the same properties as point $\mathbf{x}_m$ for $\mathbf{p}_{\alpha\beta}^{(cr)}$. Namely, $L_{\alpha\beta}(\mathbf{x}_m) = 0$, $L_{\alpha\alpha}(\mathbf{x}_{m\pm\varepsilon}) < 0$, or $L_{\beta\beta}(\mathbf{x}_{m\pm\varepsilon}) > 0$.

If the two critical points have the different properties, the sliding flow between two different critical points is discussed later. The $L_{\alpha\beta}$-functions of flows are $L_{\alpha\beta}(\mathbf{x}_{k_1}) = 0$ and $L_{\alpha\alpha}(\mathbf{x}_{k_1\pm\varepsilon}) < 0$ for point $\mathbf{x}_{k_1}$ but $L_{\alpha\beta}(\mathbf{x}_{k_2}) = 0$ and $L_{\beta\beta}(\mathbf{x}_{k_2\pm\varepsilon}) < 0$ for point $\mathbf{x}_{k_2}$. From the $L_{\alpha\beta}$-function of flows, Theorems 2.14, 2.16, and 2.18 can be restated.

Theorem 2.20. *For a discontinuous dynamical system in (2.1), there is a point $\mathbf{x}(t_m) = \mathbf{x}_m \in [\mathbf{x}_{m_1}, \mathbf{x}_{m_2}] \subset \overrightarrow{\partial\Omega}_{ij}$ at time t_m between two adjacent domains Ω_α ($\alpha = i,j$). For an arbitrarily small $\varepsilon > 0$, there are two time intervals (i.e., $[t_{m-\varepsilon}, t_m)$ and $(t_m, t_{m+\varepsilon}]$), and $\mathbf{x}^{(i)}(t_{m-}) = \mathbf{x}_m = \mathbf{x}^{(j)}(t_{m\pm})$. The flows $\mathbf{x}^{(i)}(t)$ and $\mathbf{x}^{(j)}(t)$ are $C^{r_i}_{[t_{m+\varepsilon}, t_m)}$- and $C^{r_j}_{[t_{m-\varepsilon}, t_{m+\varepsilon}]}$-continuous ($r_\alpha \geq 2$, $\alpha = i,j$) for time t, respectively. The sliding bifurcation of the flow $\mathbf{x}^{(i)}(t) \cup \mathbf{x}^{(j)}(t)$ at point $(\mathbf{x}_m, t_m)$ from $\overrightarrow{\partial\Omega}_{ij}$ to $\widetilde{\partial\Omega}_{ij}$ occurs if and only if*

$$L_{ij}(\mathbf{x}_m, t_m, \mathbf{p}_i, \mathbf{p}_j) = 0, \tag{2.70}$$

$$\mathbf{n}_{\partial\Omega_{ij}}^{\mathrm{T}} \cdot \mathbf{F}^{(i)}(t_{m-}) \neq 0 \quad \text{and} \quad \alpha = i,j. \tag{2.71}$$

Proof. Applying the L_{ij}-functions of flows in Definition 2.17 to Theorem 2.14, the foregoing theorem can be easily proved. ■

Theorem 2.21. *For a discontinuous dynamical system in (2.1), there is a point $\mathbf{x}(t_m) = \mathbf{x}_m \in [\mathbf{x}_{m_1}, \mathbf{x}_{m_2}] \subset \overrightarrow{\partial\Omega}_{ij}$ at time t_m between two adjacent domains Ω_α ($\alpha = i,j$). For an arbitrarily small $\varepsilon > 0$, there are two time intervals (i.e., $[t_{m-\varepsilon}, t_m)$ and $(t_m, t_{m+\varepsilon}]$), and $\mathbf{x}^{(i)}(t_{m\pm}) = \mathbf{x}_m = \mathbf{x}^{(j)}(t_{m+})$. The flows $\mathbf{x}^{(i)}(t)$ and $\mathbf{x}^{(j)}(t)$ are $C^{r_i}_{[t_{m-\varepsilon}, t_{m+\varepsilon}]}$- and $C^{r_j}_{[t_{m+\varepsilon}, t_m)}$-continuous ($r_\alpha \geq 2$, $\alpha = i,j$) for time t, respectively. The source bifurcation of the flow $\mathbf{x}^{(i)}(t) \cup \mathbf{x}^{(j)}(t)$ at point $(\mathbf{x}_m, t_m)$ from $\overrightarrow{\partial\Omega}_{ij}$ to $\widetilde{\partial\Omega}_{ij}$ occurs if and only if*

$$L_{ij}(\mathbf{x}_m, t_m, \mathbf{p}_i, \mathbf{p}_j) = 0, \tag{2.72}$$

$$\mathbf{n}_{\partial\Omega_{ij}}^{\mathrm{T}} \cdot \mathbf{F}^{(j)}(t_{m+}) \neq 0 \quad \text{and} \quad L_{ii}(\mathbf{x}_{m\pm\varepsilon}, t_{m\pm\varepsilon}, \mathbf{p}_i) < 0. \tag{2.73}$$

Proof. Applying the L-function of flows in Definition 2.17 to Theorem 2.16, this theorem can be easily proved. ■

Theorem 2.22. *For a discontinuous dynamical system in (2.1), there is a point* $\mathbf{x}(t_m) = \mathbf{x}_m \in [\mathbf{x}_{m_1}, \mathbf{x}_{m_2}] \subset \overrightarrow{\partial\Omega}_{ij}$ *at time* t_m *between two adjacent domains* Ω_α $(\alpha = i, j)$*. For an arbitrarily small* $\varepsilon > 0$*, there are two time intervals (i.e.,* $[t_{m-\varepsilon}, t_m)$ *and* $(t_m, t_{m+\varepsilon}]$*), and* $\mathbf{x}^{(i)}(t_{m\mp}) = \mathbf{x}_m = \mathbf{x}^{(j)}(t_{m\pm})$*. The flows* $\mathbf{x}^{(\alpha)}(t)$ $(\alpha = i, j)$ *are* $C^{r_\alpha}_{[t_{m-\varepsilon}, t_{m+\varepsilon}]}$*-continuous* $(r_\alpha \geq 2)$ *for time* t*. The switching bifurcation of the flow* $\mathbf{x}^{(i)}(t) \cup \mathbf{x}^{(j)}(t)$ *at point* $(\mathbf{x}_m, t_m)$ *from* $\overrightarrow{\partial\Omega}_{ij}$ *to* $\overleftarrow{\partial\Omega}_{ij}$ *occurs if and only if*

$$L_{ij}(\mathbf{x}_m, t_m, \mathbf{p}_i, \mathbf{p}_j) = 0, \tag{2.74}$$

$$\mathbf{n}^{\mathrm{T}}_{\partial\Omega_{ij}} \cdot \mathbf{F}^{(\alpha)}(t_{m\pm}) = 0 \quad \text{and} \quad L_{\alpha\alpha}(\mathbf{x}_{m\pm\varepsilon}, t_{m\pm\varepsilon}, \mathbf{p}_\alpha) < 0 \quad (\alpha = i, j). \tag{2.75}$$

Proof. Applying the L-function of flows in Definition 2.17 to Theorem 2.18, the theorem can be easily proved. ■

Remark. For the $(n-1)$-dimensional *plane* boundary $\partial\Omega_{ij}$, the second conditions in (2.71), (2.73), and (2.75) in Theorems 2.20–2.22 can be replaced by (2.56), (2.61), and (2.65) in Theorems 2.15, 2.17, and 2.19, respectively.

For the passable flow at $\mathbf{x}(t_m) \equiv \mathbf{x}_m \in [\mathbf{x}_{m_1}, \mathbf{x}_{m_2}] \subset \overrightarrow{\partial\Omega}_{ij}$, consider the time interval $[t_{m_1}, t_{m_2}]$ for $[\mathbf{x}_{m_1}, \mathbf{x}_{m_2}]$ on the boundary, and the L-functions of flows (i.e., L_{ij}) for $t_m \in [t_{m_1}, t_{m_2}]$ and $\mathbf{x}_m \in [\mathbf{x}_{m_1}, \mathbf{x}_{m_2}]$ is also positive, i.e., $L_{ij}(\mathbf{x}_m, t_m, \mathbf{p}_i, \mathbf{p}_j) > 0$. To determine the switching bifurcation, the local minimum of $L_{ij}(\mathbf{x}_m, t_m, \mathbf{p}_i, \mathbf{p}_j)$ is introduced. Because $\mathbf{x}_m$ is a vector function of time t_m, the two total derivatives of $L_{ij}(\mathbf{x}_m, t_m, \mathbf{p}_i, \mathbf{p}_j)$ are introduced, i.e.,

$$\begin{aligned} DL_{ij}(\mathbf{x}_m, t_m, \mathbf{p}_i, \mathbf{p}_j) = {} & \nabla L_{ij}(\mathbf{x}_m, t_m, \mathbf{p}_i, \mathbf{p}_j).\mathbf{F}^{(0)}_{ij}(\mathbf{x}_m, t_m) \\ & + \frac{\partial L_{ij}(\mathbf{x}_m, t_m, \mathbf{p}_i, \mathbf{p}_j)}{\partial t_m}, \end{aligned} \tag{2.76}$$

$$D^k L_{ij}(\mathbf{x}_m, t_m, \mathbf{p}_i, \mathbf{p}_j) = D^{k-1}(DL_{ij}(\mathbf{x}_m, t_m, \mathbf{p}_i, \mathbf{p}_j)) \quad \text{for} \quad k = 1, 2, \ldots. \tag{2.77}$$

Thus, the local minimum of $L_{ij}(\mathbf{x}_m, t_m, \mathbf{p}_i, \mathbf{p}_j)$ is determined by

$$D^k L_{ij}(\mathbf{x}_m, t_m, \mathbf{p}_i, \mathbf{p}_j) = 0 \quad (k = 1, 2, \ldots, 2l-1), \tag{2.78}$$

$$D^{2l} L_{ij}(\mathbf{x}_m, t_m, \mathbf{p}_i, \mathbf{p}_j) > 0. \tag{2.79}$$

Definition 2.18. For a discontinuous dynamical system in (2.1), there is a point $\mathbf{x}(t_m) = \mathbf{x}_m \in [\mathbf{x}_{m_1}, \mathbf{x}_{m_2}] \subset \overrightarrow{\partial\Omega}_{ij}$ at time t_m between two adjacent domains Ω_α $(\alpha = i, j)$. For an arbitrarily small $\varepsilon > 0$, there are two time intervals (i.e., $[t_{m-\varepsilon}, t_m)$ and

$(t_m, t_{m+\varepsilon}]$). $\mathbf{x}^{(i)}(t_{m\mp}) = \mathbf{x}_m = \mathbf{x}^{(j)}(t_{m\pm})$. The flows $\mathbf{x}^{(\alpha)}(t)$ are $C^{r_\alpha}_{[t_{m-\varepsilon}, t_{m+\varepsilon}]}$-continuous ($r_\alpha \geq 2l$, $\alpha = i, j$) for time t. *The local minimum set* of $L_{ij}(\mathbf{x}_m, t_m, \mathbf{p}_i, \mathbf{p}_j)$ is defined by

$$
{}_{\min}L_{ij}(t_m) = \left\{ L_{ij}(\mathbf{x}_m, t_m, \mathbf{p}_i, \mathbf{p}_j) \left| \begin{array}{l} \forall t_m \in [t_{m_1}, t_{m_2}], \exists \mathbf{x}_m \in [\mathbf{x}_{m_1}, \mathbf{x}_{m_2}], \\ \text{so that } D^k L_{ij}(\mathbf{x}_m, t_m, \mathbf{p}_i, \mathbf{p}_j) = 0 \\ \text{for } k = 1, 2, \cdots 2l - 1 \text{ and} \\ D^{2l} L_{ij}(\mathbf{x}_m, t_m, \mathbf{p}_i, \mathbf{p}_j) > 0. \end{array} \right. \right\} \tag{2.80}
$$

From the local minimum set of $L_{ij}(\mathbf{x}_m, t_m, \mathbf{p}_i, \mathbf{p}_j)$, the corresponding global minimum can be determined as follows.

Definition 2.19. For a discontinuous dynamical system in (2.1), there is a point $\mathbf{x}(t_m) = \mathbf{x}_m \in [\mathbf{x}_{m_1}, \mathbf{x}_{m_2}] \subset \overrightarrow{\partial\Omega}_{ij}$ at time t_m between two adjacent domains Ω_α ($\alpha = i, j$). For an arbitrarily small $\varepsilon > 0$, there are two time intervals (i.e., $[t_{m-\varepsilon}, t_m)$ and $(t_m, t_{m+\varepsilon}]$), and $\mathbf{x}^{(i)}(t_{m\mp}) = \mathbf{x}_m = \mathbf{x}^{(j)}(t_{m\pm})$. The flows $\mathbf{x}^{(\alpha)}(t)$ ($\alpha = i, j$) are $C^{r_\alpha}_{[t_{m-\varepsilon}, t_{m+\varepsilon}]}$-continuous ($r_\alpha \geq 2l$, $\alpha = i, j$) for time t, respectively. The global minimum set of $L_{ij}(\mathbf{x}_m, t_m, \mu_i, \mu_j)$ is defined by

$$
{}_{G\min}L_{ij}(t_m) = \min_{t_m \in [t_{m_1}, t_{m_2}]} \left\{ \begin{array}{l} {}_{\min}L_{ij}(t_m), L_{ij}(\mathbf{x}_{m_1}, t_{m_1}, \mathbf{p}_i, \mathbf{p}_j), \\ L_{ij}(\mathbf{x}_{m_2}, t_{m_2}, \mathbf{p}_i, \mathbf{p}_j). \end{array} \right\} \tag{2.81}
$$

From the foregoing definition, Theorems 2.20–2.22 can be expressed through the global minimum of $L_{ij}(\mathbf{x}_m, t_m, \mathbf{p}_i, \mathbf{p}_j)$. So the following corollaries can be achieved, which give the conditions for onsets of switching bifurcations.

Corollary 2.1. *For a discontinuous dynamical system in (2.1), there is a point* $\mathbf{x}(t_m) = \mathbf{x}_m \in [\mathbf{x}_{m_1}, \mathbf{x}_{m_2}] \subset \overrightarrow{\partial\Omega}_{ij}$ *for time* t_m *between two adjacent domains* Ω_α ($\alpha = i, j$). *For an arbitrarily small* $\varepsilon > 0$, *there are two time intervals (i.e.,* $[t_{m-\varepsilon}, t_m)$ *and* $(t_m, t_{m+\varepsilon}]$*), and* $\mathbf{x}^{(i)}(t_{m-}) = \mathbf{x}_m = \mathbf{x}^{(j)}(t_{m\pm})$. *The flows* $\mathbf{x}^{(i)}(t)$ *and* $\mathbf{x}^{(j)}(t)$ *are* $C^{r}_{[t_{m+\varepsilon}, t_m)}$- *and* $C^{r}_{[t_{m-\varepsilon}, t_{m+\varepsilon}]}$-*continuous* ($r_\alpha \geq 2l$, $\alpha = i, j$) *for time t, respectively. The necessary and sufficient conditions for the sliding bifurcation onset of the flow* $\mathbf{x}^{(i)}(t) \cup \mathbf{x}^{(j)}(t)$ *at point* $(\mathbf{x}_m, t_m)$ *on the boundary* $\overrightarrow{\partial\Omega}_{ij}$ *are*

$$
{}_{G\min}L_{ij}(t_m) = 0, \tag{2.82}
$$

$$
\mathbf{n}^{\mathrm{T}}_{\partial\Omega_{ij}} \cdot \mathbf{F}^{(i)}(t_{m-}) \neq 0 \quad \text{and} \quad L_{jj}(\mathbf{x}_{m\pm\varepsilon}, t_{m\pm\varepsilon}, \mathbf{p}_j) < 0. \tag{2.83}
$$

Proof. $L_{ij}(\mathbf{x}_m, t_m, \mathbf{p}_i, \mathbf{p}_j)$ replaced by its global minimum in Theorem 2.20 gives this corollary. This corollary is proved. ■

Corollary 2.2. *For a discontinuous dynamical system in (2.1), there is a point* $\mathbf{x}(t_m) = \mathbf{x}_m \in [\mathbf{x}_{m_1}, \mathbf{x}_{m_2}] \subset \overrightarrow{\partial\Omega}_{ij}$ *for time* t_m *between two adjacent domains* Ω_α

$(\alpha = i,j)$. For an arbitrarily small $\varepsilon > 0$, there are two time intervals (i.e., $[t_{m-\varepsilon}, t_m)$ and $(t_m, t_{m+\varepsilon}]$), and $\mathbf{x}^{(i)}(t_{m\pm}) = \mathbf{x}_m = \mathbf{x}^{(j)}(t_{m+})$. The flows $\mathbf{x}^{(i)}(t)$ and $\mathbf{x}^{(j)}(t)$ are $C^{r_i}_{[t_{m-\varepsilon}, t_{m+\varepsilon}]}$- and $C^{r_j}_{[t_{m+\varepsilon}, t_m)}$-continuous ($r_\alpha \geq 2l$, $\alpha = i,j$) for time t, respectively. The necessary and sufficient conditions for the source bifurcation onset of the flow $\mathbf{x}^{(i)}(t) \cup \mathbf{x}^{(j)}(t)$ at point $(\mathbf{x}_m, t_m)$ on the boundary $\overrightarrow{\partial\Omega}_{ij}$ are

$$_{G\min}L_{ij}(t_m) = 0, \tag{2.84}$$

$$\mathbf{n}^{\mathrm{T}}_{\partial\Omega_{ij}} \cdot \mathbf{F}^{(j)}(t_{m+}) \neq 0 \quad \text{and} \quad L_{ii}(\mathbf{x}_{m\pm\varepsilon}, t_{m\pm\varepsilon}, \mathbf{p}_i) < 0. \tag{2.85}$$

Proof. $L_{ij}(\mathbf{x}_m, t_m, \mathbf{p}_i, \mathbf{p}_j)$ replaced by its global minimum in Theorem 2.21 gives this corollary. ■

Corollary 2.3. *For a discontinuous dynamical system in (2.1), there is a point $\mathbf{x}(t_m) = \mathbf{x}_m \in [\mathbf{x}_{m_1}, \mathbf{x}_{m_2}] \subset \overrightarrow{\partial\Omega}_{ij}$ for time t_m between two adjacent domains Ω_α $(\alpha = i,j)$. For an arbitrarily small $\varepsilon > 0$, there are two time intervals (i.e., $[t_{m-\varepsilon}, t_m)$ and $(t_m, t_{m+\varepsilon}]$), and $\mathbf{x}^{(i)}(t_{m\mp}) = \mathbf{x}_m = \mathbf{x}^{(j)}(t_{m\pm})$. The flows $\mathbf{x}^{(\alpha)}(t)$ are $C^{r_\alpha}_{[t_{m-\varepsilon}, t_{m+\varepsilon}]}$-continuous ($r_\alpha \geq 2l$) for time t. The necessary and sufficient conditions for the switching bifurcation onset of the flow $\mathbf{x}^{(i)}(t) \cup \mathbf{x}^{(j)}(t)$ at point $(\mathbf{x}_m, t_m)$ on the boundary $\overrightarrow{\partial\Omega}_{ij}$ are*

$$_{G\min}L_{ij}(t_m) = 0, \tag{2.86}$$

$$\mathbf{n}^{\mathrm{T}}_{\partial\Omega_{ij}} \cdot \mathbf{F}^{(t)}(t_{m\pm}) = 0 \quad \text{and} \quad L_{\alpha\alpha}(\mathbf{x}_{m\pm\varepsilon}, t_{m\pm\varepsilon}, \mathbf{p}_\alpha) < 0 \quad \text{for} \quad \alpha = i,j. \tag{2.87}$$

Proof. $L_{ij}(\mathbf{x}_m, t_m, \mathbf{p}_i, \mathbf{p}_j)$ replaced by its global minimum in Theorem 2.22 gives this corollary. ■

2.6 Switching Bifurcations of Nonpassable Flows

The onset and vanishing of the sliding and source flows on the boundary were discussed. The fragmentations of the sliding and source flows on the boundary are of great interest in this section. This kind of bifurcation is still a switching bifurcation. The definitions for such fragmentation bifurcations of flows on the nonpassable boundary are similar to the switching bifurcations of flows from the semipassable boundary to nonpassable boundary. The necessary and sufficient conditions for the fragmentation bifurcation from a nonpassable flow to a passable flow on the boundary are quite similar to the sliding and source bifurcations from a passable flow to a nonpassable flow.

Definition 2.20. For a discontinuous dynamical system in (2.1), there is a point $\mathbf{x}(t_m) = \mathbf{x}_m \in [\mathbf{x}_{m_1}, \mathbf{x}_{m_2}] \subset \widetilde{\partial\Omega}_{ij}$ at time t_m between two adjacent domains Ω_α ($\alpha = i, j$). For an arbitrarily small $\varepsilon > 0$, there are two time intervals (i.e., $[t_{m-\varepsilon}, t_m)$ and $(t_m, t_{m+\varepsilon}]$), and $\mathbf{x}^{(i)}(t_{m-}) = \mathbf{x}_m = \mathbf{x}^{(j)}(t_{m\mp})$. The flows $\mathbf{x}^{(i)}(t)$ and $\mathbf{x}^{(j)}(t)$ are $C^{r_i}_{[t_{m+\varepsilon}, t_m)}$- and $C^{r_j}_{[t_{m-\varepsilon}, t_{m+\varepsilon}]}$-continuous ($r_\alpha \geq 1$, $\alpha = i, j$) for time t, respectively. The tangential bifurcation of the flow $\mathbf{x}^{(j)}(t)$ *at point* $(\mathbf{x}_m, t_m)$ on the boundary $\widetilde{\partial\Omega}_{ij}$ is termed *the switching bifurcation* of a nonpassable flow *of the first kind* from $\widetilde{\partial\Omega}_{ij}$ to $\overrightarrow{\partial\Omega}_{ij}$ (or simply called *the sliding fragmentation bifurcation*) if

$$\mathbf{n}^{\mathrm{T}}_{\partial\Omega_{ij}} \cdot \dot{\mathbf{x}}^{(j)}(t_{m\mp}) = 0 \quad \text{and} \quad \mathbf{n}^{\mathrm{T}}_{\partial\Omega_{ij}} \cdot \dot{\mathbf{x}}^{(i)}(t_{m-}) \neq 0, \tag{2.88}$$

$$\text{either} \left. \begin{array}{l} \mathbf{n}^{\mathrm{T}}_{\partial\Omega_{ij}} \cdot [\mathbf{x}^{(i)}(t_{m-}) - \mathbf{x}^{(i)}(t_{m-\varepsilon})] > 0 \\ \mathbf{n}^{\mathrm{T}}_{\partial\Omega_{ij}} \cdot [\mathbf{x}^{(j)}(t_{m-}) - \mathbf{x}^{(j)}(t_{m-\varepsilon})] < 0 \\ \mathbf{n}^{\mathrm{T}}_{\partial\Omega_{ij}} \cdot [\mathbf{x}^{(j)}(t_{m+\varepsilon}) - \mathbf{x}^{(j)}(t_{m+})] > 0 \end{array} \right\} \text{ for } \mathbf{n}_{\partial\Omega_{ij}} \to \Omega_j \tag{2.89}$$

$$\text{or} \quad \left. \begin{array}{l} \mathbf{n}^{\mathrm{T}}_{\partial\Omega_{ij}} \cdot [\mathbf{x}^{(i)}(t_{m-}) - \mathbf{x}^{(i)}(t_{m-\varepsilon})] < 0 \\ \mathbf{n}^{\mathrm{T}}_{\partial\Omega_{ij}} \cdot [\mathbf{x}^{(j)}(t_{m-}) - \mathbf{x}^{(j)}(t_{m-\varepsilon})] > 0 \\ \mathbf{n}^{\mathrm{T}}_{\partial\Omega_{ij}} \cdot [\mathbf{x}^{(j)}(t_{m+\varepsilon}) - \mathbf{x}^{(j)}(t_{m+})] < 0 \end{array} \right\} \text{ for } \mathbf{n}_{\partial\Omega_{ij}} \to \Omega_i. \tag{2.90}$$

Definition 2.21. For a discontinuous dynamical system in (2.1), there is a point $\mathbf{x}(t_m) = \mathbf{x}_m \in [\mathbf{x}_{m_1}, \mathbf{x}_{m_2}] \subset \widehat{\partial\Omega}_{ij}$ at time t_m between two adjacent domains Ω_α ($\alpha = i, j$). For an arbitrarily small $\varepsilon > 0$, there are two time intervals (i.e., $[t_{m-\varepsilon}, t_m)$ and $(t_m, t_{m+\varepsilon}]$), and $\mathbf{x}^{(i)}(t_{m\pm}) = \mathbf{x}_m = \mathbf{x}^{(j)}(t_{m+})$. The flows $\mathbf{x}^{(i)}(t)$ and $\mathbf{x}^{(j)}(t)$ are $C^{r_i}_{[t_{m-\varepsilon}, t_{m+\varepsilon}]}$- and $C^{r_j}_{[t_{m+\varepsilon}, t_m)}$-continuous ($r_\alpha \geq 1$, $\alpha = i, j$) for time t, respectively. The tangential bifurcation of flows $\mathbf{x}^{(i)}(t)$ and $\mathbf{x}^{(j)}(t)$ *at point* $(\mathbf{x}_m, t_m)$ on the boundary $\widehat{\partial\Omega}_{ij}$ is termed *the switching bifurcation* of a nonpassable flow *of the second kind* from $\widehat{\partial\Omega}_{ij}$ to $\overrightarrow{\partial\Omega}_{ij}$ (or simply called *the source fragmentation bifurcation*) if

$$\mathbf{n}^{\mathrm{T}}_{\partial\Omega_{ij}} \cdot \dot{\mathbf{x}}^{(i)}(t_{m\pm}) = 0 \quad \text{and} \quad \mathbf{n}^{\mathrm{T}}_{\partial\Omega_{ij}} \cdot \dot{\mathbf{x}}^{(j)}(t_{m+}) \neq 0, \tag{2.91}$$

$$\text{either} \left. \begin{array}{l} \mathbf{n}^{\mathrm{T}}_{\partial\Omega_{ij}} \cdot [\mathbf{x}^{(i)}(t_{m-}) - \mathbf{x}^{(i)}(t_{m-\varepsilon})] > 0 \\ \mathbf{n}^{\mathrm{T}}_{\partial\Omega_{ij}} \cdot [\mathbf{x}^{(i)}(t_{m+\varepsilon}) - \mathbf{x}^{(i)}(t_{m+})] < 0 \\ \mathbf{n}^{\mathrm{T}}_{\partial\Omega_{ij}} \cdot [\mathbf{x}^{(j)}(t_{m+\varepsilon}) - \mathbf{x}^{(j)}(t_{m+})] > 0 \end{array} \right\} \text{ for } \mathbf{n}_{\partial\Omega_{ij}} \to \Omega_j \tag{2.92}$$

$$\text{or} \quad \left. \begin{array}{l} \mathbf{n}^{\mathrm{T}}_{\partial\Omega_{ij}} \cdot [\mathbf{x}^{(i)}(t_{m-}) - \mathbf{x}^{(i)}(t_{m-\varepsilon})] < 0 \\ \mathbf{n}^{\mathrm{T}}_{\partial\Omega_{ij}} \cdot [\mathbf{x}^{(i)}(t_{m+\varepsilon}) - \mathbf{x}^{(i)}(t_{m+}) > 0] \\ \mathbf{n}^{\mathrm{T}}_{\partial\Omega_{ij}} \cdot [\mathbf{x}^{(j)}(t_{m+\varepsilon}) - \mathbf{x}^{(j)}(t_{m+})] < 0 \end{array} \right\} \text{ for } \mathbf{n}_{\partial\Omega_{ij}} \to \Omega_i. \tag{2.93}$$

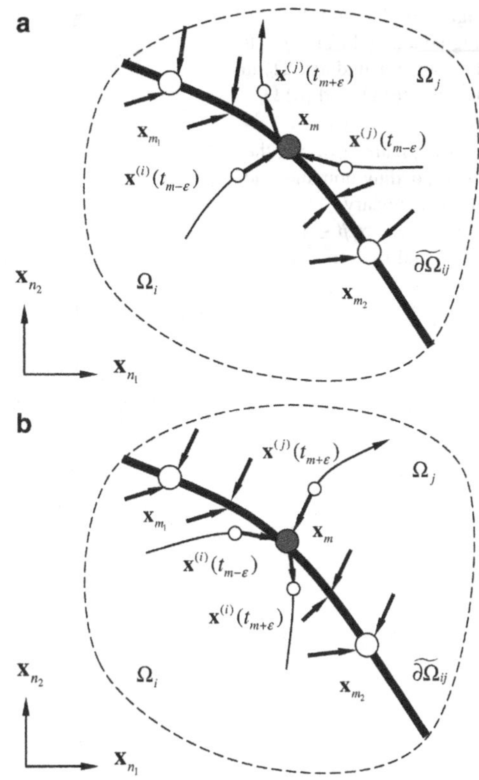

Fig. 2.9 The sliding fragmentation bifurcation to the sink boundary $\widetilde{\partial\Omega}_{ij}$ in domain: **(a)** Ω_j and **(b)** Ω_i. Four points $\mathbf{x}^{(\alpha)}(t_{m\pm\varepsilon})$, $\mathbf{x}^{(\beta)}(t_{m-\varepsilon})$, and $\mathbf{x}_m$ lie in the corresponding domains and on the boundary $\partial\Omega_{ij}$, respectively. $\alpha, \beta \in \{i, j\}$ but $\alpha \neq \beta$ and $n_1 + n_2 = n$

For the fragmentation bifurcation of the nonpassable flow on the boundary, the vector fields near the sink and source boundaries are sketched in Figs. 2.9 and 2.10, respectively. The switching from the sink or source flow to the semipassable flow has two possibilities. Therefore, the conditions in Definitions 2.20 and 2.21 have been changed accordingly. Before the fragmentation bifurcation of the nonpassable flow occurs on the boundary, the flow $\mathbf{x}^{(\alpha)}(t)$ ($\alpha \in \{i, j\}$) exists for $t \in [t_{m-\varepsilon}, t_{m-})$ or $t \in (t_{m+}, t_{m+\varepsilon}]$ on the sink or source boundary. Only the sliding flow exists on such a boundary. After the fragmentation bifurcation occurs, the sliding flow on the boundary will split into at least two portions of the sliding and semipassable motions. This phenomenon is called *the fragmentation of the sliding flow on the boundary*, which can help one easily understand the sliding dynamics on the boundary. In addition, for the nonpassable boundary, if flows on both sides of the nonpassable boundary possess the local singularity at the boundary, the nonpassable flow of the first kind switches into the nonpassable flow of the second kind, and vice versa. The local singularity of such switchability is similar to the switching between the two semipassable flows on the boundary, and the corresponding definition of the switching bifurcation is given as follows.

Definition 2.22. For a discontinuous dynamical system in (2.1), there is a point $\mathbf{x}(t_m) = \mathbf{x}_m \in [\mathbf{x}_{m_1}, \mathbf{x}_{m_2}] \subset \widetilde{\partial\Omega}_{ij}$ (or $\widehat{\partial\Omega}_{ij}$) at time t_m between two adjacent domains

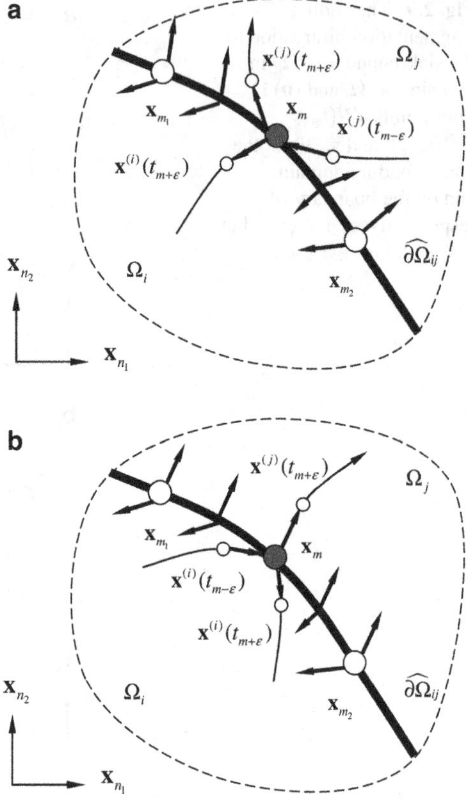

Fig. 2.10 The source fragmentation bifurcation to the source boundary $\widehat{\partial\Omega}_{ij}$ in domain: (**a**) Ω_j and (**b**) Ω_i. Four points $\mathbf{x}^{(\alpha)}(t_{m\pm\varepsilon})$, $\mathbf{x}^{(\beta)}(t_{m+\varepsilon})$, and $\mathbf{x}_m$ lie in the corresponding domains and on the boundary $\partial\Omega_{ij}$, respectively. $\alpha, \beta \in \{i,j\}$ but $\alpha \neq \beta$ and $n_1 + n_2 = n$

Ω_α $(\alpha = i,j)$. For an arbitrarily small $\varepsilon > 0$, there are two time intervals (i.e., $[t_{m-\varepsilon}, t_m)$ and $(t_m, t_{m+\varepsilon}]$), and $\mathbf{x}^{(\alpha)}(t_{m\pm}) = \mathbf{x}_m$. The flows $\mathbf{x}^{(\alpha)}(t)$ $(\alpha = i,j)$ are $C^{r_\alpha}_{[t_{m-\varepsilon},t_{m+\varepsilon}]}$-continuous $(r_\alpha \geq 1)$ for time t. The tangential bifurcation of the flow $\mathbf{x}^{(i)}(t)$ and $\mathbf{x}^{(j)}(t)$ at point $(\mathbf{x}_m, t_m)$ on the boundary $\widetilde{\partial\Omega}_{ij}$ (or $\widehat{\partial\Omega}_{ij}$) is termed the switching bifurcation of a nonpassable flow from $\widetilde{\partial\Omega}_{ij}$ to $\widehat{\partial\Omega}_{ij}$ (or $\widehat{\partial\Omega}_{ij}$ to $\widetilde{\partial\Omega}_{ij}$) if

$$\mathbf{n}^{\mathrm{T}}_{\partial\Omega_{ij}} \cdot \dot{\mathbf{x}}^{(\alpha)}(t_{m\pm}) = 0 \quad \text{for} \quad \alpha = i,j, \tag{2.94}$$

$$\text{either} \left.\begin{array}{l} \mathbf{n}^{\mathrm{T}}_{\partial\Omega_{ij}} \cdot [\mathbf{x}^{(i)}(t_{m-}) - \mathbf{x}^{(i)}(t_{m-\varepsilon})] > 0 \text{ and} \\ \mathbf{n}^{\mathrm{T}}_{\partial\Omega_{ij}} \cdot [\mathbf{x}^{(i)}(t_{m+\varepsilon}) - \mathbf{x}^{(i)}(t_{m+})] < 0 \\ \mathbf{n}^{\mathrm{T}}_{\partial\Omega_{ij}} \cdot [\mathbf{x}^{(j)}(t_{m-}) - \mathbf{x}^{(j)}(t_{m-\varepsilon})] < 0 \text{ and} \\ \mathbf{n}^{\mathrm{T}}_{\partial\Omega_{ij}} \cdot [\mathbf{x}^{(j)}(t_{m+\varepsilon}) - \mathbf{x}^{(j)}(t_{m+})] > 0 \end{array}\right\} \text{for } \mathbf{n}_{\partial\Omega_{ij}} \to \Omega_j \tag{2.95}$$

$$\text{or} \left.\begin{array}{l} \mathbf{n}^{\mathrm{T}}_{\partial\Omega_{ij}} \cdot [\mathbf{x}^{(i)}(t_{m-}) - \mathbf{x}^{(i)}(t_{m-\varepsilon})] < 0 \text{ and} \\ \mathbf{n}^{\mathrm{T}}_{\partial\Omega_{ij}} \cdot [\mathbf{x}^{(i)}(t_{m+\varepsilon}) - \mathbf{x}^{(i)}(t_{m+})] > 0 \\ \mathbf{n}^{\mathrm{T}}_{\partial\Omega_{ij}} \cdot [\mathbf{x}^{(j)}(t_{m-}) - \mathbf{x}^{(j)}(t_{m-\varepsilon})] > 0 \text{ and} \\ \mathbf{n}^{\mathrm{T}}_{\partial\Omega_{ij}} \cdot [\mathbf{x}^{(j)}(t_{m+\varepsilon}) - \mathbf{x}^{(j)}(t_{m+})] < 0 \end{array}\right\} \text{for } \mathbf{n}_{\partial\Omega_{ij}} \to \Omega_i. \tag{2.96}$$

Theorem 2.23. *For a discontinuous dynamical system in (2.1), there is a point* $\mathbf{x}(t_m) = \mathbf{x}_m \in [\mathbf{x}_{m_1}, \mathbf{x}_{m_2}] \subset \widetilde{\partial\Omega}_{ij}$ *at time* t_m *between two adjacent domains* Ω_α $(\alpha = i,j)$*. For an arbitrarily small* $\varepsilon > 0$*, there are two time intervals (i.e.,* $[t_{m-\varepsilon}, t_m)$ *and* $(t_m, t_{m+\varepsilon}]$*), and* $\mathbf{x}^{(i)}(t_{m-}) = \mathbf{x}_m = \mathbf{x}^{(j)}(t_{m\pm})$*. The flows* $\mathbf{x}^{(i)}(t)$ *and* $\mathbf{x}^{(j)}(t)$ *are* $C^{r_i}_{[t_{m-\varepsilon}, t_m)}$*- and* $C^{r_j}_{[t_{m-\varepsilon}, t_{m+\varepsilon}]}$*-continuous* $(r_\alpha \geq 1,\ \alpha = i,j)$ *for time t, respectively. The sliding fragmentation bifurcation of the flow* $\mathbf{x}^{(i)}(t) \cup \mathbf{x}^{(j)}(t)$ *at point* $(\mathbf{x}_m, t_m)$ *from* $\widetilde{\partial\Omega}_{ij}$ *to* $\overrightarrow{\partial\Omega}_{ij}$ *occurs if and only if*

$$\mathbf{n}^{\mathrm{T}}_{\partial\Omega_{ij}} \cdot \mathbf{F}^{(j)}(t_{m\pm}) = 0 \quad \text{and} \quad \mathbf{n}^{\mathrm{T}}_{\partial\Omega_{ij}} \cdot \mathbf{F}^{(i)}(t_{m-}) \neq 0, \tag{2.97}$$

$$\left.\begin{array}{lll} \text{either } \mathbf{n}^{\mathrm{T}}_{\partial\Omega_{ij}} \cdot \mathbf{F}^{(i)}(t_{m-}) > 0 & \text{for } \mathbf{n}_{\partial\Omega_{ij}} \to \Omega_j \\ \text{or } \quad \mathbf{n}^{\mathrm{T}}_{\partial\Omega_{ij}} \cdot \mathbf{F}^{(i)}(t_{m-}) < 0 & \text{for } \mathbf{n}_{\partial\Omega_{ij}} \to \Omega_i, \end{array}\right\} \tag{2.98}$$

$$\begin{array}{ll} \text{either} & \left.\begin{array}{l} \mathbf{n}^{\mathrm{T}}_{\partial\Omega_{ij}} \cdot \mathbf{F}^{(j)}(t_{m-\varepsilon}) < 0 \\ \mathbf{n}^{\mathrm{T}}_{\partial\Omega_{ij}} \cdot \mathbf{F}^{(j)}(t_{m+\varepsilon}) > 0 \end{array}\right\} \text{ for } \mathbf{n}_{\partial\Omega_{ij}} \to \Omega_j \\ \text{or} & \left.\begin{array}{l} \mathbf{n}^{\mathrm{T}}_{\partial\Omega_{ij}} \cdot \mathbf{F}^{(j)}(t_{m-\varepsilon}) > 0 \\ \mathbf{n}^{\mathrm{T}}_{\partial\Omega_{ij}} \cdot \mathbf{F}^{(j)}(t_{m+\varepsilon}) < 0 \end{array}\right\} \text{ for } \mathbf{n}_{\partial\Omega_{ij}} \to \Omega_i. \end{array} \tag{2.99}$$

Proof. Following the proof procedures of Theorems 2.8 and 2.9, the above theorem can be easily proved. ■

Theorem 2.24. *For a discontinuous dynamical system in (2.1), there is a point* $\mathbf{x}(t_m) = \mathbf{x}_m \in [\mathbf{x}_{m_1}, \mathbf{x}_{m_2}] \subset \widetilde{\partial\Omega}_{ij}$ *at time* t_m *on the* $(n-1)$*-dimensional plane boundary* $\partial\Omega_{ij}$ *between two adjacent domains* Ω_α $(\alpha = i,j)$*. For an arbitrarily small* $\varepsilon > 0$*, there are two time intervals (i.e.,* $[t_{m-\varepsilon}, t_m)$ *and* $(t_m, t_{m+\varepsilon}]$*), and* $\mathbf{x}^{(i)}(t_{m-}) = \mathbf{x}_m = \mathbf{x}^{(j)}(t_{m\pm})$*. The flows* $\mathbf{x}^{(i)}(t)$ *and* $\mathbf{x}^{(j)}(t)$ *are* $C^{r_i}_{[t_{m-\varepsilon}, t_m)}$*- and* $C^{r_j}_{[t_{m-\varepsilon}, t_{m+\varepsilon}]}$*-continuous* $(r_\alpha \geq 2,\ \alpha = i,j)$ *for time t, respectively. The sliding fragmentation bifurcation of the flow* $\mathbf{x}^{(i)}(t) \cup \mathbf{x}^{(j)}(t)$ *at point* $(\mathbf{x}_m, t_m)$ *from* $\widetilde{\partial\Omega}_{ij}$ *to* $\overrightarrow{\partial\Omega}_{ij}$ *occurs if and only if*

$$\mathbf{n}^{\mathrm{T}}_{\partial\Omega_{ij}} \cdot \mathbf{F}^{(j)}(t_{m\pm}) = 0 \quad \text{and} \quad \mathbf{n}^{\mathrm{T}}_{\partial\Omega_{ij}} \cdot \mathbf{F}^{(i)}(t_{m-}) \neq 0, \tag{2.100}$$

$$\begin{array}{ll} \text{either} & \left.\begin{array}{l} \mathbf{n}^{\mathrm{T}}_{\partial\Omega_{ij}} \cdot \mathbf{F}^{(i)}(t_{m-}) > 0 \\ \mathbf{n}^{\mathrm{T}}_{\partial\Omega_{ij}} \cdot D\mathbf{F}^{(j)}(t_{m\pm}) < 0 \end{array}\right\} \text{ for } \mathbf{n}_{\partial\Omega_{ij}} \to \Omega_j \\ \text{or} & \left.\begin{array}{l} \mathbf{n}^{\mathrm{T}}_{\partial\Omega_{ij}} \cdot \mathbf{F}^{(i)}(t_{m-}) < 0 \\ \mathbf{n}^{\mathrm{T}}_{\partial\Omega_{ij}} \cdot D\mathbf{F}^{(j)}(t_{m\pm}) < 0 \end{array}\right\} \text{ for } \mathbf{n}_{\partial\Omega_{ij}} \to \Omega_i. \end{array} \tag{2.101}$$

Proof. Following the proof procedures of Theorems 2.10 and 2.11, the above theorem can be easily proved. ■

Theorem 2.25. *For a discontinuous dynamical system in (2.1), there is a point* $\mathbf{x}(t_m) = \mathbf{x}_m \in [\mathbf{x}_{m_1}, \mathbf{x}_{m_2}] \subset \widehat{\partial\Omega}_{ij}$ *for time* t_m *between two adjacent domains* Ω_α $(\alpha = i,j)$*. For an arbitrarily small* $\varepsilon > 0$*, there are two time intervals (i.e.,* $[t_{m-\varepsilon}, t_m)$ *and* $(t_m, t_{m+\varepsilon}]$*), and* $\mathbf{x}^{(i)}(t_{m\pm}) = \mathbf{x}_m = \mathbf{x}^{(j)}(t_{m+})$*. The flows* $\mathbf{x}^{(i)}(t)$ *and* $\mathbf{x}^{(j)}(t)$ *are* $C^{r_i}_{[t_{m-\varepsilon}, t_{m+\varepsilon}]}$*- and* $C^{r_j}_{(t_m, t_{m+\varepsilon}]}$*-continuous* $(r_\alpha \geq 1,\ \alpha = i,j)$ *for time t, respectively. The source fragmentation bifurcation of the flow* $\mathbf{x}^{(i)}(t) \cup \mathbf{x}^{(j)}(t)$ *at point* $(\mathbf{x}_m, t_m)$ *on the boundary* $\widehat{\partial\Omega}_{ij}$ *occurs if and only if*

$$\mathbf{n}^{\mathrm{T}}_{\partial\Omega_{ij}} \cdot \mathbf{F}^{(i)}(t_{m\pm}) = 0 \quad \text{and} \quad \mathbf{n}^{\mathrm{T}}_{\partial\Omega_{ij}} \cdot \mathbf{F}^{(j)}(t_{m+}) \neq 0, \tag{2.102}$$

$$\begin{aligned} &\text{either } \mathbf{n}^{\mathrm{T}}_{\partial\Omega_{ij}} \cdot \mathbf{F}^{(j)}(t_{m+}) > 0 \quad \text{for } \mathbf{n}_{\partial\Omega_{ij}} \to \Omega_j \\ &\text{or} \qquad \mathbf{n}^{\mathrm{T}}_{\partial\Omega_{ij}} \cdot \mathbf{F}^{(j)}(t_{m+}) < 0 \quad \text{for } \mathbf{n}_{\partial\Omega_{ij}} \to \Omega_i, \end{aligned} \tag{2.103}$$

$$\begin{aligned} &\text{either} \left.\begin{aligned} \mathbf{n}^{\mathrm{T}}_{\partial\Omega_{ij}} \cdot \mathbf{F}^{(i)}(t_{m-\varepsilon}) > 0 \\ \mathbf{n}^{\mathrm{T}}_{\partial\Omega_{ij}} \cdot \mathbf{F}^{(i)}(t_{m+\varepsilon}) < 0 \end{aligned}\right\} \text{for } \mathbf{n}_{\partial\Omega_{ij}} \to \Omega_j \\ &\text{or} \left.\begin{aligned} \mathbf{n}^{\mathrm{T}}_{\partial\Omega_{ij}} \cdot \mathbf{F}^{(i)}(t_{m-\varepsilon}) < 0 \\ \mathbf{n}^{\mathrm{T}}_{\partial\Omega_{ij}} \cdot \mathbf{F}^{(i)}(t_{m+\varepsilon}) > 0 \end{aligned}\right\} \text{for } \mathbf{n}_{\partial\Omega_{ij}} \to \Omega_i. \end{aligned} \tag{2.104}$$

Proof. Following the proof procedures of Theorems 2.8 and 2.9, the above theorem can be easily proved. ■

Theorem 2.26. *For a discontinuous dynamical system in (2.1), there is a point* $\mathbf{x}(t_m) = \mathbf{x}_m \in [\mathbf{x}_{m_1}, \mathbf{x}_{m_2}] \subset \widehat{\partial\Omega}_{ij}$ *at time* t_m *on the* $(n-1)$*-dimensional plane boundary* $\partial\Omega_{ij}$ *between two adjacent domains* Ω_α $(\alpha = i,j)$*. For an arbitrarily small* $\varepsilon > 0$*, there are two time intervals (i.e.,* $[t_{m-\varepsilon}, t_m)$ *and* $(t_m, t_{m+\varepsilon}]$*), and* $\mathbf{x}^{(i)}(t_{m\pm}) = \mathbf{x}_m = \mathbf{x}^{(j)}(t_{m+})$*. The flows* $\mathbf{x}^{(i)}(t)$ *and* $\mathbf{x}^{(j)}(t)$ *are* $C^{r_i}_{[t_{m-\varepsilon}, t_{m+\varepsilon}]}$ *and* $C^{r_j}_{(t_m, t_{m+\varepsilon}]}$ *continuous* $(r_\alpha \geq 1,\ \alpha = i,j)$ *for time t, respectively. The source fragmentation bifurcation of the flow* $\mathbf{x}^{(i)}(t) \cup \mathbf{x}^{(j)}(t)$ *at point* $(\mathbf{x}_m, t_m)$ *from* $\widehat{\partial\Omega}_{ij}$ *to* $\overrightarrow{\partial\Omega}_{ij}$ *occurs if and only if*

$$\mathbf{n}^{\mathrm{T}}_{\partial\Omega_{ij}} \cdot \mathbf{F}^{(i)}(t_{m\pm}) = 0 \quad \text{and} \quad \mathbf{n}^{\mathrm{T}}_{\partial\Omega_{ij}} \cdot \mathbf{F}^{(j)}(t_{m+}) \neq 0, \tag{2.105}$$

$$\begin{aligned} &\text{either} \left.\begin{aligned} \mathbf{n}^{\mathrm{T}}_{\partial\Omega_{ij}} \cdot \mathbf{F}^{(j)}(t_{m+}) > 0 \\ \mathbf{n}^{\mathrm{T}}_{\partial\Omega_{ij}} \cdot D\mathbf{F}^{(i)}(t_{m\pm}) < 0 \end{aligned}\right\} \text{for } \mathbf{n}_{\partial\Omega_{ij}} \to \Omega_j \\ &\text{or} \left.\begin{aligned} \mathbf{n}^{\mathrm{T}}_{\partial\Omega_{ij}} \cdot \mathbf{F}^{(j)}(t_{m+}) < 0 \\ \mathbf{n}^{\mathrm{T}}_{\partial\Omega_{ij}} \cdot D\mathbf{F}^{(i)}(t_{m\pm}) < 0 \end{aligned}\right\} \text{for } \mathbf{n}_{\partial\Omega_{ij}} \to \Omega_i. \end{aligned} \tag{2.106}$$

Proof. Following the proof procedures in Theorems 2.10 and 2.11, the above theorem can be easily proved. ■

Theorem 2.27. *For a discontinuous dynamical system in (2.1), there is a point* $\mathbf{x}(t_m) = \mathbf{x}_m \in [\mathbf{x}_{m_1}, \mathbf{x}_{m_2}] \subset \widetilde{\partial\Omega}_{ij}$ *(or* $\widehat{\partial\Omega}_{ij}$*) at time* t_m *between two adjacent domains* Ω_α $(\alpha = i, j)$*. For an arbitrarily small* $\varepsilon > 0$*, there are two time intervals (i.e.,* $[t_{m-\varepsilon}, t_m)$ *and* $(t_m, t_{m+\varepsilon}]$*), and* $\mathbf{x}^{(i)}(t_{m\pm}) = \mathbf{x}_m = \mathbf{x}^{(j)}(t_{m\pm})$*. The flows* $\mathbf{x}^{(\alpha)}(t)$ $(\alpha = i, j)$ *are* $C^{r_\alpha}_{[t_{m-\varepsilon}, t_{m+\varepsilon}]}$*-continuous* $(r_\alpha \geq 1)$ *for time t. The switching bifurcation of the flow at point* $(\mathbf{x}_m, t_m)$ *from* $\widetilde{\partial\Omega}_{ij}$ *to* $\widehat{\partial\Omega}_{ij}$ *(or* $\widehat{\partial\Omega}_{ij}$ *to* $\widetilde{\partial\Omega}_{ij}$*) occurs if and only if*

$$\mathbf{n}^{\mathrm{T}}_{\partial\Omega_{ij}} \cdot \mathbf{F}^{(\alpha)}(t_{m\pm}) = 0, \tag{2.107}$$

$$\begin{aligned}
&\text{either} && \left.\begin{aligned} \mathbf{n}^{\mathrm{T}}_{\partial\Omega_{ij}} \cdot \mathbf{F}^{(\alpha)}(t_{m-\varepsilon}) > 0 \\ \mathbf{n}^{\mathrm{T}}_{\partial\Omega_{ij}} \cdot \mathbf{F}^{(\alpha)}(t_{m+\varepsilon}) < 0 \end{aligned}\right\} && \text{for } \mathbf{n}_{\partial\Omega_{ij}} \to \Omega_\beta \\
&\text{or} && \left.\begin{aligned} \mathbf{n}^{\mathrm{T}}_{\partial\Omega_{ij}} \cdot \mathbf{F}^{(\alpha)}(t_{m-\varepsilon}) < 0 \\ \mathbf{n}^{\mathrm{T}}_{\partial\Omega_{ij}} \cdot \mathbf{F}^{(\alpha)}(t_{m+\varepsilon}) > 0 \end{aligned}\right\} && \text{for } \mathbf{n}_{\partial\Omega_{ij}} \to \Omega_\alpha,
\end{aligned} \tag{2.108}$$

with $\alpha, \beta = i, j$ *but* $\alpha \neq \beta$*.*

Proof. Following the proof procedures of Theorems 2.8 and 2.9, the above theorem can be easily proved. ■

Theorem 2.28. *For a discontinuous dynamical system in (2.1), there is a point* $\mathbf{x}(t_m) = \mathbf{x}_m \in [\mathbf{x}_{m_1}, \mathbf{x}_{m_2}] \subset \widetilde{\partial\Omega}_{ij}$ *(or* $\widehat{\partial\Omega}_{ij}$*) at time* t_m *on the* $(n-1)$*-dimensional plane boundary* $\partial\Omega_{ij}$ *between two adjacent domains* Ω_α $(\alpha = i, j)$*. For an arbitrarily small* $\varepsilon > 0$*, there are two time intervals (i.e.,* $[t_{m-\varepsilon}, t_m)$ *and* $(t_m, t_{m+\varepsilon}]$*), and* $\mathbf{x}^{(\alpha)}(t_{m\pm}) = \mathbf{x}_m$*. The flows* $\mathbf{x}^{(\alpha)}(t)$ $(\alpha = i, j)$ *are* $C^{r_\alpha}_{[t_{m-\varepsilon}, t_{m+\varepsilon}]}$*-continuous* $(r_\alpha \geq 2)$ *for time t. The switching bifurcation of the flow at point* $(\mathbf{x}_m, t_m)$ *from* $\widetilde{\partial\Omega}_{ij}$ *to* $\widehat{\partial\Omega}_{ij}$ *(or* $\widehat{\partial\Omega}_{ij}$ *to* $\widetilde{\partial\Omega}_{ij}$*) exists if and only if*

$$\mathbf{n}^{\mathrm{T}}_{\partial\Omega_{ij}} \cdot \mathbf{F}^{(\alpha)}(t_{m\pm}) = 0 \quad \text{for} \quad \alpha = i, j, \tag{2.109}$$

$$\begin{aligned}
&\text{either} && \left.\begin{aligned} \mathbf{n}^{\mathrm{T}}_{\partial\Omega_{ij}} \cdot D\mathbf{F}^{(i)}(t_{m\pm}) < 0 \\ \mathbf{n}^{\mathrm{T}}_{\partial\Omega_{ij}} \cdot D\mathbf{F}^{(j)}(t_{m\pm}) > 0 \end{aligned}\right\} && \text{for } \mathbf{n}_{\partial\Omega_{ij}} \to \Omega_j, \\
&\text{or} && \left.\begin{aligned} \mathbf{n}^{\mathrm{T}}_{\partial\Omega_{ij}} \cdot D\mathbf{F}^{(i)}(t_{m\pm}) > 0 \\ \mathbf{n}^{\mathrm{T}}_{\partial\Omega_{ij}} \cdot D\mathbf{F}^{(j)}(t_{m\pm}) < 0 \end{aligned}\right\} && \text{for } \mathbf{n}_{\partial\Omega_{ij}} \to \Omega_i.
\end{aligned} \tag{2.110}$$

Proof. Following the proof procedures of Theorems 2.10 and 2.11, the above theorem can be easily proved. ■

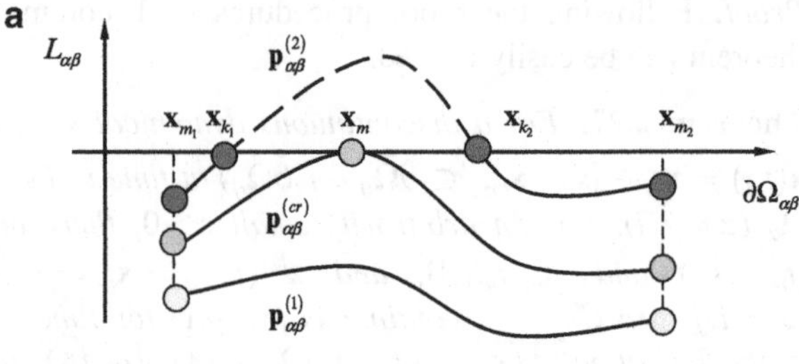

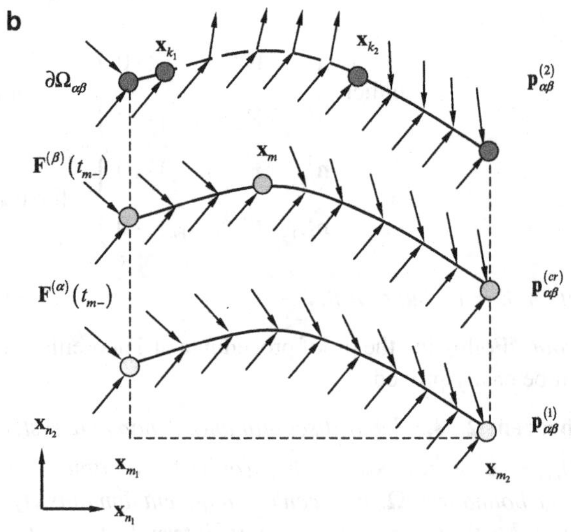

Fig. 2.11 (**a**) The *L*-function of flows ($L_{\alpha\beta}$) and (**b**) the vector fields between two points $\mathbf{x}_{m_1}$ and $\mathbf{x}_{m_2}$ on the boundary $\widetilde{\partial\Omega}_{\alpha\beta}$. The point $\mathbf{x}_m$ for $\mathbf{p}_{\alpha\beta}^{(\mathrm{cr})}$ is the critical point for the switching bifurcation. Two points $\mathbf{x}_{k_1}$ and $\mathbf{x}_{k_2}$ are the starting and vanishing of the passable flow on the boundary $\widetilde{\partial\Omega}_{\alpha\beta}$. The *dashed* and *solid curves* represent $L_{\alpha\beta}>0$ and $L_{\alpha\beta}\leqslant 0$, respectively $(n_1+n_2=n)$

Similarly, the $L_{\alpha\beta}$-functions of flows varying with $\mathbf{p}_{ij}\in\{\mu_\alpha\}_{\alpha\in\{i,j\}}$ is used to discuss the switching of the nonpassable flow from $\overline{\partial\Omega}_{\alpha\beta}$ to $\overrightarrow{\partial\Omega}_{\alpha\beta}$. The $L_{\alpha\beta}$-functions of a nonpassable flow to the boundary is $L_{\alpha\beta}<0$ with varying boundary location. The $L_{\alpha\beta}$-function for flows between two points $\mathbf{x}_{m_1}$ and $\mathbf{x}_{m_2}$ on the sink boundary $\widetilde{\partial\Omega}_{\alpha\beta}$ are sketched in Fig. 2.11 for $\mathbf{p}_{\alpha\beta}$ between $\mathbf{p}_{\alpha\beta}^{(1)}$ and $\mathbf{p}_{\alpha\beta}^{(2)}$. The $L_{\alpha\beta}$-function is sketched in Fig. 2.11a, and the corresponding vector fields varying with system parameters on the boundary $\partial\Omega_{\alpha\beta}$ are illustrated in Fig. 2.11b. $\mathbf{F}^{(\alpha)}(t_{m-})$ and $\mathbf{F}^{(\beta)}(t_{m\pm})$ are limits of the vector fields to the boundary $\partial\Omega_{\alpha\beta}$ in domains Ω_α and Ω_β, respectively. The boundary relative to the nonpassable flows with $L_{\alpha\beta}<0$ is a sink boundary. There is a specific value $\mathbf{p}_{\alpha\beta}^{(\mathrm{cr})}$ between $\mathbf{p}_{\alpha\beta}^{(1)}$ and $\mathbf{p}_{\alpha\beta}^{(2)}$. For this specific value, a point $\mathbf{x}_m$ on the sink boundary can be found for the sliding fragmentation bifurcation on the boundary. Two points $\mathbf{x}_{k_1}$ and $\mathbf{x}_{k_2}$ are onset and vanishing points of the passable flow on the boundary $\partial\Omega_{\alpha\beta}$ for $\mathbf{p}_{\alpha\beta}$. The dashed and solid curves represent $L_{\alpha\beta}>0$ and $L_{\alpha\beta}\leq 0$, respectively. For $\mathbf{p}_{\alpha\beta}$ varying from $\mathbf{p}_{\alpha\beta}^{(1)}\to\mathbf{p}_{\alpha\beta}^{(\mathrm{cr})}$, the $L_{\alpha\beta}$-functions of flows for $\mathbf{x}\in(\mathbf{x}_{m_1},\mathbf{x}_{m_2})$ on the boundary is negative (i.e., $L_{\alpha\beta}<0$). Therefore, the boundary $\partial\Omega_{\alpha\beta}$ is nonpassable. For $\mathbf{p}_{\alpha\beta}$ varying from $\mathbf{p}_{\alpha\beta}^{(cr)}\to\mathbf{p}_{\alpha\beta}^{(2)}$,

$L_{\alpha\beta} < 0$ are for $\mathbf{x} \in [\mathbf{x}_{m_1}, \mathbf{x}_{k_1}) \cup (\mathbf{x}_{k_2}, \mathbf{x}_{m_2}]$, and $L_{\alpha\beta} > 0$ are for $\mathbf{x} \in (\mathbf{x}_{k_1}, \mathbf{x}_{k_2})$. From (2.67), the flow for the portion of $\mathbf{x} \in (\mathbf{x}_{k_1}, \mathbf{x}_{k_2})$ boundary with $L_{\alpha\beta} > 0$ is semipassable. For $\mathbf{p}_{\alpha\beta}$ varying from $\mathbf{p}_{\alpha\beta}^{(1)} \to \mathbf{p}_{\alpha\beta}^{(2)}$, the point $(\mathbf{x}_m, \mathbf{p}_{\alpha\beta}^{(\mathrm{cr})})$ on the boundary $\partial\Omega_{\alpha\beta}$ is the onset point of the semipassable flow on the boundary. The sliding flow on the boundary will be fragmentized. However, for $\mathbf{p}_{\alpha\beta}$ varying from $\mathbf{p}_{\alpha\beta}^{(2)} \to \mathbf{p}_{\alpha\beta}^{(1)}$, the sliding fragmentation disappears at such a point. At three critical points $(\mathbf{x}_m, \mathbf{x}_{k_1}, \mathbf{x}_{k_2})$, $L_{\alpha\beta} = 0$. The flow at the critical points $(\mathbf{x}_{k_1}, \mathbf{x}_{k_2})$ has the same properties as at the critical point $\mathbf{x}_m$. If the two critical points have the different properties, the sliding flow between the two different critical points is discussed later. From the $L_{\alpha\beta}$-function of flows, the criteria for the sliding fragmentation bifurcation can be given as similar to Theorems 2.22, 2.25, and 2.27. Thus, the corresponding bifurcation conditions are stated herein.

Theorem 2.29. *For a discontinuous dynamical system in (2.1), there is a point $\mathbf{x}(t_m) = \mathbf{x}_m \in [\mathbf{x}_{m_1}, \mathbf{x}_{m_2}] \subset \widetilde{\partial\Omega}_{ij}$ at time t_m between two adjacent domains Ω_α $(\alpha = i, j)$. For an arbitrarily small $\varepsilon > 0$, there are two time intervals (i.e., $[t_{m-\varepsilon}, t_m)$ and $(t_m, t_{m+\varepsilon}]$), and $\mathbf{x}^{(i)}(t_{m-}) = \mathbf{x}_m = \mathbf{x}^{(j)}(t_{m\pm})$. The flows $\mathbf{x}^{(i)}(t)$ and $\mathbf{x}^{(j)}(t)$ are $C^{r_i}_{[t_{m-\varepsilon}, t_m)}$- and $C^{r_j}_{[t_{m-\varepsilon}, t_{m+\varepsilon}]}$-continuous $(r_\alpha \geq 2,\ \alpha = i, j)$ for time t, respectively. The sliding fragmentation bifurcation of the flow $\mathbf{x}^{(i)}(t) \cup \mathbf{x}^{(j)}(t)$ at point $(\mathbf{x}_m, t_m)$ on the boundary $\widetilde{\partial\Omega}_{ij}$ exists if and only if*

$$L_{ij}(\mathbf{x}_m, t_m, \mathbf{p}_i, \mathbf{p}_j) = 0, \tag{2.111}$$

$$\mathbf{n}^{\mathrm{T}}_{\partial\Omega_{ij}} \cdot \mathbf{F}^{(i)}(t_{m-}) \neq 0 \quad \text{and} \quad L_{jj}(\mathbf{x}_{m\pm\varepsilon}, t_{m\pm\varepsilon}, \mathbf{p}_j) < 0. \tag{2.112}$$

Proof. Applying the L-function of flows in Definition 2.17 to Theorem 2.22, the foregoing theorem can be easily proved. ■

Theorem 2.30. *For a discontinuous dynamical system in (2.1), there is a point $\mathbf{x}(t_m) = \mathbf{x}_m \in [\mathbf{x}_{m_1}, \mathbf{x}_{m_2}] \subset \widehat{\partial\Omega}_{ij}$ at time t_m between two adjacent domains Ω_α $(\alpha = i, j)$. For an arbitrarily small $\varepsilon > 0$, there are two time intervals (i.e., $[t_{m-\varepsilon}, t_m)$ and $(t_m, t_{m+\varepsilon}]$), and $\mathbf{x}^{(i)}(t_{m\pm}) = \mathbf{x}_m = \mathbf{x}^{(j)}(t_{m+})$. The flows $\mathbf{x}^{(i)}(t)$ and $\mathbf{x}^{(j)}(t)$ are $C^{r_i}_{[t_{m-\varepsilon}, t_{m+\varepsilon}]}$- and $C^{r_j}_{(t_m, t_{m+\varepsilon}]}$-continuous $(r_\alpha \geq 2,\ \alpha = i, j)$ for time t, respectively. The source fragmentation bifurcation of the flow $\mathbf{x}^{(i)}(t) \cup \mathbf{x}^{(j)}(t)$ at point $(\mathbf{x}_m, t_m)$ on the boundary $\widehat{\partial\Omega}_{ij}$ occurs if and only if*

$$L_{ij}(\mathbf{x}_m, t_m, \mathbf{p}_i, \mathbf{p}_j) = 0, \tag{2.113}$$

$$\mathbf{n}^{\mathrm{T}}_{\partial\Omega_{ij}} \cdot \mathbf{F}^{(j)}(t_{m+}) \neq 0 \quad \text{and} \quad L_{jj}(\mathbf{x}_{m\pm\varepsilon}, t_{m\pm\varepsilon}, \mathbf{p}_j) < 0. \tag{2.114}$$

Proof. Applying the L_{ij} and L_{jj}-functions of flows in Definition 2.17 to Theorem 2.25, the foregoing theorem can be easily proved. ■

Theorem 2.31. *For a discontinuous dynamical system in (2.1), there is a point* $\mathbf{x}(t_m) = \mathbf{x}_m \in [\mathbf{x}_{m_1}, \mathbf{x}_{m_2}] \subset \widetilde{\partial\Omega}_{ij}$ *(or* $\widehat{\partial\Omega}_{ij}$*) for time* t_m *between two adjacent domains* Ω_α $(\alpha = i,j)$*. For an arbitrarily small* $\varepsilon > 0$*, there are two time intervals (i.e.,* $[t_{m-\varepsilon}, t_m)$ *and* $(t_m, t_{m+\varepsilon}]$*), and* $\mathbf{x}^{(i)}(t_{m\pm}) = \mathbf{x}_m = \mathbf{x}^{(j)}(t_{m\pm})$*. The flows* $\mathbf{x}^{(\alpha)}(t)$ $(\alpha = i,j)$ *are* $C^{r_\alpha}_{[t_{m-\varepsilon}, t_{m+\varepsilon}]}$*-continuous* $(r_\alpha \geq 2)$ *for time* t*. The switching bifurcation of the flow at point* $(\mathbf{x}_m, t_m)$ *from* $\widetilde{\partial\Omega}_{ij}$ *to* $\widehat{\partial\Omega}_{ij}$ *(or* $\widehat{\partial\Omega}_{ij}$ *to* $\widetilde{\partial\Omega}_{ij}$*) occurs if and only if*

$$L_{ij}(\mathbf{x}_m, t_m, \mathbf{p}_i, \mathbf{p}_j) = 0, \tag{2.115}$$

$$\mathbf{n}^{\mathrm{T}}_{\partial\Omega_{ij}} \cdot \mathbf{F}^{(\alpha)}(t_{m\pm}) = 0 \quad \text{and} \quad L_{\alpha\alpha}(\mathbf{x}_{m\pm\varepsilon}, t_{m\pm\varepsilon}, \mathbf{p}_\alpha) < 0 \quad \text{for} \quad \alpha = i,j. \tag{2.116}$$

Proof. Applying the L_{ij} and $L_{\alpha\alpha}$-functions of flows in Definition 2.17 to Theorem 2.27, the foregoing theorem can be easily proved. ■

For the nonpassable flow at $\mathbf{x}(t_m) = \mathbf{x}_m \in [\mathbf{x}_{m_1}, \mathbf{x}_{m_2}] \subset \widetilde{\partial\Omega}_{ij}$ (or $\widehat{\partial\Omega}_{ij}$), consider a time interval $[t_{m_1}, t_{m_2}]$ for $[\mathbf{x}_{m_1}, \mathbf{x}_{m_2}]$ on the boundary, for $t_m \in [t_{m_1}, t_{m_2}]$ and $\mathbf{x}_m \in [\mathbf{x}_{m_1}, \mathbf{x}_{m_2}]$, $L_{ij}(\mathbf{x}_m, t_m, \mathbf{p}_i, \mathbf{p}_j) < 0$. To determine the switching bifurcation, the local maximum of $L_{ij}(\mathbf{x}_m, t_m, \mathbf{p}_i, \mathbf{p}_j)$ can be determined. With (2.78) and (2.79), the local maximum of $L_{ij}(\mathbf{x}_m, t_m, \mathbf{p}_i, \mathbf{p}_j)$ is determined by

$$D^k L_{ij}(\mathbf{x}_m, t_m, \mathbf{p}_i, \mathbf{p}_j) = 0 \quad (k = 0, 1, 2, \ldots, 2l-1), \tag{2.117}$$

$$D^{2l} L_{ij}(\mathbf{x}_m, t_m, \mathbf{p}_i, \mathbf{p}_j) < 0. \tag{2.118}$$

Definition 2.23. For a discontinuous dynamical system in (2.1), there is a point $\mathbf{x}(t_m) = \mathbf{x}_m \in [\mathbf{x}_{m_1}, \mathbf{x}_{m_2}] \subset \widetilde{\partial\Omega}_{ij}$ (or $\widehat{\partial\Omega}_{ij}$) at time t_m between two adjacent domains Ω_α $(\alpha = i,j)$. For an arbitrarily small $\varepsilon > 0$, there are two time intervals (i.e., $[t_{m-\varepsilon}, t_m)$ and $(t_m, t_{m+\varepsilon}]$), and $\mathbf{x}^{(i)}(t_{m\pm}) = \mathbf{x}_m = \mathbf{x}^{(j)}(t_{m\mp})$. The flows $\mathbf{x}^{(i)}(t)$ and $\mathbf{x}^{(j)}(t)$ are $C^r_{[t_{m-\varepsilon}, t_{m+\varepsilon}]}$-continuous $(r \geq 2l)$ for time t. The local maximum set of $L_{ij}(\mathbf{x}_m, t_m, \mathbf{p}_i, \mathbf{p}_j)$ is defined by

$$_{\max}L_{ij}(t_m) = \left\{ L_{ij}(\mathbf{x}_m, t_m, \mathbf{p}_i, \mathbf{p}_j) \left| \begin{array}{l} \forall t_m \in [t_{m_1}, t_{m_2}], \exists \mathbf{x}_m \in [\mathbf{x}_{m_1}, \mathbf{x}_{m_2}], \\ \text{so that } D^k L_{ij}(\mathbf{x}_m, t_m, \mathbf{p}_i, \mathbf{p}_j) = 0 \\ \text{for } k = \{1, 2, \ldots 2l-1\} \text{ and} \\ D^{2l} L_{ij}(\mathbf{x}_m, t_m, \mathbf{p}_i, \mathbf{p}_j) < 0. \end{array} \right. \right\} \tag{2.119}$$

From the local maximum set of $L_{ij}(\mathbf{x}_m, t_m, \mathbf{p}_i, \mathbf{p}_j)$, the corresponding global maximum can be determined as follows.

Definition 2.24. For a discontinuous dynamical system in (2.1), there is a point $\mathbf{x}(t_m) = \mathbf{x}_m \in [\mathbf{x}_{m_1}, \mathbf{x}_{m_2}] \subset \widetilde{\partial\Omega}_{ij}$ (or $\widehat{\partial\Omega}_{ij}$) at time t_m between two adjacent domains Ω_α $(\alpha = i,j)$. For an arbitrarily small $\varepsilon > 0$, there are two time intervals (i.e., $[t_{m-\varepsilon}, t_m)$ and $(t_m, t_{m+\varepsilon}]$), and $\mathbf{x}^{(i)}(t_{m\pm}) = \mathbf{x}_m = \mathbf{x}^{(j)}(t_{m\mp})$. The global maximum set of $L_{ij}(\mathbf{x}_m, t_m, \mathbf{p}_i, \mathbf{p}_j)$ is defined by

$$
{}_{G\max}L_{ij}(t_m) = \max_{t_m \in [t_{m_1}, t_{m_2}]} \left\{ \begin{array}{l} {}_{\max}L_{ij}(t_m), L_{ij}(\mathbf{x}_{m_1}, t_{m_1}, \mathbf{p}_i, \mathbf{p}_j), \\ L_{ij}(\mathbf{x}_{m_2}, t_{m_2}, \mathbf{p}_i, \mathbf{p}_j). \end{array} \right\} \tag{2.120}
$$

From the foregoing definition, Theorems 2.22, 2.25, and 2.27 can be expressed through the global minimum of $L_{ij}(\mathbf{x}_m, t_m, \mathbf{p}_i, \mathbf{p}_j)$. So the following corollaries can be achieved, which give the condition of sliding fragmentation bifurcation.

Corollary 2.4. *For a discontinuous dynamical system in (2.1), there is a point $\mathbf{x}(t_m) = \mathbf{x}_m \in [\mathbf{x}_{m_1}, \mathbf{x}_{m_2}] \subset \widetilde{\partial\Omega}_{ij}$ at time t_m between two adjacent domains Ω_α ($\alpha = i,j$). For an arbitrarily small $\varepsilon > 0$, there are two time intervals (i.e., $[t_{m-\varepsilon}, t_m)$ and $(t_m, t_{m+\varepsilon}]$), and $\mathbf{x}^{(i)}(t_{m-}) = \mathbf{x}_m = \mathbf{x}^{(j)}(t_{m\pm})$. The flows $\mathbf{x}^{(i)}(t)$ and $\mathbf{x}^{(j)}(t)$ are $C^{r_i}_{[t_{m-\varepsilon}, t_m)}$- and $C^{r_j}_{[t_{m-\varepsilon}, t_{m+\varepsilon}]}$-continuous ($r_\alpha \geq 2l$, $\alpha = i,j$) for time t, respectively. The sliding fragmentation bifurcation of the flow $\mathbf{x}^{(i)}(t) \cup \mathbf{x}^{(j)}(t)$ at point $(\mathbf{x}_m, t_m)$ on the boundary $\widetilde{\partial\Omega}_{ij}$ occurs if and only if*

$$
{}_{G\max}L_{ij}(t_m) = 0, \tag{2.121}
$$

$$
\mathbf{n}^{\mathrm{T}}_{\partial\Omega_{ij}} \cdot \mathbf{F}^{(i)}(t_{m-}) \neq 0 \quad \text{and} \quad L_{jj}(\mathbf{x}_{m\pm\varepsilon}, t_{m\pm\varepsilon}, \mathbf{p}_j) < 0. \tag{2.122}
$$

Proof. In Theorem 2.29, replacing $L_{ij}(\mathbf{x}_m, t_m, \mathbf{p}_i, \mathbf{p}_j)$ by its global maximum value ${}_{G\max}L_{ij}(t_m)$ gives the above corollary. ∎

Corollary 2.5. *For a discontinuous dynamical system in (2.1), there is a point $\mathbf{x}(t_m) = \mathbf{x}_m \in [\mathbf{x}_{m_1}, \mathbf{x}_{m_2}] \subset \widehat{\partial\Omega}_{ij}$ at time t_m between two adjacent domains Ω_α ($\alpha = i,j$). For an arbitrarily small $\varepsilon > 0$, there are two time intervals (i.e., $[t_{m-\varepsilon}, t_m)$ and $(t_m, t_{m+\varepsilon}]$), and $\mathbf{x}^{(i)}(t_{m\pm}) = \mathbf{x}_m = \mathbf{x}^{(j)}(t_{m+})$. The flows $\mathbf{x}^{(i)}(t)$ and $\mathbf{x}^{(j)}(t)$ are $C^{r_i}_{[t_{m-\varepsilon}, t_{m+\varepsilon}]}$- and $C^{r_j}_{(t_m, t_{m+\varepsilon}]}$-continuous ($r_\alpha \geq 2l$, $\alpha = i,j$) for time t, respectively. The source fragmentation bifurcation of the flow $\mathbf{x}^{(i)}(t) \cup \mathbf{x}^{(j)}(t)$ at point $(\mathbf{x}_m, t_m)$ on the boundary $\widehat{\partial\Omega}_{ij}$ occurs if and only if*

$$
{}_{G\max}L_{ij}(t_m) = 0, \tag{2.123}
$$

$$
\mathbf{n}^{\mathrm{T}}_{\partial\Omega_{ij}} \cdot \mathbf{F}^{(i)}(t_{m+}) \neq 0 \quad \text{and} \quad L_{jj}(\mathbf{x}_{m\pm\varepsilon}, t_{m\pm\varepsilon}, \mathbf{p}_j) < 0. \tag{2.124}
$$

Proof. In Theorem 2.30, replacing $L_{ij}(\mathbf{x}_m, t_m, \mathbf{p}_i, \mathbf{p}_j)$ by the global maximum ${}_{G\max}L_{ij}(t_m)$ gives the above corollary. ∎

Corollary 2.6. *For a discontinuous dynamical system in (2.1), there is a point $\mathbf{x}(t_m) = \mathbf{x}_m \in [\mathbf{x}_{m_1}, \mathbf{x}_{m_2}] \subset \widetilde{\partial\Omega}_{ij}$ (or $\widehat{\partial\Omega}_{ij}$) at time t_m between two adjacent domains Ω_α ($\alpha = i,j$). For an arbitrarily small $\varepsilon > 0$, there are two time intervals (i.e., $[t_{m-\varepsilon}, t_m)$ and $(t_m, t_{m+\varepsilon}]$), and $\mathbf{x}^{(i)}(t_{m\pm}) = \mathbf{x}_m = \mathbf{x}^{(j)}(t_{m\pm})$. The flows $\mathbf{x}^{(\alpha)}(t)$*

$(\alpha = i,j)$ are $C^{r_\alpha}_{[t_{m-\varepsilon},t_{m+\varepsilon}]}$-continuous $(r_\alpha \geq 2l)$ for time t. The switching bifurcation of the flow at point $(\mathbf{x}_m, t_m)$ from $\overrightarrow{\partial\Omega}_{ij}$ to $\overleftarrow{\partial\Omega}_{ij}$ (or $\overleftarrow{\partial\Omega}_{ij}$ to $\overrightarrow{\partial\Omega}_{ij}$) occurs if and only if

$$_{G\max}L_{ij}(t_m) = 0, \tag{2.125}$$

$$\mathbf{n}^{\mathrm{T}}_{\partial\Omega_{ij}}.\mathbf{F}^{(\alpha)}(t_{m\pm}) = 0 \quad \text{and} \quad L_{\alpha\alpha}(\mathbf{x}_{m\pm\varepsilon}, t_{m\pm\varepsilon}, \mathbf{p}_\alpha) < 0 \quad \text{for} \quad \alpha = i,j. \tag{2.126}$$

Proof. In Theorem 2.31, replacing $L_{ij}(\mathbf{x}_m, t_m, \boldsymbol{\mu}_i, \boldsymbol{\mu}_j)$ through its global maximum $_{G\max}L_{ij}(t_m)$ gives the above corollary. ■

References

Luo, A.C.J., 2005, A theory for non-smooth dynamic systems on the connectable domains, *Communications in Nonlinear Science and Numerical Simulation*, **10**, pp. 1–55.

Luo, A.C.J., 2006, *Singularity and Dynamics on Discontinuous Vector Fields*, Elsevier: Amsterdam.

Luo, A.C.J., 2011, *Discontinuous Dynamical Systems*, HEP-Springer: Heidelberg.

Chapter 3
Friction-Induced Oscillators

In this chapter, the passability of a flow to the separation boundary for two friction-induced oscillators is presented for a better understanding of cutting mechanism in manufacturing. The friction-induced oscillator with a constant velocity belt is presented first. Through this practical example, the theory for flow singularity and passability in discontinuous dynamical systems is understandable. To further build the basic concepts and knowledge of discontinuous dynamical systems for cutting dynamics, the friction-induced oscillator with a time-varying velocity belt is addressed. From the theory of discontinuous dynamical systems, the analytical conditions for stick and grazing motions to the velocity boundary are presented, and intuitive illustrations are used to help one understand the physical meaning of the mathematical conditions.

3.1 Constant Velocity Boundary

To help one understand the above concepts, as in Luo and Gegg (2006a, b, c, 2007), consider a periodically forced oscillator consisting of a mass (m), a spring of stiffness (k), and a damper of viscous damping coefficient (r), as shown in Fig. 3.1. The grazing flow for this problem was discussed. This oscillator also rests on the horizontal belt surface traveling with a constant speed (V). The absolute coordinate system (x, t) is for the mass. Consider a periodical force $Q_0 \cos \Omega t$ exerted on the mass where Q_0 and Ω are excitation strength and frequency, respectively. Since the mass contacts the moving belt with friction, the mass can move along or rest on the belt surface. Once the nonstick motion exists, a kinetic friction force shown in Fig. 3.1b is described as

$$\overline{F}_f(\dot{x}) \begin{cases} = \mu_k F_N, & \dot{x} \in [V, \infty), \\ \in [-\mu_k F_N, \mu_k F_N], & \dot{x} = V, \\ = -\mu_k F_N, & \dot{x} \in (-\infty, V], \end{cases} \tag{3.1}$$

B.C. Gegg et al., *Machine Tool Vibrations and Cutting Dynamics*,
DOI 10.1007/978-1-4419-9801-9_3, © Springer Science+Business Media, LLC 2011

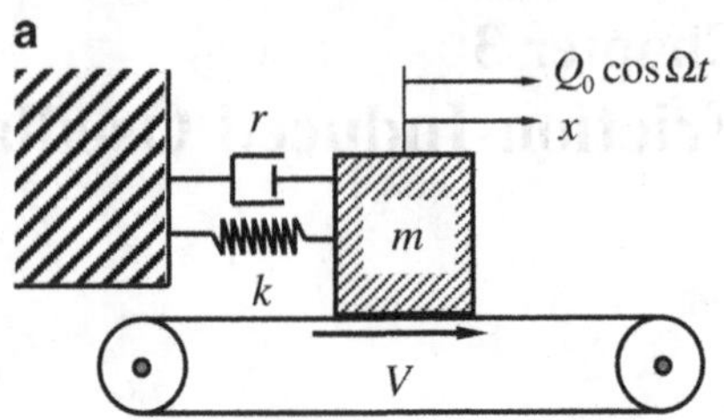

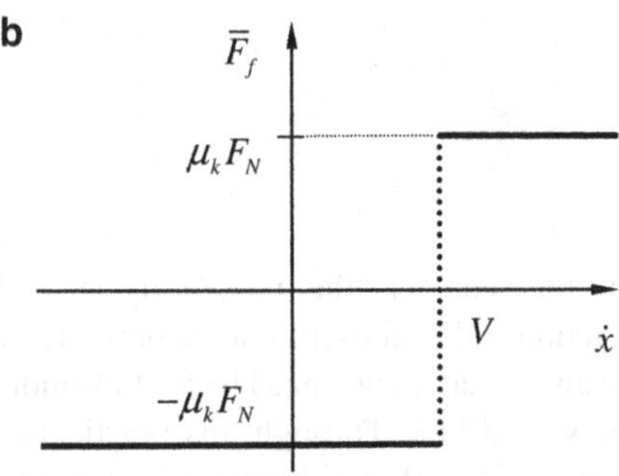

Fig. 3.1 (**a**) Schematic of mechanical model and (**b**) friction force

where $\dot{x} \triangleq \mathrm{d}x/\mathrm{d}t$, μ_k and F_N are friction coefficient and a normal force to the contact surface, respectively. For the model in Fig. 3.1, the normal force is $F_\mathrm{N} = mg$ where g is the gravitational acceleration.

For the mass moving with the same speed of the belt surface, the nonfriction forces per unit mass in the x-direction is defined as

$$F_s = A_0 \cos \Omega t - 2dV - cx, \quad \text{for } \dot{x} = V, \tag{3.2}$$

where $A_0 = Q_0/m, d = r/2m$, and $c = k/m$. This force cannot overcome the friction force during the stick motion (i.e., $|F_s| \leq F_f$ and $F_f = \mu_k F_\mathrm{N}/m$). Therefore, the mass does not have any relative motion to the belt. No acceleration exists, i.e.,

$$\ddot{x} = 0, \quad \text{for } \dot{x} = V. \tag{3.3}$$

If $|F_s| > F_f$, the nonfriction force will overcome the static friction force on the mass and the nonstick motion will appear. For the nonstick motion, the total force acting on the mass is

$$F = A_0 \cos \Omega t - F_f \,\mathrm{sgn}(\dot{x} - V) - 2d\dot{x} - cx, \quad \text{for } \dot{x} \neq V, \tag{3.4}$$

where $\mathrm{sgn}(\cdot)$ is the sign function. Therefore, the equation of the nonstick motion for this oscillator with friction is

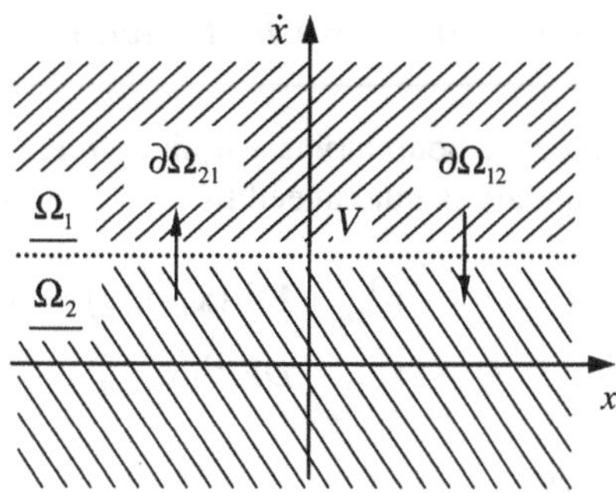

Fig. 3.2 Domain partitions in phase plane

$$\ddot{x} + 2d\dot{x} + cx = A_0 \cos \Omega t - F_f \mathrm{sgn}(\dot{x} - V), \quad \text{for } \dot{x} \neq V. \tag{3.5}$$

Since the friction force is dependent on the direction of the relative velocity, the phase plane is partitioned into two regions in which the motion is described through the continuous dynamical systems, as shown in Fig. 3.2. The two regions are expressed by Ω_α ($\alpha \in \{1, 2\}$). The following vectors are introduced as

$$\mathbf{x} \triangleq (x, \dot{x})^{\mathrm{T}} \equiv (x, y)^{\mathrm{T}} \quad \text{and} \quad \mathbf{F} \triangleq (y, F)^{\mathrm{T}}. \tag{3.6}$$

The mathematical description of the regions and boundary are

$$\left.\begin{aligned} &\Omega_1 = \{(x, y)|y \in (V, \infty)\}, \ \Omega_2 = \{(x, y)|y \in (-\infty, V)\}, \\ &\partial\Omega_{12} = \{(x, y)|\varphi_{12}(x, y) \equiv y - V = 0\}, \\ &\partial\Omega_{21} = \{(x, y)|\varphi_{21}(x, y) \equiv y - V = 0\}. \end{aligned}\right\} \tag{3.7}$$

The boundary $\partial\Omega_{\alpha\beta}$ gives a motion from Ω_α to Ω_β ($\alpha, \beta \in \{1, 2\}$ and $\alpha \neq \beta$). The equations of motion in (3.3) and (3.4) can be described as

$$\dot{\mathbf{x}} = \mathbf{F}^{(\alpha)}(\mathbf{x}, t) \text{ in } \Omega_\alpha (\alpha \in \{1, 2\}), \tag{3.8}$$

where

$$\mathbf{F}^{(\alpha)}(\mathbf{x}, t) = (y, F_\alpha(\mathbf{x}, \Omega t))^{\mathrm{T}}, \tag{3.9}$$

$$F_\alpha(\mathbf{x}, \Omega t) = A_0 \cos \Omega t - b_\alpha - 2d_\alpha y - c_\alpha x. \tag{3.10}$$

Note that $b_1 = \mu g$, $b_2 = -\mu g$, $d_\alpha = d$, and $c_\alpha = c$ for the model in Fig. 3.1.

3.1.1 Grazing Phenomena

Since the boundary is a straight *line*, from Theorem 2.11, the grazing motion to the boundary is guaranteed by

$$\left.\begin{aligned} &\mathbf{n}^{\mathrm{T}}_{\partial\Omega_{\alpha\beta}}\cdot\mathbf{F}^{(\alpha)}(\mathbf{x}_m,t_{m\pm})=0,\quad \alpha\in\{1,2\},\\ &\mathbf{n}^{\mathrm{T}}_{\partial\Omega_{\alpha\beta}}\cdot D\mathbf{F}^{(1)}(\mathbf{x}_m,t_{m\pm})>0,\quad \mathbf{n}^{\mathrm{T}}_{\partial\Omega_{\alpha\beta}}\cdot D\mathbf{F}^{(2)}(\mathbf{x}_m,t)<0,\end{aligned}\right\}\tag{3.11}$$

where

$$D\mathbf{F}^{(\alpha)}(\mathbf{x},t)=(F_\alpha(\mathbf{x},t),\nabla F_\alpha(\mathbf{x},t)\cdot\mathbf{F}^{(\alpha)}(\mathbf{x},t)+\partial_t F_\alpha(\mathbf{x},t))^{\mathrm{T}}.\tag{3.12}$$

where $\nabla=(\partial/\partial x,\ \partial/\partial y)^{\mathrm{T}}$ is the Hamilton operator and $\partial_t=\partial/\partial t$. The time t_m represents the time for the motion on the velocity boundary. $t_{m\pm}=t_m\pm 0$ reflects the responses on the regions rather than boundary. Using the third and fourth equations of (3.7), the normal vector of the boundary is

$$\mathbf{n}_{\partial\Omega_{12}}=\mathbf{n}_{\partial\Omega_{21}}=(0,1)^{\mathrm{T}}.\tag{3.13}$$

Therefore,

$$\left.\begin{aligned} &\mathbf{n}^{\mathrm{T}}_{\partial\Omega_{\alpha\beta}}\cdot\mathbf{F}^{(\alpha)}(\mathbf{x},t)=F_\alpha(\mathbf{x},\Omega t),\\ &\mathbf{n}^{\mathrm{T}}_{\partial\Omega_{\alpha\beta}}\cdot D\mathbf{F}^{(\alpha)}(\mathbf{x},t)=\nabla F_\alpha(\mathbf{x},\Omega t)\cdot\mathbf{F}^{(\alpha)}(\mathbf{x},t)+\partial_t F_\alpha(\mathbf{x},t).\end{aligned}\right\}\tag{3.14}$$

From (3.13) and (3.14), the necessary and sufficient conditions for grazing motions are from Theorem 2.9:

$$F_\alpha(\mathbf{x}_m,\Omega t_{m\pm})=0,\quad F_\alpha(\mathbf{x}_m,\Omega t_{m-\varepsilon})\times F_\alpha(\mathbf{x}_m,\Omega t_{m+\varepsilon})<0.\tag{3.15}$$

or more accurately,

$$\left.\begin{aligned} &F_\alpha(\mathbf{x}_m,\Omega t_{m\pm})=0,\\ &F_1(\mathbf{x}_m,\Omega t_{m-\varepsilon})<0,\quad F_1(\mathbf{x}_m,\Omega t_{m+\varepsilon})>0,\\ &F_2(\mathbf{x}_m,\Omega t_{m-\varepsilon})>0,\quad F_2(\mathbf{x}_m,\Omega t_{m+\varepsilon})<0.\end{aligned}\right\}\tag{3.16}$$

However, from Theorem 2.11, the necessary and sufficient conditions for grazing is given by

$$\left.\begin{aligned} &F_\alpha(\mathbf{x}_m,\Omega t_{m\pm})=0,\\ &\nabla F_\alpha(\mathbf{x}_m,\Omega t_{m\pm})\cdot\mathbf{F}^{(\alpha)}(t_{m\pm})+\partial_t F_\alpha(\mathbf{x}_m,\Omega t_{m\pm})\begin{cases}>0, & \text{for } \alpha=1,\\ <0, & \text{for } \alpha=2.\end{cases}\end{aligned}\right\}\tag{3.17}$$

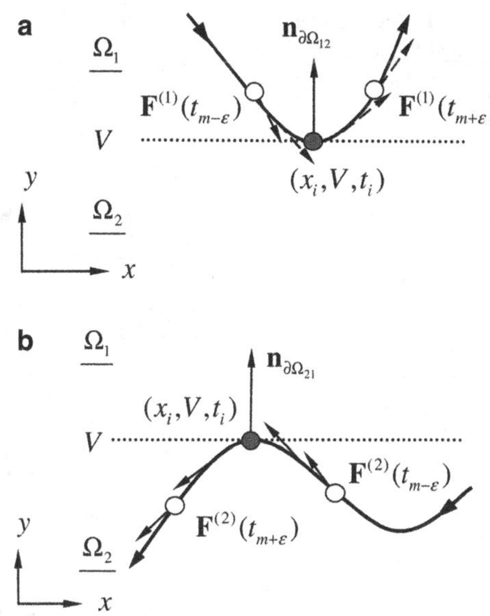

Fig. 3.3 Vector fields for grazing motion in Ω_α $(\alpha = 1, 2)$ and $V > 0$

A sketch of grazing motions in domain Ω_α $(\alpha = 1, 2)$ is illustrated in Fig. 3.3a, b. The grazing conditions are also presented, and the vector fields in Ω_1 and Ω_2 are expressed by the dashed and solid arrow-lines, respectively. The condition in (3.17) for the grazing motion in Ω_α is presented by the vector fields of $\mathbf{F}^{(\alpha)}(t)$. In addition to $F_\alpha(\mathbf{x}_m, t_{m\pm}) = 0$, the sufficient condition also requires $F_1(\mathbf{x}_m, \Omega t_{m-\varepsilon}) < 0$ and $F_1(\mathbf{x}_m, \Omega t_{m+\varepsilon}) > 0$ in domain Ω_1; and $F_2(\mathbf{x}_m, \Omega t_{m-\varepsilon}) > 0$ and $F_2(\mathbf{x}_m, \Omega t_{m+\varepsilon}) < 0$ in domain Ω_2. For detailed discussion, the reader can refer to Luo and Gegg (2006a, b).

After grazing motion, the sliding motion will appear. Direct integration of (3.3) with initial condition (t_i, x_i, V) gives the sliding motion, i.e.,

$$x = V \times (t - t_i) + x_i. \tag{3.18}$$

Substitution of (3.18) into (3.10) gives the forces for a small δ-neighborhood of the stick motion ($\delta \to 0$) in the two domains Ω_α $(\alpha \in \{1, 2\})$, i.e.,

$$F_\alpha(\mathbf{x}_m, \Omega t_{m-}) = -2d_\alpha V - c_\alpha[V \times (t_m - t_i) + x_i] + A_0 \cos \Omega t_m - b_\alpha. \tag{3.19}$$

For the nonstick motion, select the initial condition on the velocity boundary (i.e., $\dot{x}_i = V$), and then the coefficients of the solution which are described in Appendix A, $C_k^{(\alpha)}(x_i, \dot{x}_i, t_i) \triangleq C_k^{(\alpha)}(x_i, t_i)$ for $k = 1, 2$. The basic solutions in Appendix A are used for the construction of mappings. In phase plane, the trajectories in Ω_α starting and ending at the velocity boundary (i.e., from $\partial\Omega_{\beta\alpha}$ to $\partial\Omega_{\alpha\beta}$) are illustrated in Fig. 3.4. The starting and ending points for mappings P_α in Ω_α are (x_i, V, t_i) and (x_{i+1}, V, t_{i+1}), respectively. The stick mapping is P_0. Define the switching planes as

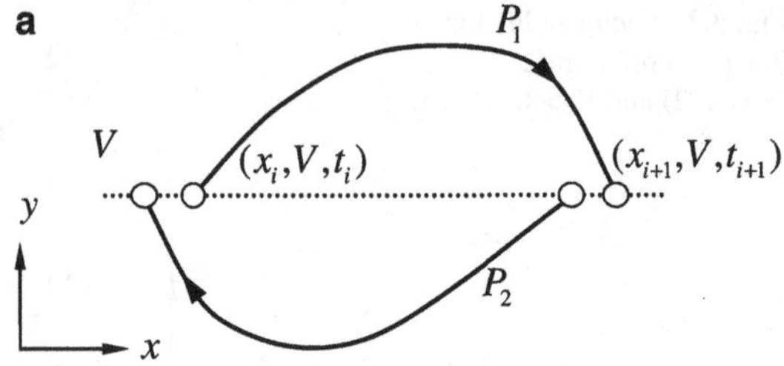

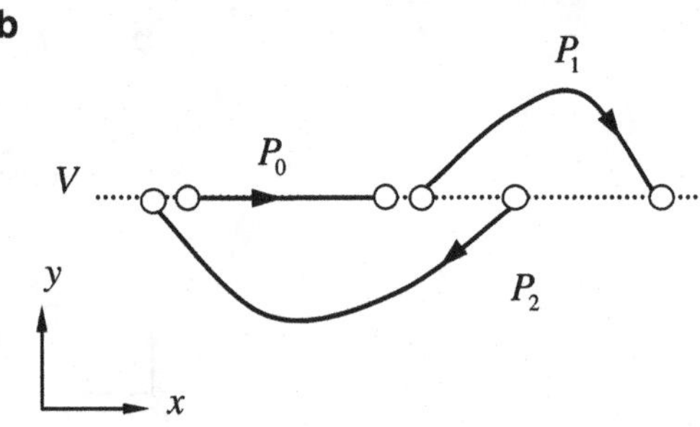

Fig. 3.4 Regular mappings (P_1 and P_2) and stick mapping P_0

$$\left.\begin{aligned} \Xi^0 &= \{(x_i, \Omega t_i) | \dot{x}_i(t_i) = V\}, \\ \Xi^1 &= \{(x_i, \Omega t_i) | \dot{x}_i(t_i) = V^+\}, \\ \Xi^2 &= \{(x_i, \Omega t_i) | \dot{x}_i(t_i) = V^-\}, \end{aligned}\right\} \tag{3.20}$$

where $V^- = \lim_{\delta \to 0} (V - \delta)$ and $V^+ = \lim_{\delta \to 0} (V + \delta)$ for arbitrarily small $\delta > 0$. Thus,

$$P_1 : \Xi^1 \to \Xi^1, \quad P_2 : \Xi^2 \to \Xi^2, \quad P_0 : \Xi^0 \to \Xi^0. \tag{3.21}$$

From the foregoing two equations, we have

$$\left.\begin{aligned} P_0 &: (x_i, V, t_i) \to (x_{i+1}, V, t_{i+1}), \\ P_1 &: (x_i, V^+, t_i) \to (x_{i+1}, V^+, t_{i+1}), \\ P_2 &: (x_i, V^-, t_i) \to (x_{i+1}, V^-, t_{i+1}). \end{aligned}\right\} \tag{3.22}$$

The governing equations for P_0 with $\alpha \in \{1, 2\}$ are

$$\begin{aligned} &- x_{i+1} + V \times (t_{i+1} - t_i) + x_i = 0, \\ &2d_\alpha V + c_\alpha [V \times (t_{i+1} - t_i) + x_i] - A_0 \cos \Omega t_{i+1} + b_\alpha = 0. \end{aligned} \tag{3.23}$$

The mapping P_0 described the starting and ending of the stick motion, the disappearance of stick motion requires $F_\alpha(\mathbf{x}_{i+1}, \Omega t_{i+1}) = 0$. This chapter will not use the stick mapping to discuss the grazing flow, which is presented herein as a generic mapping. For sliding motion, the stick mapping will be used, and the reader can refer to Luo and Gegg (2006a, b, 2007) for more information. From this problem, the two domains Ω_α ($\alpha = 1$ or 2) are unbounded. The flows of the dynamical systems on the corresponding domains should be bounded from Hypothesis (H2.1)–(H2.3) in nonsmooth dynamical systems. Therefore, for nonstick motion, there are three possible stable motions in the two domains Ω_α ($\alpha \in \{1,2\}$), and the governing equations of mapping P_α ($\alpha \in \{1,2\}$) are obtained from the displacement and velocity responses for the three cases of motions in Appendix A. Therefore, the governing equations of mapping P_α ($\alpha \in \{0,1,2\}$) can be expressed by

$$\begin{aligned} f_1^{(\alpha)}(x_i, \Omega t_i, x_{i+1}, \Omega t_{i+1}) &= 0, \\ f_2^{(\alpha)}(x_i, \Omega t_i, x_{i+1}, \Omega t_{i+1}) &= 0. \end{aligned} \tag{3.24}$$

If the grazing for the two nonstick mappings occurs at the final state (x_{i+1}, V, t_{i+1}), from (3.17), the grazing conditions based on mappings are obtained, i.e.,

$$\left.\begin{aligned} &F_\alpha(x_{i+1}, V, \Omega t_{i+1}) = 0, \\ &\nabla F_\alpha(\mathbf{x}_{i+1}, \Omega t_{i+1}) \cdot \mathbf{F}^{(\alpha)}(t_{i+1}) + \frac{\partial F_\alpha(\mathbf{x}_{i+1}, \Omega t_{i+1})}{\partial t} \begin{cases} > 0 & \text{for } \alpha = 1, \\ < 0 & \text{for } \alpha = 2. \end{cases} \end{aligned}\right\} \tag{3.25}$$

With (3.12), the grazing condition becomes

$$\left.\begin{aligned} &A_0 \cos \Omega t_{i+1} - b_\alpha - 2d_\alpha V - c_\alpha x_{i+1} = 0, \\ &- c_\alpha V - A_0 \Omega \sin \Omega t_{i+1} \begin{cases} > 0 & \text{for } \alpha = 1, \\ < 0 & \text{for } \alpha = 2. \end{cases} \end{aligned}\right\} \tag{3.26}$$

To ensure the initial switching sets to be passable, from Luo (2005, 2006), the initial switching sets of mapping P_α ($\alpha \in \{1,2\}$) should satisfy the following condition as in Luo and Gegg (2006a).

$$\left.\begin{aligned} F_1(x_i, V^-, \Omega t_i) < 0 \text{ and } F_2(x_i, V^-, \Omega t_i) < 0 \quad \text{for } \Omega_1 \to \Omega_2, \\ F_1(x_i, V^+, \Omega t_i) > 0 \text{ and } F_2(x_i, V^-, \Omega t_i) > 0 \quad \text{for } \Omega_2 \to \Omega_1. \end{aligned}\right\} \tag{3.27}$$

To make sure motions relative to mappings P_α ($\alpha = 1,2$) exist, the initial switching force product $F_1 \times F_2$ at the boundary should be nonnegative. The comprehensive discussion of the foregoing condition can be referred to Luo and Gegg (2006c). The condition of (3.27) guarantees that the flow relative to the initial switching sets of mapping P_α ($\alpha \in \{1,2\}$) is passable on the discontinuous boundary (i.e., $y_i = V$). The force product for the initial switching sets is also illustrated to ensure the nonstick mapping exists. The force conditions for the final switching sets of mapping P_α ($\alpha \in \{1,2\}$) is presented in (3.15). However, the equivalent grazing conditions based on (3.17) give the inequality condition in (3.26).

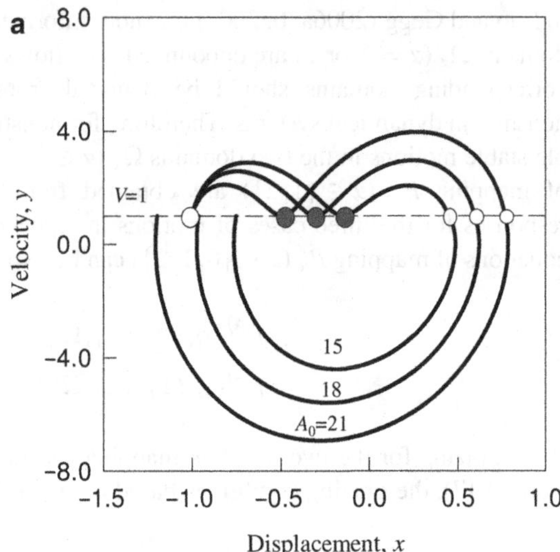

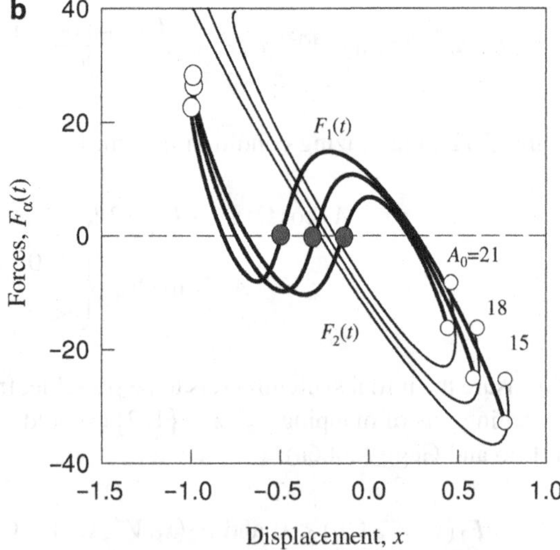

Fig. 3.5 Grazing motion of mapping P_1 for $A_0 = 15,\ 18,\ 21$: (**a**) phase trajectory, (**b**) forces distribution along displacement, (**c**) velocity time-history, and (**d**) force distribution on velocity ($\Omega = 8,\ V = 1,\ d_1 = 1,\ d_2 = 0,\ b_1 = -b_2 = 3$, and $c_1 = c_2 = 30$). The initial conditions are $(x_i, y_i) = (-1., 1.)$ and $\Omega t_i \approx 1.3617,\ 1.6958,\ 1.4830$, accordingly

To illustrate the analytical prediction of the grazing motions, the motion responses of the oscillator will be demonstrated through time-history responses and trajectories in phase plane. The grazing strongly depends on the force responses in this discontinuous dynamical system. The force responses are presented to illustrate the force criteria for the grazing motions in such a friction-induced oscillator. The starting and grazing points of mapping P_α ($\alpha \in \{1, 2\}$) are represented by the large, hollow and dark-solid circular symbols, respectively. The switching points from domain α to β ($\alpha, \beta \in \{1, 2\},\ \alpha \neq \beta$) are depicted by smaller circular symbols. In Fig. 3.5, phase trajectories, forces distribution along displacement, velocity time-history and forces distribution on velocity are presented for the grazing motion of mapping P_1. The parameters ($\Omega = 8,\ V = 1,$

Fig. 3.5 (continued)

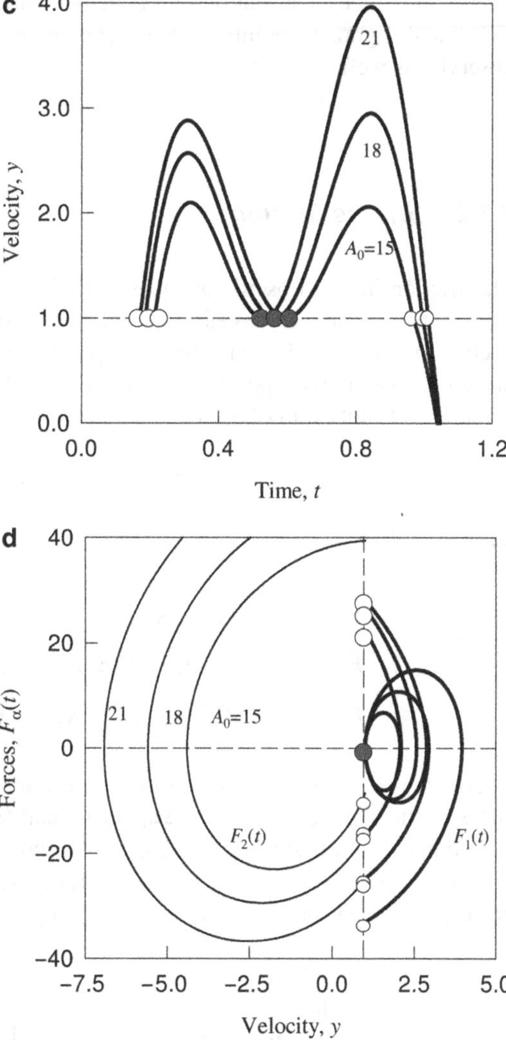

and $b_1 = -b_2 = 3$) plus the initial conditions $(x_i, y_i) = (-1., 1.)$, and $(\Omega t_i \approx 1.3617, 1.6958, 1.4830)$ corresponding to $(A_0 = 15,\ 18,\ 21)$ are adopted. In phase plane, the three grazing trajectories are tangential to the discontinuous boundary (i.e., $y = V$), which are seen in Fig. 3.5a. In Fig. 3.5b, the thick- and thin-solid curves represent the forces $F_1(t)$ and $F_2(t)$, respectively. From the force distribution along displacement, the force $F_1(t)$ has a sign change from negative to positive. This indicates that the grazing conditions in (3.15) are satisfied. The forces $F_1(t)$ and $F_2(t)$ at the switching points from domain Ω_1 to Ω_2 have a jump with the same sign, which satisfies (3.16). In the velocity time-history plot, the velocity curves are tangential to the discontinuous boundary (see Fig. 3.5c).

Finally, the forces distributions along velocity are presented in Fig. 3.5d. The force $F_1(t)$ at the grazing points is zero. The force jump from domain Ω_1 to Ω_2 is observed as well.

3.1.2 Sliding Motion

The friction-induced oscillator in (3.2)–(3.5) is considered herein as an example to demonstrate how to develop the analytical conditions for sliding motion. The vectors in (3.8) are adopted. The corresponding subdomains and boundary in (3.9) are used. The singular points are at $(\pm\infty, V)$. To include the sliding motion, the equations of motion in (3.3) and (3.5) can be described as

$$\dot{\mathbf{x}} = \mathbf{F}^{(\lambda)}(\mathbf{x}, t), \quad \lambda \in \{0, \alpha\}, \tag{3.28}$$

where

$$\left.\begin{aligned} &\mathbf{F}^{(\alpha)}(\mathbf{x}, t) = (y, F_\alpha(\mathbf{x}, \Omega t))^{\mathrm{T}} \text{ in } \Omega_\alpha \quad (\alpha \in \{1,2\}),\\ &\mathbf{F}^{(0)}_{\alpha\beta}(\mathbf{x}, t) = (V, 0)^{\mathrm{T}} \text{ on } \widetilde{\partial\Omega}_{\alpha\beta} \quad (\alpha, \beta \in \{1,2\}, \alpha \neq \beta),\\ &\mathbf{F}^{(0)}_{\alpha\beta}(\mathbf{x}, t) \in [\mathbf{F}^{(\alpha)}(\mathbf{x}, t), \mathbf{F}^{(\beta)}(\mathbf{x}, t)] \quad \text{on } \overrightarrow{\partial\Omega}_{\alpha\beta}. \end{aligned}\right\} \tag{3.29}$$

Notice that $F_\alpha(\mathbf{x}, \Omega t)$ is given in (3.10). For $\widetilde{\partial\Omega}_{\alpha\beta}$, the critical initial and final states of the sliding motion are $(\Omega t_{\mathrm{c}}, x_{\mathrm{c}}, V)$ and $(\Omega t_{\mathrm{f}}, x_{\mathrm{f}}, V)$, accordingly. Consider the sliding motion starting at $(\Omega t_i, x_i, V)$ and ending at $(\Omega t_{i+1}, x_{i+1}, V)$, where $(\Omega t_{i+1}, x_{i+1}, V) \triangleq (\Omega t_{\mathrm{f}}, x_{\mathrm{f}}, V)$. From Theorem 2.4, the sliding motion on the boundary is guaranteed for $t_m \in [t_i, t_{i+1}) \subseteq [t_{\mathrm{c}}, t_{\mathrm{f}})$ by

$$\begin{aligned} &\text{either} \quad \left.\begin{aligned} &\mathbf{n}^{\mathrm{T}}_{\partial\Omega_{\alpha\beta}} \cdot \mathbf{F}^{(\alpha)}(\mathbf{x}_m, t_{m-}) < 0\\ &\mathbf{n}^{\mathrm{T}}_{\partial\Omega_{\alpha\beta}} \cdot \mathbf{F}^{(\beta)}(\mathbf{x}_m, t_{m-}) > 0 \end{aligned}\right\} \quad \text{for } \mathbf{n}_{\partial\Omega_{\alpha\beta}} \to \Omega_\alpha\\ &\text{or} \quad \left.\begin{aligned} &\mathbf{n}^{\mathrm{T}}_{\partial\Omega_{\alpha\beta}} \cdot \mathbf{F}^{(\alpha)}(\mathbf{x}_m, t_{m-}) > 0\\ &\mathbf{n}^{\mathrm{T}}_{\partial\Omega_{\alpha\beta}} \cdot \mathbf{F}^{(\beta)}(\mathbf{x}_m, t_{m-}) < 0 \end{aligned}\right\} \quad \text{for } \mathbf{n}_{\partial\Omega_{\alpha\beta}} \to \Omega_\beta. \end{aligned} \tag{3.30}$$

In other words, one can obtain

$$[\mathbf{n}^{\mathrm{T}}_{\partial\Omega_{\alpha\beta}} \cdot \mathbf{F}^{(\alpha)}(\mathbf{x}_m, t_{m-})] \times [\mathbf{n}^{\mathrm{T}}_{\partial\Omega_{\alpha\beta}} \cdot \mathbf{F}^{(\beta)}(\mathbf{x}_m, t_{m-})] < 0. \tag{3.31}$$

For a boundary $\overrightarrow{\partial\Omega}_{\alpha\beta}$ with nonzero measure, the starting and ending states of the passable motion are $(\Omega t_{\mathrm{s}}, x_{\mathrm{s}}, V)$ and $(\Omega t_{\mathrm{e}}, x_{\mathrm{e}}, V)$ accordingly. From Theorem 2.2, the nonsliding motion (or called passable motion) to the boundary in Luo (2005, 2006a) is guaranteed for $t_m \in (t_s, t_e)$ by

$$
\begin{aligned}
&\text{either} \quad \left.\begin{aligned} \mathbf{n}^{\mathrm{T}}_{\partial\Omega_{\alpha\beta}} \cdot \mathbf{F}^{(\alpha)}(\mathbf{x}_m, t_{m-}) < 0 \\ \mathbf{n}^{\mathrm{T}}_{\partial\Omega_{\alpha\beta}} \cdot \mathbf{F}^{(\beta)}(\mathbf{x}_m, t_{m+}) < 0 \end{aligned}\right\} \quad \text{for } \mathbf{n}_{\partial\Omega_{\alpha\beta}} \to \Omega_\alpha \\
&\text{or} \quad \left.\begin{aligned} \mathbf{n}^{\mathrm{T}}_{\partial\Omega_{\alpha\beta}} \cdot \mathbf{F}^{(\alpha)}(\mathbf{x}_m, t_{m-}) > 0 \\ \mathbf{n}^{\mathrm{T}}_{\partial\Omega_{\alpha\beta}} \cdot \mathbf{F}^{(\beta)}(\mathbf{x}_m, t_{m+}) > 0 \end{aligned}\right\} \quad \text{for } \mathbf{n}_{\partial\Omega_{\alpha\beta}} \to \Omega_\beta.
\end{aligned} \tag{3.32}
$$

In other words,

$$[\mathbf{n}^{\mathrm{T}}_{\partial\Omega_{\alpha\beta}} \cdot \mathbf{F}^{(\alpha)}(\mathbf{x}_m, t_{m-})] \times [\mathbf{n}^{\mathrm{T}}_{\partial\Omega_{\alpha\beta}} \cdot \mathbf{F}^{(\beta)}(\mathbf{x}_m, t_{m+})] > 0. \tag{3.33}$$

For the boundary switching from $\overrightarrow{\partial\Omega}_{\alpha\beta}$ to $\widetilde{\partial\Omega}_{\alpha\beta}$, one has $t_e = t_c$, otherwise, $t_f = t_s$. Suppose the starting state is a switching state of the sliding motion, then from Theorem 2.15, the switching condition of the sliding motion from $\overrightarrow{\partial\Omega}_{\alpha\beta}$ to $\widetilde{\partial\Omega}_{\alpha\beta}$ at $t_m = t_c$ is

$$
\left.\begin{aligned}
&\text{either } \mathbf{n}^{\mathrm{T}}_{\partial\Omega_{\alpha\beta}} \cdot \mathbf{F}^{(\alpha)}(\mathbf{x}_m, t_{m-}) < 0, \quad \text{for } \mathbf{n}_{\partial\Omega_{\alpha\beta}} \to \Omega_\alpha \\
&\text{or} \quad \mathbf{n}^{\mathrm{T}}_{\partial\Omega_{\alpha\beta}} \cdot \mathbf{F}^{(\alpha)}(\mathbf{x}_m, t_{m-}) > 0, \quad \text{for } \mathbf{n}_{\partial\Omega_{\alpha\beta}} \to \Omega_\beta, \\
&\mathbf{n}^{\mathrm{T}}_{\partial\Omega_{\alpha\beta}} \cdot \mathbf{F}^{(\beta)}(\mathbf{x}_m, t_{m\pm}) = 0, \\
&\mathbf{n}^{\mathrm{T}}_{\partial\Omega_{\alpha\beta}} \cdot D\mathbf{F}^{(\beta)}(\mathbf{x}_m, t_{m\pm}) > 0, \quad \text{for } \mathbf{n}_{\partial\Omega_{\alpha\beta}} \to \Omega_\beta, \\
&\mathbf{n}^{\mathrm{T}}_{\partial\Omega_{\alpha\beta}} \cdot D\mathbf{F}^{(\beta)}(\mathbf{x}_m, t_{m\pm}) < 0, \quad \text{for } \mathbf{n}_{\partial\Omega_{\alpha\beta}} \to \Omega_\alpha.
\end{aligned}\right\} \tag{3.34}
$$

From Theorem 2.24, the sliding motion vanishing from $\widetilde{\partial\Omega}_{\alpha\beta}$ to $\overrightarrow{\partial\Omega}_{\alpha\beta}$ and going into the domain Ω_β at $t_m = t_f$ requires:

$$
\left.\begin{aligned}
&\text{either } \mathbf{n}^{\mathrm{T}}_{\partial\Omega_{\alpha\beta}} \cdot \mathbf{F}^{(\alpha)}(\mathbf{x}_m, t_{m-}) < 0, \quad \text{for } \mathbf{n}_{\partial\Omega_{\alpha\beta}} \to \Omega_\alpha \\
&\text{or} \quad \mathbf{n}^{\mathrm{T}}_{\partial\Omega_{\alpha\beta}} \cdot \mathbf{F}^{(\alpha)}(\mathbf{x}_m, t_{m-}) > 0, \quad \text{for } \mathbf{n}_{\partial\Omega_{\alpha\beta}} \to \Omega_\beta, \\
&\mathbf{n}^{\mathrm{T}}_{\partial\Omega_{\alpha\beta}} \cdot \mathbf{F}^{(\beta)}(\mathbf{x}_m, t_{m\mp}) = 0, \\
&\mathbf{n}^{\mathrm{T}}_{\partial\Omega_{\alpha\beta}} \cdot D\mathbf{F}^{(\beta)}(\mathbf{x}_m, t_{m\mp}) > 0, \quad \text{for } \mathbf{n}_{\partial\Omega_{\alpha\beta}} \to \Omega_\beta, \\
&\mathbf{n}^{\mathrm{T}}_{\partial\Omega_{\alpha\beta}} \cdot D\mathbf{F}^{(\beta)}(\mathbf{x}_m, t_{m\mp}) < 0, \quad \text{for } \mathbf{n}_{\partial\Omega_{\alpha\beta}} \to \Omega_\alpha.
\end{aligned}\right\} \tag{3.35}
$$

From (3.34) and (3.35), the switching conditions for the sliding motion to the passable motion and vice versa are summarized as

$$\left.\begin{array}{l}[\mathbf{n}_{\partial\Omega_{\alpha\beta}}^{\mathrm{T}}\cdot\mathbf{F}^{(\alpha)}(\mathbf{x}_m,t_{m-})]\times[\mathbf{n}_{\partial\Omega_{\alpha\beta}}^{\mathrm{T}}\cdot\mathbf{F}^{(\beta)}(\mathbf{x}_m,t_{m\pm})]=0.\\ \mathbf{n}_{\partial\Omega_{\alpha\beta}}^{\mathrm{T}}\cdot D\mathbf{F}^{(\beta)}(\mathbf{x}_m,t_{m\pm})>0,\quad \text{for }\mathbf{n}_{\partial\Omega_{\alpha\beta}}\to\Omega_\beta,\\ \mathbf{n}_{\partial\Omega_{\alpha\beta}}^{\mathrm{T}}\cdot D\mathbf{F}^{(\beta)}(\mathbf{x}_m,t_{m\pm})<0,\quad \text{for }\mathbf{n}_{\partial\Omega_{\alpha\beta}}\to\Omega_\alpha.\end{array}\right\}\tag{3.36}$$

The normal vector of the boundary $\partial\Omega_{12}$ and $\partial\Omega_{21}$ is given in (3.13). Therefore, the projection of the vector field in the normal direction of the boundary is

$$\mathbf{n}_{\partial\Omega_{\alpha\beta}}^{\mathrm{T}}\cdot\mathbf{F}^{(\alpha)}(\mathbf{x},t)=\mathbf{n}_{\partial\Omega_{\beta\alpha}}^{\mathrm{T}}\cdot\mathbf{F}^{(\alpha)}(\mathbf{x},t)=F_\alpha(\mathbf{x},\Omega t).\tag{3.37}$$

which is the force for the oscillator in (3.28). For a better understanding of the force characteristic of the sliding motion, the conditions for sliding and nonsliding motions in (3.30) and (3.32) can be re-written by

$$F_1(\mathbf{x}_m,\Omega t_{m-})<0 \text{ and } F_2(\mathbf{x}_m,\Omega t_{m-})>0\quad \text{for }\widetilde{\partial\Omega}_{12}\tag{3.38}$$

and

$$\left.\begin{array}{l}F_1(\mathbf{x}_m,\Omega t_{m-})<0 \text{ and } F_2(\mathbf{x}_m,\Omega t_{m+})<0\quad \text{for }\overrightarrow{\partial\Omega}_{12},\\ F_2(\mathbf{x}_m,\Omega t_{m-})>0 \text{ and } F_1(\mathbf{x}_m,\Omega t_{m+})>0\quad \text{for }\overrightarrow{\partial\Omega}_{21}.\end{array}\right\}\tag{3.39}$$

Equations (3.34) and (3.35), respectively, give the switching conditions for the sliding motion in (3.30), i.e.,

$$\left.\begin{array}{l}F_1(\mathbf{x}_m,\Omega t_{m-})<0,\ F_2(\mathbf{x}_m,\Omega t_{m\pm})=0 \text{ and } D\mathbf{F}^{(2)}(\mathbf{x}_m,t_{m\pm})<0\\ \quad\text{for }\overrightarrow{\partial\Omega}_{12}\to\widetilde{\partial\Omega}_{12},\\ F_2(\mathbf{x}_m,\Omega t_{m-})>0,\ F_1(\mathbf{x}_m,\Omega t_{m\pm})=0\ \text{ and } D\mathbf{F}^{(1)}(\mathbf{x}_m,t_{m\pm})>0\\ \quad\text{for }\overrightarrow{\partial\Omega}_{21}\to\widetilde{\partial\Omega}_{21}\end{array}\right\}\tag{3.40}$$

at $t_m=t_c$, and

$$\left.\begin{array}{l}F_1(\mathbf{x}_m,\Omega t_{m-})<0,\ F_2(\mathbf{x}_m,\Omega t_{m\mp})=0 \text{ and } D\mathbf{F}^{(2)}(\mathbf{x}_m,t_{m\mp})<0\\ \quad\text{for }\widetilde{\partial\Omega}_{12}\to\Omega_2,\\ F_2(\mathbf{x}_m,\Omega t_{m-})>0,\ F_1(\mathbf{x}_m,\Omega t_{m\mp})=0 \text{ and } D\mathbf{F}^{(1)}(\mathbf{x}_m,t_{m\mp})>0\\ \quad\text{for }\widetilde{\partial\Omega}_{12}\to\Omega_1,\end{array}\right\}\tag{3.41}$$

at $t_m=t_f$.

A sketch of the vector field and classification of the sliding motion is illustrated in Fig. 3.6. In Fig. 3.6a, the switching condition for the sliding motion is presented through the vector fields of $\mathbf{F}^{(1)}(\mathbf{x},t)$ and $\mathbf{F}^{(2)}(\mathbf{x},t)$. The ending condition for sliding

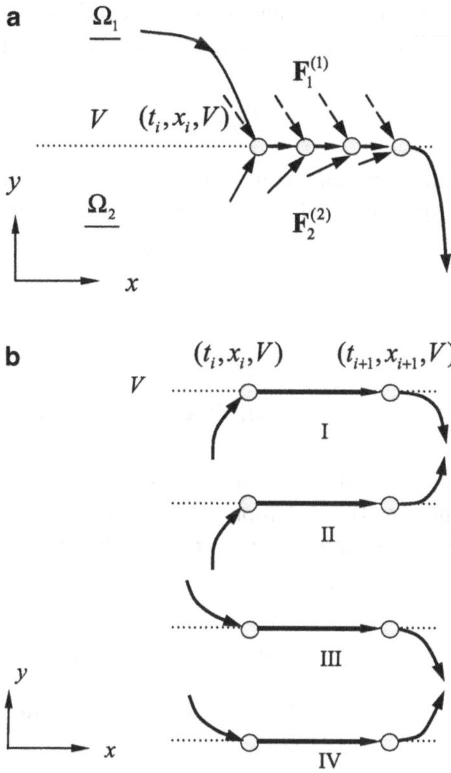

Fig. 3.6 (**a**) The vector field and (**b**) classification of sliding motions for belt speed $V > 0$

(or stick) motion along the velocity boundary is illustrated by $F_2(t_{m-}) = 0$. However, the starting point of the sliding motion may not be switching points from the possible boundary to the sliding motion boundary. When the flow arrives to the boundary, once (3.40) holds, the sliding motion along the boundary will be formed. From the switching conditions in (3.40) and (3.41), there are four possible sliding motions (I)–(IV), as shown in Fig. 3.6b. The switching condition in (3.40) is the critical condition for the formation of the sliding motion. The above switching conditions for onset, forming, and vanishing of the sliding motions along the boundary at a certain velocity are strongly dependent on the total force acting on the oscillator. Because the total force can be contributed from the linear or nonlinear continuous forces from spring and/or damper, such switching conditions can be applied to dynamical systems possessing nonlinear, continuous spring and viscous damping forces with a nonlinear friction with a C^0-discontinuity. The three generic mappings in (3.20) is governed the algebraic equations in (3.22). For nonsliding mapping, two equations are given. Two equations with four unknowns in (3.22) cannot give the unique solutions for the sliding motion. Once the initial state is given, the final state of the sliding motion is uniquely determined. Consider the switching states $(\Omega t_c, x_c, V)$ and $(\Omega t_f, x_f, V)$ as the critical and final conditions of the sliding motion, respectively. For the time interval $[t_i, t_{i+1}] \subseteq [t_c, t_f]$ of any sliding motion, the sliding motion requires $(\Omega t_{i+1}, x_{i+1}, V) \triangleq (\Omega t_f, x_f, V)$.

From (3.29) and (3.40), the initial condition of the sliding motion for all $t_i \in (t_c, t_f)$ and on the boundary $\widetilde{\partial\Omega}_{\alpha\beta}$ satisfies the following force relation

$$L_{12}(t_i) = F_1(x_i, V, \Omega t_i) \times F_2(x_i, V, \Omega t_i) < 0. \tag{3.42}$$

The foregoing equation is the product of normal vector fields in (3.30). The switching condition in (3.36) or (3.41) for the sliding motion at the critical time $t_i = t_c$ also gives the *initial force product condition*, i.e.,

$$\left.\begin{aligned} &L_{12}(t_c) = F_1(x_c, V, \Omega t_c) \times F_2(x_c, V, \Omega t_c) = 0, \\ &D\mathbf{F}^{(2)}(\mathbf{x}_c, t_{c\pm}) < 0 \quad \text{for } \overrightarrow{\partial\Omega}_{12} \to \widetilde{\partial\Omega}_{12}, \\ &D\mathbf{F}^{(1)}(\mathbf{x}_c, t_{c\pm}) > 0 \quad \text{for } \overrightarrow{\partial\Omega}_{21} \to \widetilde{\partial\Omega}_{21}. \end{aligned}\right\} \tag{3.43}$$

If the initial condition satisfies (3.42), the sliding motion is called the critical sliding motion. The condition in (3.37) or (3.43) for the sliding motion at the time $t_{i+1} = t_f$ gives the *final force product condition*, i.e.,

$$\left.\begin{aligned} &L_{12}(t_{i+1}) = F_1(x_{i+1}, V, \Omega t_{i+1}) \times F_2(x_{i+1}, V, \Omega t_{i+1}) = 0. \\ &D\mathbf{F}^{(2)}(\mathbf{x}_{i+1}, t_{(i+1)\mp}) < 0 \quad \text{for } \widetilde{\partial\Omega}_{12} \to \Omega_2, \\ &D\mathbf{F}^{(1)}(\mathbf{x}_{i+1}, t_{(i+1)\mp}) > 0 \quad \text{for } \widetilde{\partial\Omega}_{12} \to \Omega_1. \end{aligned}\right\} \tag{3.44}$$

From (3.44), it is observed that the force product on the boundary is also very significant for the friction oscillator. Once the force product of the sliding motion for $t_m \in (t_i, t_{i+1})$ changes its sign, the sliding motion for the friction-induced oscillator will vanish. So the characteristic of the force product for the sliding motion should be further discussed. To explain the force mechanism of the sliding motion, the force product for sliding motion starting at $(\Omega t_i, x_i, V)$ and ending at $(\Omega t_{i+1}, x_{i+1}, V)$ is sketched in Fig. 3.7 for given parameters with the

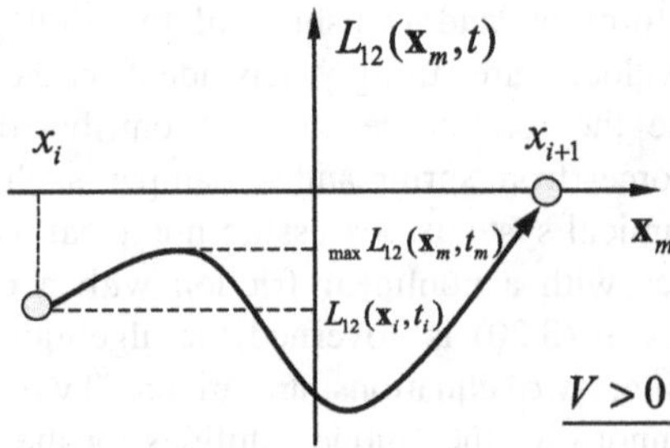

Fig. 3.7 A sketch of a force product for sliding motion under given parameters for belt speed $V > 0$. ${}_{\max}L_{12}(\mathbf{x}_m, t_m)$ and $L_{12}(\mathbf{x}_i, t_i)$ represent the local *maximum* and *initial* force products, respectively. x_i and x_{i+1} represent the switching displacements of the stating and vanishing points of the sliding motion, respectively

belt speed $V > 0$. x_i and x_{i+1} represent the switching displacements of the starting and ending points of the sliding motion, respectively.

From the local normal vector product, the local peak force product relative to the domains Ω_1 and Ω_2 is defined as

$$_{\max}L_{12}(t_m) = \left\{ L_{12}(\mathbf{x}_m,\ t_m) \left| \begin{array}{l} \forall t_m \in (t_i, t_{i+1}), \exists \mathbf{x}_m \in (\mathbf{x}_i, \mathbf{x}_{i+1}) \\ DL_{12}(\mathbf{x}, t)|_{(\mathbf{x}_m, t_m)} = 0 \text{ and} \\ D^2L_{12}(\mathbf{x}, t)|_{(\mathbf{x}_m, t_m)} < 0 \end{array} \right. \right\} \tag{3.45}$$

and from the global normal vector field product, the maximum force product for the computational convenience is defined as

$$_{\mathrm{Gmax}}L_{12}(t_k) = \max_{t_k \in (t_i, t_{i+1})} \{L_{12}(\mathbf{x}_i, t_i), {}_{\max}L_{12}(\mathbf{x}_\mathrm{m}, t_\mathrm{m})\}. \tag{3.46}$$

Using the chain rule and $\mathrm{d}x/\mathrm{d}t = V$ for the sliding motion, the above definitions are described as

$$_{\max}L_{12}(x_m) = \left\{ L_{12}(x_m) \left| \begin{array}{l} \forall x_m \in (x_i, x_{i+1}), \exists t_m = \dfrac{x_m - x_i}{V} \in [t_i, t_{i+1}], \\ \dfrac{dL_{12}(x)}{dx}\Big|_{x=x_m} = 0 \text{ and } \dfrac{d^2L_{12}(x)}{dx^2}\Big|_{x=x_m} < 0 \end{array} \right. \right\} \tag{3.47}$$

$$_{\mathrm{Gmax}}L_{12}(x_m) = \max_{x_m \in (x_i, x_{i+1})} \{L_{12}(x_i), {}_{\max}L_{12}(x_m)\}. \tag{3.48}$$

Once one of the global maximum force products is greater than zero, the sliding motion will disappear. Further, the sliding motion will be fragmentized. The corresponding critical condition is

$$L_{12}(t_m) = 0 \text{ or } {}_{\mathrm{Gmax}}L_{12}(x_m) = 0, \quad \text{for } t_m \in (t_i, t_{i+1}) \tag{3.49}$$

The foregoing condition is termed the *global maximum force product condition* of the sliding fragmentation. For simplicity, it is also called the *sliding fragmentation condition.* After fragmentation, consider the starting to ending points of the two sliding motions to be $(\Omega t_i, x_i, V)$ to $(\Omega t_{i+1}, x_{i+1}, V)$ and $(\Omega t_{i+2}, x_{i+2}, V)$ to $(\Omega t_{i+3}, x_{i+3}, V)$, respectively. It is assumed that $t_i < t_{i+1} \leq t_{i+2} < t_{i+3}$. The nonsliding motion between the two fragmentized motions requires from (3.33)

$$L_{12}(t_m) > 0 \quad \text{for } t_m \in (t_{i+1}, t_{i+2}). \tag{3.50}$$

The inverse process of the sliding motion fragmentation is the merging of the two adjacent sliding motions. If the two sliding motions merge together, the condition of the global maximum force product in (3.49) will be satisfied at the gluing point of the two sliding motions. Once the force products of the two sliding motions are monitored, the peak force product condition similar to (3.50) can be observed. Before merging of the two sliding motions, consider the fragmentized sliding motions with $(\Omega t_{i+1}, x_{i+1}, V) \neq (\Omega t_{i+2}, x_{i+2}, V)$. Extending definitions of (3.44) and (3.45) to the starting and ending points of the sliding motions, then the merging condition for the two adjacent sliding motions is

$$\left.\begin{array}{l} L_{12}(t_k) = 0 \text{ or } {}_{\text{Gmax}}L_{12}(x_k) = 0 \quad \text{for } k \in \{i+1, i+2\} \\ (\Omega t_{i+1}, x_{i+1}, V) = (\Omega t_{i+2}, x_{i+2}, V). \end{array}\right\} \tag{3.51}$$

To explain the above condition of the fragmentation and merging of the sliding motions, the corresponding force product characteristics and phase plane of the sliding motion are sketched in Fig. 3.8. The sliding, critical sliding, and sliding fragmentation motions are depicted for given parameters with belt speed $V > 0$ and excitation amplitude $A_0^{(1)} < A_0^{(2)} < A_0^{(3)}$. The sliding motion vanishes from the boundary and goes into the domain Ω_1. The displacements x_i and x_{i+1} represent the starting and vanishing switching ones of the sliding motion, respectively. The point $x_k \in (x_i, x_{i+1})$ represents the critical point for the fragmentation or merging of the sliding motion. After the fragmentation (or before the merging), the critical point x_k is split into two new points x_{i+1} and x_{i+2}. However, after fragmentation (or merging), the index of x_{i+1} (or x_{x+3}) will be shifted as x_{x+3} (or x_{i+1}). Suppose the peak force product increases with increasing excitation amplitude, then the

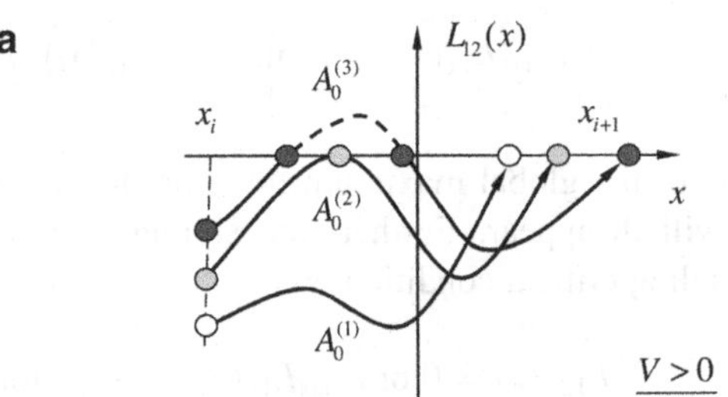

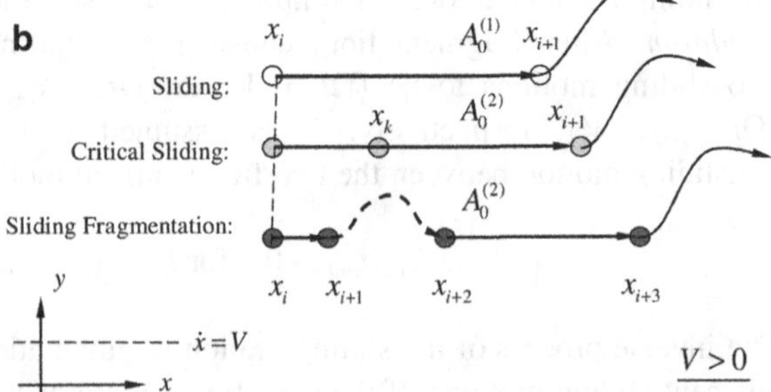

Fig. 3.8 A sketch of (**a**) force product and (**b**) phase plane for sliding, critical sliding, and sliding fragmentation motions under given parameters for belt speed $V > 0$ and $A_0^{(1)} < A_0^{(2)} < A_0^{(3)}$. The ending point of the sliding motion vanishes on the boundary and goes to the domain Ω_1

motions from the sliding to the sliding fragmentation are given for $A_0^{(1)} \xrightarrow{\text{increase}} A_0^{(3)}$. The dashed and thin-solid curves represent nonsliding motion. It is clear that the nonsliding portion of the sliding fragmentation needs $L_{12}(t) > 0$ as in (3.50) for $t \in U_n = (t_j^n, t_{j+1}^n) \subset (t_i, t_{i+1})$ where $L_{12}(t_k) = 0$ for $t_k = \{t_j^n, t_{j+1}^n\}$. However, for the sliding portion, the force products keep the relation $L_{12}(t) < 0$ for $t \in (t_i, t_{i+1}) \setminus \cup_n U_n$. Otherwise, suppose the peak force products at the potential merging points of the two sliding motion decrease with decreasing excitation, then the merging of the two sliding motion can be observed for $A_0^{(3)} \xrightarrow{\text{decreases}} A_0^{(1)}$. Whatever the merging or fragmentation of the sliding motion occur, the old sliding motion will be destroyed. Therefore, the conditions for the fragmentation and merging of the sliding motion are called the *vanishing* conditions for the sliding motion.

Consider the sliding motion with the nonzero measure of $\widetilde{\partial\Omega_{\alpha\beta}}$ with starting and ending points $(\Omega t_i, x_i, V)$ and $(\Omega t_{i+1}, x_{i+1}, V)$. Define $\delta = (\delta_{x_i}^2 + \delta_{t_i}^2)^{1/2}$ with $\delta_{x_i} = x_{i+1} - x_i$ and $\delta_{t_i} = \Omega t_{i+1} - \Omega t_i \geq 0$. Suppose the starting and ending points satisfy the initial and final force product conditions, as $\delta \to 0$, the force product condition is termed the *onset* condition for the sliding motion.

$$\left.\begin{aligned} &L_{12}(t_m) = 0 \quad \text{for } m \in \{i, i+1\}, \\ &(\Omega t_{i+1}, x_{i+1}, V) = \lim_{\delta \to 0} (\Omega t_i + \delta_{t_i}, x_i + \delta_{x_i}, V). \end{aligned}\right\} \tag{3.52}$$

The foregoing condition is also called *the onset force product condition.*

The onset condition has four possible cases from (3.40) and (3.41), as sketched in Fig. 3.9. The four cases are: $F_2(t_i) = F_2(t_{i+1}) = 0$ (Case I), $F_2(t_i) = F_1(t_{i+1}) = 0$ (Case II), $F_1(t_i) = F_2(t_{i+1}) = 0$ (Case III), and $F_1(t_i) = F_1(t_{i+1}) = 0$ (Case IV). For $\delta \neq 0$, the four sliding motions exist. The onset conditions of the sliding motion for Cases I and IV is the same as for the grazing motion. For details, one can refer to Luo and Gegg (2006c). The two onsets of the sliding motions can be called the grazing onsets. The other onset conditions based on the Cases II and III are the *inflexed* onsets for the sliding motions. In this subsection, the force product criteria for

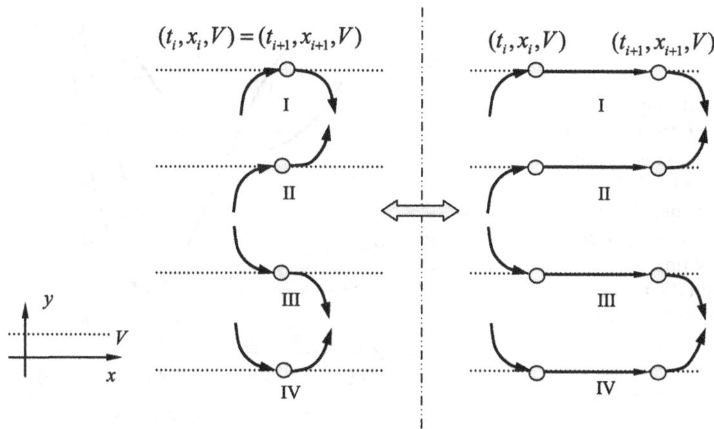

Fig. 3.9 The four onsets of the sliding motion and the corresponding sliding motion in phase plane for belt speed $V > 0$

onset, forming, and vanishing of the sliding motions along the discontinuous friction boundary at a certain velocity have been developed. From the expressions of such criteria, the achieved criteria can be applied for a nonlinear, continuous spring, and viscous damper oscillator including a nonlinear friction with a C^0-discontinuity.

Consider the parameters (i.e., $V = 1$, $\Omega = 1$, $d_1 = 1$, $d_2 = 0$, $b_1 = -b_2 = 30$, $c_1 = c_2 = 30$) with initial condition ($x_i = 3$, $y_i = 1$ and $\Omega t_i = \pi$) for demonstration. For specific excitation amplitudes $A_0 = \{58.0,\ 85.0,\ 116.1\}$, the phase trajectories of corresponding sliding motions are plotted in Fig. 3.10a. The force product versus displacement for the sliding motion is presented in Fig. 3.10b, and it is observed

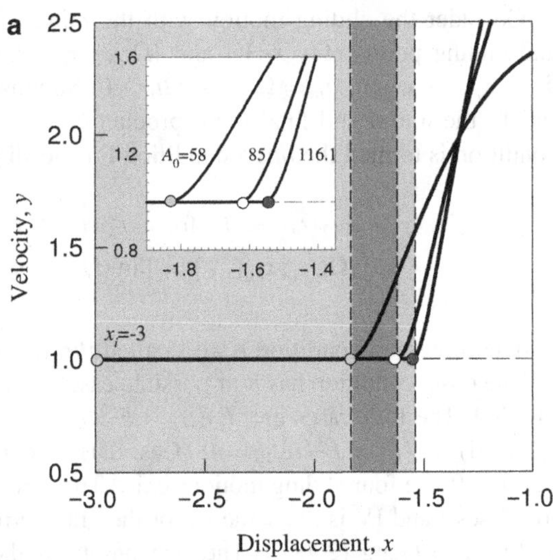

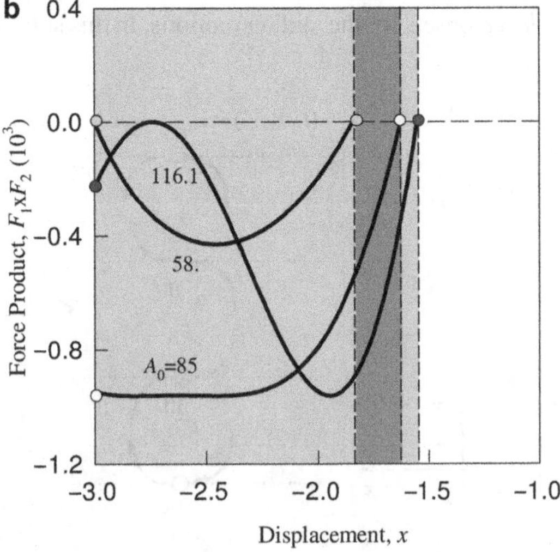

Fig. 3.10 Sliding motion vanishing on the boundary and going into the domain Ω_1 for ($A_0 = 58.0$, 85.0, 116.1): (**a**) phase plane, (**b**) force product versus displacement, (**c**) velocity time-history, (**d**) force product time-history. ($V = 1$, $\Omega = 1$, $d_1 = 1$, $d_2 = 0$, $b_1 = -b_2 = 30$, $c_1 = c_2 = 30$, $x_i = -3$, $\Omega t_i = \pi$)

Fig. 3.10 (continued)

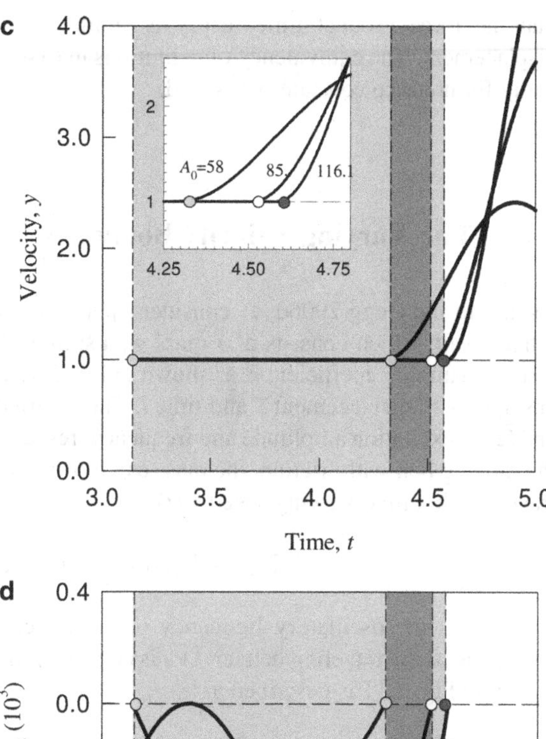

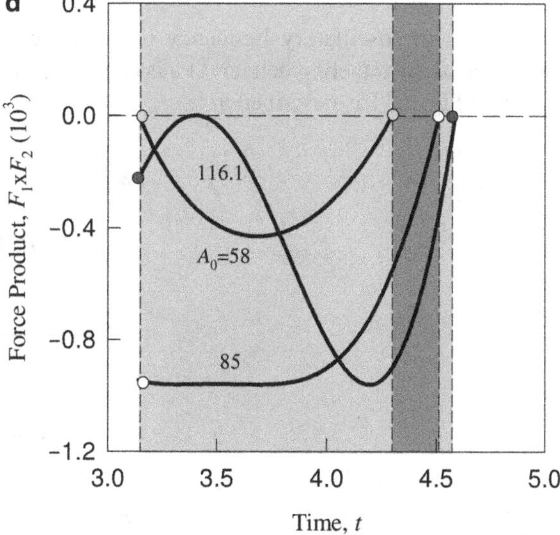

that the sliding motions satisfy $F_1 \times F_2 \leq 0$. For $A_0 = 58.0$, the initial force product is zero, it implies that $A_{0\,\min} \approx 58.0$. If $A_0 < A_{0\,\min}$, such a sliding motion cannot be observed. For $A_0 \approx 116.1$, the maximum force product is tangential to zero. This value will be the maximum value for this sliding motion $A_{0\,\max} \approx 116.1$. If $A_0 > A_{0\,\max}$, the sliding motion will be fragmentized into two parts, as discussed in Fig. 3.10. Thus, under the above parameters, the sliding motion starting at $(x_i, y_i, \Omega t_i) = (-3., 1., \pi)$ will exist for $A_{0\,\min} \leq A_0 \leq A_{0\,\max}$. Further, the velocity and force product time-histories are also presented in Fig. 3.10c, d. It is interesting that the force product time-history is similar to the force product versus the displacement. The equivalency of definitions in (3.45) and (3.47) is verified numerically for nonzero constant belt speed.

3.2 Time-Varying Velocity Boundary

As in Luo and Gegg(2006d, e), consider a periodically forced oscillator attached to a fixed wall, which consists of a mass m, a spring of stiffness k, and a damper of viscous damping coefficient r, as shown in Fig. 3.1. The coordinate system $(\overline{x}, \overline{t})$ is absolute with displacement $\overline{x}$ and time $\overline{t}$. The external force is $\overline{Q}_0 \cos \overline{\Omega}\overline{t}$ where $\overline{Q}_0$ and $\overline{\Omega}$ are excitation amplitude and frequency, respectively. Since the mass contacts the moving belt with friction, the mass can move along or rest on the belt. The belt travels with a time-varying speed $\overline{V}(\overline{t})$.

$$\overline{V}(\overline{t}) = \overline{V}_0 \cos(\omega\overline{t} + \beta) + \overline{V}_1, \tag{3.53}$$

where ω is the oscillatory frequency of the traveling belt, $\overline{V}_0$ is the oscillatory amplitude of the traveling belt, and $\overline{V}_1$ is constant. Further, the kinetic friction force shown in Fig. 3.11 is described as

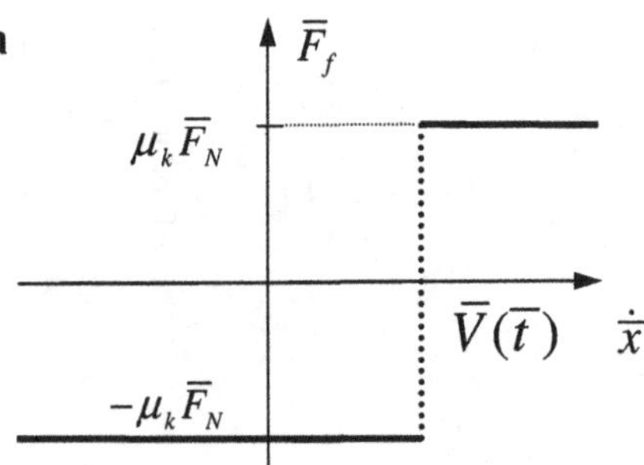

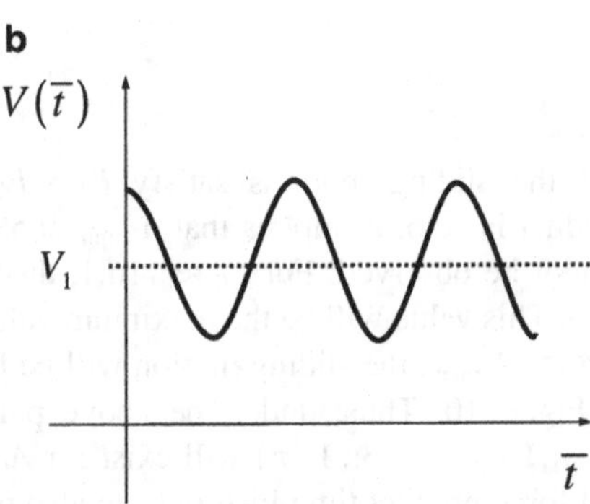

Fig. 3.11 (**a**) Friction forces and (**b**) the oscillation transport speed of the belt

$$\overline{F}_f(\dot{\overline{x}})\begin{cases} = \mu_k F_N, & \dot{\overline{x}} \in [\overline{V}(\overline{t}), \infty) \\ \in [-\mu_k F_N, \mu_k F_N], & \dot{\overline{x}} = \overline{V}(\overline{t}) \\ = -\mu_k F_N, & \dot{\overline{x}} \in (-\infty, \overline{V}(\overline{t})] \end{cases} \tag{3.54}$$

where $\dot{\overline{x}} \triangleq \mathrm{d}\overline{x}/\mathrm{d}t$, μ_k and F_N are a dynamical friction coefficient and a normal force to the contact surface, respectively.

For the mass moving with the same speed of the belt surface, the nonfriction force acting on the mass in the x-direction is defined as

$$F_s = \overline{Q}_0 \cos \overline{\Omega}\overline{t} - r\overline{V}(\overline{t}) - k\overline{x}, \quad \text{for } \dot{\overline{x}} = \overline{V}(\overline{t}). \tag{3.55}$$

For stick motion, this force cannot overcome the static friction force for stick motions (i.e., $|F_s| \le \mu_k F_\mathrm{N}$). Thus, the mass does not have any relative motion to the belt. In other words, the relative acceleration should be zero, i.e.,

$$\ddot{\overline{x}} = \dot{\overline{V}}(t) = -\overline{V}_0\omega \sin(\omega\overline{t} + \beta), \quad \text{for } \dot{\overline{x}} = \overline{V}(\overline{t}). \tag{3.56}$$

If $|F_s| > \mu_k F_\mathrm{N}$, the nonfriction force will overcome the static friction force on the mass and the nonstick motion will appear. For the nonstick motion, the total force acting on the mass is

$$F = \overline{Q}_0 \cos \overline{\Omega}\overline{t} - \mu_k F_\mathrm{N} \mathrm{sgn}(\dot{\overline{x}} - \overline{V}) - r\dot{\overline{x}} - k\overline{x}, \quad \text{for } \dot{\overline{x}} \ne \overline{V}, \tag{3.57}$$

where $\mathrm{sgn}(\cdot)$ is the sign function. Therefore, the equation of the nonstick motion for this oscillator with the dry-friction is

$$\ddot{\overline{x}} + 2\overline{d}\dot{\overline{x}} + \overline{c}\overline{x} = \overline{A}_0 \cos \overline{\Omega}\overline{t} - \tilde{F}_f \mathrm{sgn}(\dot{\overline{x}} - \overline{V}(\overline{t})), \quad \text{for } \dot{\overline{x}} \ne \overline{V}(\overline{t}), \tag{3.58}$$

where $\overline{A}_0 = \overline{Q}_0/m$, $\overline{d} = r/2m$, $\overline{c} = k/m$, and $\tilde{F}_f = \mu_k F_\mathrm{N}/m$. Introduce nondimensional frequency and time. Thus,

$$\left.\begin{array}{l} \Omega = \overline{\Omega}/\omega,\ t = \omega\overline{t},\ d = \overline{d}/\omega,\ c = \overline{c}/\omega^2,\ A_0 = \overline{A}_0/\omega^2, \\ F_f = \tilde{F}_f/\omega^2,\ V_0 = \overline{V}_0/\omega,\ V_1 = \overline{V}_1/\omega,\ V = V/\omega, x = \overline{x}, \end{array}\right\} \tag{3.59}$$

$$V(t) = V_0 \cos(t + \beta) + V_1. \tag{3.60}$$

The phase constant β is used to synchronize the periodic force input and the velocity discontinuity after the modulus of time has been computed. The integration of (3.53) gives for $t > t_i$

$$X(t) = V_0[\sin(t + \beta) - \sin(t_i + \beta)] + V_1 \times (t - t_i) + x_i, \tag{3.61}$$

which is the displacement response of the periodically time-varying traveling belt. For $t = t_i$, $X(t_i) = x(t_i) \equiv x_i$.

Introduce the relative displacement, velocity, and acceleration as

$$\left.\begin{aligned} z(t) &= x(t) - X(t),\\ \dot{z}(t) &= \dot{x}(t) - V(t),\\ \ddot{z}(t) &= \ddot{x}(t) - \dot{V}(t). \end{aligned}\right\} \tag{3.62}$$

The nonfriction force in (3.55) for $\dot{z} = 0$ becomes

$$F_s = A_0 \cos \Omega t - 2d[V(t) + \dot{z}(t)] - c[X(t) + z(t)] - \dot{V}(t). \tag{3.63}$$

Since the mass does not have any relative motion to the vibrating belt, the relative acceleration is zero, i.e.,

$$\ddot{x} = \dot{V}(t) = -V_0 \sin(t + \beta), \quad \text{for } \dot{x} = V(t). \tag{3.64}$$

or

$$\ddot{z} = 0, \quad \text{for } \dot{z} = 0. \tag{3.65}$$

For nonstick motion, the equation of this oscillator with friction becomes: for $\dot{x} \neq V(t)$

$$\ddot{x} + 2d\dot{x} + cx = A_0 \cos \Omega t - F_f \operatorname{sgn}(\dot{x} - V(t)), \tag{3.66}$$

in the absolute frame, or for $\dot{z} \neq 0$

$$\begin{aligned} \ddot{z}(t) + 2d\dot{z}(t) + cz(t) = {} & A_0 \cos \Omega t - F_f \operatorname{sgn}(\dot{z}(t)) \\ & - 2dV(t) - cX(t) - \dot{V}(t), \end{aligned} \tag{3.67}$$

in the relative frame.

3.2.1 Analytical Conditions

In this section, the sufficient and necessary conditions for passable, sliding, and grazing motions to the time-varying boundary are discussed. The corresponding physical interpretation is given for a better understanding of mechanism for such motions.

3.2.1.1 Equations of Motion

Because of the discontinuity, the phase plane partition for this friction oscillator is sketched in the absolute and relative frames in Fig. 3.12. In the absolute frame, the separation boundary in phase plane is a curve determined by eliminating the parameter t in two parameter equations $X(t)$ and $V(t)$ in (3.60) and (3.61). For

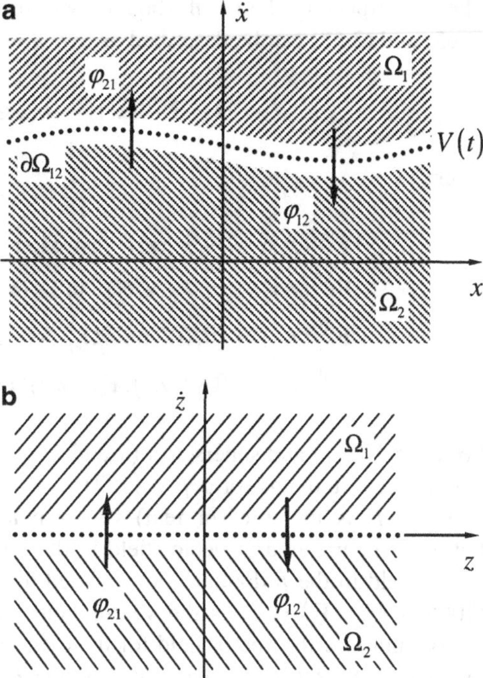

Fig. 3.12 Phase plane partition in (**a**) absolute and (**b**) relative phase planes for time-varying boundary

$0 < V_0 < V_1$, the separation boundary is the prolate trochoid, as shown in Fig. 4.3a. Similarly, the separation boundary for $V_0 > V_1 > 0$ is the curtate trochoid. However, the boundary velocity in the relative frame is zero in Fig. 3.12b. In the relative frame, the criteria for motion to pass through the discontinuous boundary or not can be easily obtained from the theory of discontinuous dynamical systems as in Chap. 2. Therefore, the criteria for such a system will be developed based on the relative frame.

In phase plane, we define

$$\mathbf{z} \triangleq (z, \dot{z})^{\mathrm{T}} \equiv (z, v)^{\mathrm{T}} \text{ and } \mathbf{F} \triangleq (v, F)^{\mathrm{T}}. \tag{3.68}$$

The corresponding regions and boundary are

$$\left.\begin{aligned} \Omega_1 &= \{(z, v) | v \in (0, \infty)\}, \\ \Omega_2 &= \{(z, v) | v \in (-\infty, 0)\}, \\ \partial\Omega_{\alpha\beta} &= \{(z, v) | \varphi_{\alpha\beta}(z, v) = v = 0\} \end{aligned}\right\} \tag{3.69}$$

or

$$\left.\begin{aligned} \Omega_1 &= \{(x, \dot{x}) | \dot{x} - V > 0\}, \\ \Omega_2 &= \{(x, \dot{x}) | \dot{x} - V < 0\}, \\ \partial\Omega_{\alpha\beta} &= \{(x, \dot{x}) | \varphi_{\alpha\beta}(x, \dot{x}) = \dot{x} - V(t) = 0\}. \end{aligned}\right\} \tag{3.70}$$

The subscripts $(\cdot)_{\alpha\beta}$ defines the boundary from Ω_α to Ω_β. The equations of motion in (3.65) and (3.67) can be described as

$$\dot{\mathbf{z}} = \mathbf{F}_\lambda^{(\kappa)}(\mathbf{z}, t) \quad (\kappa, \lambda \in \{0, 1, 2\}) \tag{3.71}$$

where

$$\left.\begin{aligned}
&\mathbf{F}_\alpha^{(\alpha)}(\mathbf{z}, t) = (v, F_\alpha(\mathbf{z}, t))^{\mathrm{T}} \text{ in } \Omega_\alpha(\alpha \in \{1, 2\}),\\
&\mathbf{F}_\alpha^{(\beta)}(\mathbf{z}, t) = (v, F_\beta(\mathbf{z}, t))^{\mathrm{T}} \text{ in } \Omega_\alpha(\alpha \neq \beta \in \{1, 2\});\\
&\mathbf{F}_0^{(0)}(\mathbf{z}, t) = (0, 0)^{\mathrm{T}} \text{ on } \partial\Omega_{\alpha\beta} \text{ for stick},\\
&\mathbf{F}_0^{(0)}(\mathbf{z}, t) \in [\mathbf{F}_\alpha^{(\alpha)}(\mathbf{z}, t), \mathbf{F}_\beta^{(\beta)}(\mathbf{z}, t)] \text{ on } \partial\Omega_{\alpha\beta} \text{ for nonstick.}
\end{aligned}\right\} \tag{3.72}$$

For the subscript and superscript (κ and λ) with nonzero values, they represent the two adjacent domains for $\alpha, \beta \in \{1, 2\}$.

$\mathbf{F}_\alpha^{(\alpha)}(\mathbf{z}, t)$ is the true (or real) vector field in the α-domain. $\mathbf{F}_\alpha^{(\beta)}(\mathbf{z}, t)$ is the fictitious (or imaginary) vector field in the α-domain, which is determined by the vector field in the β-domain. For detailed discussion, the reader can refer to Luo (2005, 2006). $\mathbf{F}_0^{(0)}(\mathbf{z}, t)$ is the vector field on the separation boundary, and the discontinuity of the vector field for the entire system is presented through such an expression. $F_\alpha(\mathbf{z}, t)$ is the scalar force in the α-domain. For the system in (3.67), we have the forces in the two domains as

$$\begin{aligned}
F_\alpha(\mathbf{z}, t) = {} & A_0 \cos \Omega t - b_\alpha - 2d_\alpha[V(t) + \dot{z}(t)]\\
& - c_\alpha[X(t) + z(t)] - \dot{V}(t).
\end{aligned} \tag{3.73}$$

Note that $b_1 = -b_2 = \mu g/\omega^2, d_\alpha = d,$ and $c_\alpha = c$ for the model in Fig. 3.1.

From (3.61), with increasing time t, the belt displacement will increase. However, the oscillator vibrates in the vicinity of equilibrium. The friction is dependent on the relative velocity between the oscillator and the belt. When the nonstick motion of the oscillator arrives to the discontinuous boundary, the particle of the belt that contacts the mass with the same velocity and displacement is different from the initial particle. To understand the stick and nonstick motions, the particle switching on the surface of the oscillating belt in the absolute phase plane is sketched in Fig. 3.13. The particles (p_1, p_2, p_3) on the belts are represented by white, light gray, dark gray circular symbols, respectively. The dark circular symbol is the oscillator location. Because the oscillating velocity boundary is trochoid, when time $t \to \infty$, the selected belt particle will move to the infinity. However, the oscillator will move around a certain location. To help one understand the particle switching on the translating belt, consider the oscillator with a belt particle p_1 at the same location at time t_i. Once the oscillator velocity is greater than the belt, the oscillator will move faster than the belt particle p_1 for $t \in (t_i, t_{i+1})$. When the oscillator has the same speed as the belt, the oscillator with the belt particle p_2 has

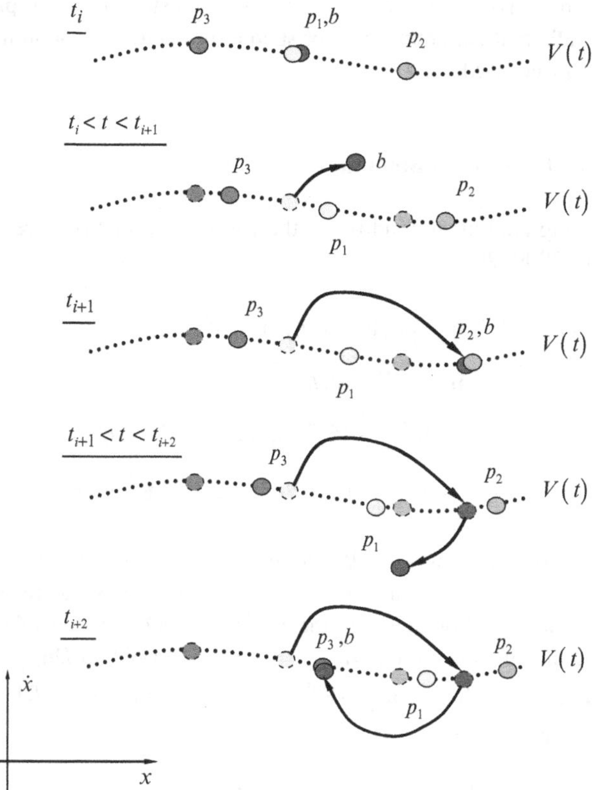

Fig. 3.13 Particles switching on the oscillating belt surface in absolute phase plane once the oscillator has the same speed as the oscillating belt at a moment t_i. The particles (p_1, p_2, p_3) on the belts are represented by *white, light gray, dark gray circular symbols*, respectively. The *dark circular symbol* is the oscillator location

the same location at t_{i+1}. Similarly, once the oscillator velocity is lower than the belt speed, the oscillator will move more slowly than the belt particle p_2 or move backwards with negative velocity for $t \in (t_{i+1}, t_{i+2})$. When the oscillator velocity is the same as the belt speed at time t_{i+2}, the oscillator with the belt particle p_3 has the same location. Such a belt particle switching can help one understand the motion mechanisms in such an oscillator. This motion switching phenomena can exist extensively in mechanical systems. If the oscillator has the same speed with the translating belt in a time interval instead of a time point, then both the oscillator and the belt will move together during this time interval. In mathematics, such a motion is called *a sliding flow*, but from a physics point of view, this motion is called *a stick motion*. In Luo (2005, 2006), the motion approaching the velocity boundary is called *a passable motion* only if the motion can pass through the boundary and into another adjacent domain in phase space. Otherwise, such a motion is termed *a nonpassable motion*. For nonpassable motion, the sink and source motions to the boundary can be classified. The stick motion is a kind of sink motion, which belongs to one of the nonpassable motions. If one is interested in this topic except for Chap. 2, the mathematical definitions for such motions to the discontinuous boundary can refer to Luo (2005, 2006).

3.2.1.2 Passable Flows to Boundary

Before presenting analytical conditions, the following functions are introduced in Luo (2008a, b, 2009, 2011)

$$\begin{aligned} G^{(0,\alpha)}(\mathbf{z},t_{m\pm}) &= \mathbf{n}^{\mathrm{T}}_{\partial\Omega_{\alpha\beta}} \cdot \left[\mathbf{F}^{(\alpha)}_{\alpha}(\mathbf{z},t_{m\pm}) - \mathbf{F}^{(0)}_{\alpha\beta}(\mathbf{z},t_{m\pm})\right] \\ &= \mathbf{n}^{\mathrm{T}}_{\partial\Omega_{\alpha\beta}} \cdot \mathbf{F}^{(\alpha)}_{\alpha}(\mathbf{z},t_{m\pm}), \\ G^{(1,\alpha)}(\mathbf{z},t_{m\pm}) &= 2D\mathbf{n}^{\mathrm{T}}_{\partial\Omega_{\alpha\beta}} \cdot \left[\mathbf{F}^{(\alpha)}_{\alpha}(\mathbf{z},t_{m\pm}) - \mathbf{F}^{(0)}_{\alpha\beta}(\mathbf{z},t_{m\pm})\right] \\ &\quad + \mathbf{n}^{\mathrm{T}}_{\partial\Omega_{\alpha\beta}} \cdot \left[D\mathbf{F}^{(\alpha)}_{\alpha}(\mathbf{z},t_{m\pm}) - D\mathbf{F}^{(0)}_{\alpha\beta}(\mathbf{z},t_{m\pm})\right], \end{aligned} \tag{3.74}$$

where $D = \dot{z}\partial/\partial z + \dot{w}\partial/\partial w + \partial/\partial t$. Notice that t_m represents the time for the motion on the velocity boundary and $t_{m\pm} = t_m \pm 0$ reflects the responses on the regions rather than the boundary. If the boundary $\partial\Omega_{\alpha\beta}$ in the relative frame is a line independent of time t (i.e., $D\mathbf{n}^{\mathrm{T}}_{\partial\Omega_{\alpha\beta}}=0$). One can obtain $D\mathbf{n}^{\mathrm{T}}_{\partial\Omega_{\alpha\beta}}\cdot\mathbf{F}^{(0)}_{\alpha\beta}(\mathbf{z},t_{m\pm})+\mathbf{n}^{\mathrm{T}}_{\partial\Omega_{\alpha\beta}}\cdot D\mathbf{F}^{(0)}_{\alpha\beta}(\mathbf{z},t_{m\pm})=0$ due to $\mathbf{n}^{\mathrm{T}}_{\partial\Omega_{\alpha\beta}}\cdot\mathbf{F}^{(0)}_{\alpha\beta}(\mathbf{z},t_{m\pm})=0$. Thus, $\mathbf{n}^{\mathrm{T}}_{\partial\Omega_{\alpha\beta}}\cdot D\mathbf{F}^{(0)}_{\alpha\beta}(\mathbf{z},t_{m\pm})=0$. Equation (3.74) reduces to

$$G^{(1,\alpha)}(\mathbf{z}_m,t_{m\pm}) = \mathbf{n}^{\mathrm{T}}_{\partial\Omega_{\alpha\beta}} \cdot D\mathbf{F}^{(\alpha)}_{\alpha}(\mathbf{z}_m,t_{m\pm}). \tag{3.75}$$

For a general case, (3.74) instead of (3.75) will be used. Notice that $\mathbf{F}^{(0)}_{\alpha\beta}(\mathbf{z},t) = (0,0)^{\mathrm{T}}$.

For the boundary $\overrightarrow{\partial\Omega}_{\alpha\beta}$ with $\mathbf{n}^{\mathrm{T}}_{\partial\Omega_{\alpha\beta}} \to \Omega_\alpha$, the necessary and sufficient conditions for the nonstick motion (or called passable motion) to boundary are given from Luo (2009, 2011)

$$\left.\begin{aligned} &\left.\begin{aligned} G^{(0,\alpha)}(\mathbf{z}_m,t_{m-}) &= \mathbf{n}^{\mathrm{T}}_{\partial\Omega_{\alpha\beta}} \cdot \mathbf{F}^{(\alpha)}_{\alpha}(\mathbf{z}_m,t_{m-}) < 0 \\ G^{(0,\beta)}(\mathbf{z}_m,t_{m+}) &= \mathbf{n}^{\mathrm{T}}_{\partial\Omega_{\alpha\beta}} \cdot \mathbf{F}^{(\beta)}_{\beta}(\mathbf{z}_m,t_{m+}) < 0 \end{aligned}\right\} \text{from } \Omega_\alpha \to \Omega_\beta, \\ &\left.\begin{aligned} G^{(0,\beta)}(\mathbf{z}_m,t_{m-}) &= \mathbf{n}^{\mathrm{T}}_{\partial\Omega_{\beta\alpha}} \cdot \mathbf{F}^{(\beta)}_{\beta}(\mathbf{z}_m,t_{m-}) > 0 \\ G^{(0,\alpha)}(\mathbf{z}_m,t_{m+}) &= \mathbf{n}^{\mathrm{T}}_{\partial\Omega_{\beta\alpha}} \cdot \mathbf{F}^{(\alpha)}_{\alpha}(\mathbf{z}_m,t_{m+}) > 0 \end{aligned}\right\} \text{from } \Omega_\beta \to \Omega_\alpha \end{aligned}\right\} \tag{3.76}$$

or

$$\begin{aligned} L_{\alpha\beta}(t_{m-}) &= G^{(0,\alpha)}(\mathbf{z}_m,t_{m-}) \times G^{(0,\beta)}(\mathbf{z}_m,t_{m+}), \\ &= \left[\mathbf{n}^{\mathrm{T}}_{\partial\Omega_{\alpha\beta}} \cdot \mathbf{F}^{(\alpha)}_{\alpha}(\mathbf{z}_m,t_{m-})\right] \times \left[\mathbf{n}^{\mathrm{T}}_{\partial\Omega_{\alpha\beta}} \cdot \mathbf{F}^{(\beta)}_{\beta}(\mathbf{z}_m,t_{m+})\right] > 0. \end{aligned} \tag{3.77}$$

Note that $\mathbf{n}_{\partial\Omega_{\alpha\beta}} \to \Omega_\alpha$ means that the normal direction of $\partial\Omega_{\alpha\beta}$ points to domain Ω_α. The G-function gives the same formulas as obtained from Chap. 2. Using the third equation of (3.69), the normal vector of the boundary $\partial\Omega_{12}$ or $\partial\Omega_{21}$ is

$$\mathbf{n}_{\partial\Omega_{12}} = \mathbf{n}_{\partial\Omega_{21}} = (0,1)^{\mathrm{T}}. \tag{3.78}$$

Therefore, we have

$$\left.\begin{aligned} G^{(0,\alpha)}(\mathbf{z},t) &= \mathbf{n}^{\mathrm{T}}_{\partial\Omega_{\alpha\beta}} \cdot \mathbf{F}^{(\alpha)}_{\alpha}(\mathbf{z},t) = F_\alpha(\mathbf{z},t), \quad \alpha \in \{1,2\}, \\ G^{(1,\alpha)}(\mathbf{z},t) &= \mathbf{n}^{\mathrm{T}}_{\partial\Omega_{\alpha\beta}} \cdot D\mathbf{F}^{(\alpha)}_{\alpha}(\mathbf{z},t) = \nabla F_\alpha(\mathbf{z},t) \cdot \mathbf{F}^{(\alpha)}_{\alpha}(\mathbf{z},t) + \partial_t F_\alpha(\mathbf{z},t). \end{aligned}\right\} \tag{3.79}$$

For this problem, the normal projection of the vector field of the oscillator to the separation boundary is the force. The conditions in (3.76) or (3.77) give

$$\left.\begin{aligned} &\left.\begin{aligned} G^{(0,\alpha)}(\mathbf{z}_m,t_{m-}) &= F_\alpha(\mathbf{z}_m,t_{m-}) < 0 \\ G^{(0,\beta)}(\mathbf{z}_m,t_{m+}) &= F_\beta(\mathbf{z}_m,t_{m+}) < 0 \end{aligned}\right\} \text{ from } \Omega_\alpha \to \Omega_\beta, \\ &\left.\begin{aligned} G^{(0,\beta)}(\mathbf{z}_m,t_{m-}) &= F_\beta(\mathbf{z}_m,t_{m-}) > 0 \\ G^{(0,\alpha)}(\mathbf{z}_m,t_{m+}) &= F_\alpha(\mathbf{z}_m,t_{m+}) > 0 \end{aligned}\right\} \text{ from } \Omega_\beta \to \Omega_\alpha. \end{aligned}\right\} \tag{3.80}$$

$$\begin{aligned} L_{\alpha\beta}(\mathbf{z}_m,t_{m\pm}) &= G^{(0,\alpha)}(\mathbf{z}_m,t_{m-}) \times G^{(0,\beta)}(\mathbf{z}_m,t_{m+}) \\ &= F_\alpha(\mathbf{z}_m,t_{m-}) \times F_\beta(\mathbf{z}_m,t_{m+}) > 0. \end{aligned} \tag{3.81}$$

The aforementioned conditions can be illustrated through the vector fields in the two domains, as shown in Fig. 3.14. In Fig. 3.14a, the vector fields in the absolute frame are presented. The normal vector $\mathbf{n}_{\partial\Omega_{12}}$ is changed with time, and both the corresponding direction and magnitude vary with time and location. Thus, the corresponding normal component of vector fields will change with normal vector. However, in the relative frame, the normal vector of the boundary is constant. With time varying, the normal vector is invariant, and the vector fields and normal components of the two vectors are sketched in Fig. 3.14b. The passable flow on the boundary requires two normal components of two vector fields are of the same direction.

3.2.1.3 Sliding Flows on Boundary

Consider the nonpassable boundary $\widetilde{\partial\Omega}_{\alpha\beta}$ and passable boundary $\overrightarrow{\partial\Omega}_{\alpha\beta}$ on the velocity boundary in Fig. 3.15. For a nonpassable boundary $\widetilde{\partial\Omega}_{\alpha\beta}$ possessing nonzero measure in phase space, the critical initial and final states of the sliding flow are $(\Omega t_{\mathrm{c}}, x_{\mathrm{c}}, V_{\mathrm{c}})$ and $(\Omega t_{\mathrm{f}}, x_{\mathrm{f}}, V_{\mathrm{f}})$, respectively. Consider a sliding flow starting at $(\Omega t_i, x_i, V_i)$ and ending at $(\Omega t_{i+1}, x_{i+1}, V_{i+1})$, where $(\Omega t_{i+1}, x_{i+1}, V_{i+1}) \triangleq (\Omega t_{\mathrm{f}}, x_{\mathrm{f}}, V_{\mathrm{f}})$.

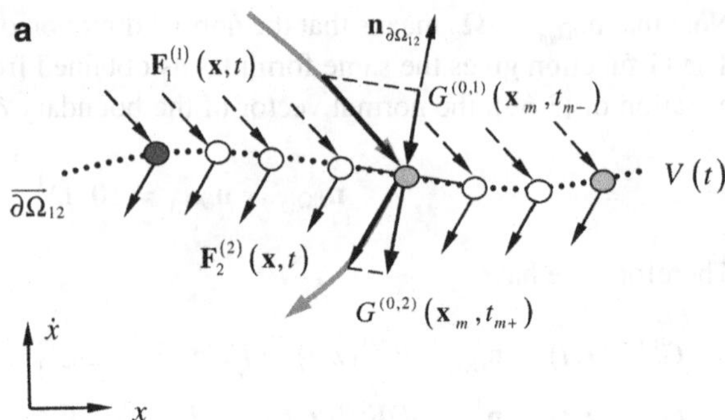

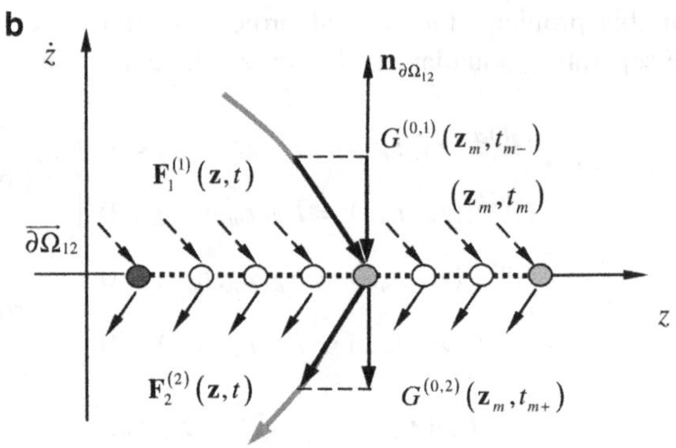

Fig. 3.14 Vector fields in the vicinity of the time-varying boundary for passable flows (**a**) in the absolute frame and (**b**) in the relative frame

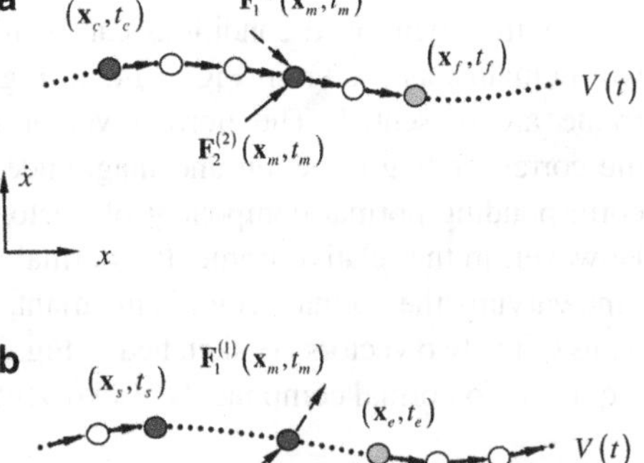

Fig. 3.15 (**a**) Nonpassable boundary $\widetilde{\partial\Omega}_{12}$ with sliding flows and (**b**) passable boundary $\overrightarrow{\partial\Omega}_{12}$ with passable flows in the absolute frame

From Luo (2009, 2011), the sliding flow (or called the stick motion in physics) through the real flow is guaranteed for $t_m \in [t_i, t_{i+1}) \subseteq [t_c, t_f)$ by

$$\begin{aligned} L_{\alpha\beta}(t_{m-}) &= G^{(0,\alpha)}(\mathbf{z}_m, t_{m-}) \times G^{(0,\beta)}(\mathbf{z}_m, t_{m-}) \\ &= \left[\mathbf{n}_{\partial\Omega_{\alpha\beta}}^{\mathrm{T}} \cdot \mathbf{F}_{\alpha}^{(\alpha)}(\mathbf{z}_m, t_{m-})\right] \times \left[\mathbf{n}_{\partial\Omega_{\alpha\beta}}^{\mathrm{T}} \cdot \mathbf{F}_{\beta}^{(\beta)}(\mathbf{z}_m, t_{m+})\right] < 0. \end{aligned} \quad (3.82)$$

For the boundary $\partial\Omega_{\alpha\beta}$ with $\mathbf{n}^{\mathrm{T}}_{\partial\Omega_{\alpha\beta}} \to \Omega_{\alpha}$, the necessary and sufficient conditions for the stick motion (or called nonpassable motion) to boundary are given from Luo (2009, 2011), i.e.,

$$\left.\begin{aligned} G^{(0,\alpha)}(\mathbf{z}_m, t_{m-}) &= \mathbf{n}^{\mathrm{T}}_{\partial\Omega_{\alpha\beta}} \cdot \mathbf{F}^{(\alpha)}_{\alpha}(\mathbf{z}_m, t_{m-}) < 0, \\ G^{(0,\beta)}(\mathbf{z}_m, t_{m-}) &= \mathbf{n}^{\mathrm{T}}_{\partial\Omega_{\alpha\beta}} \cdot \mathbf{F}^{(\beta)}_{\beta}(\mathbf{z}_m, t_{m-}) > 0. \end{aligned}\right\} \tag{3.83}$$

For a boundary $\overrightarrow{\partial\Omega}_{\alpha\beta}$ with the nonzero measure, the starting and ending states of the passable motion on the boundary are accordingly $(\Omega t_{\mathrm{s}}, z_{\mathrm{s}}, V_{\mathrm{s}})$ and $(\Omega t_{\mathrm{e}}, z_{\mathrm{e}}, V_{\mathrm{e}})$, respectively, as shown in Fig. 3.14b. From $\overrightarrow{\partial\Omega}_{\alpha\beta}$ to $\widetilde{\partial\Omega}_{\alpha\beta}$, we have $t_{\mathrm{e}} = t_{\mathrm{c}}$. Otherwise, $t_{\mathrm{f}} = t_{\mathrm{s}}$. Suppose the starting state is a switching state of the sliding flow, the switching condition of the sliding flows from $\overrightarrow{\partial\Omega}_{\alpha\beta}$ to $\widetilde{\partial\Omega}_{\alpha\beta}$ at $t_m = t_{\mathrm{c}}$ is for $\mathbf{n}_{\partial\Omega_{\alpha\beta}} \to \Omega_{\alpha}$,

$$\left.\begin{aligned} G^{(0,\alpha)}(\mathbf{z}_m, t_{m-}) &= \mathbf{n}^{\mathrm{T}}_{\partial\Omega_{\alpha\beta}} \cdot \mathbf{F}^{(\alpha)}_{\alpha}(\mathbf{z}_m, t_{m-}) < 0, \\ G^{(0,\beta)}(\mathbf{z}_m, t_{m\pm}) &= \mathbf{n}^{\mathrm{T}}_{\partial\Omega_{\alpha\beta}} \cdot \mathbf{F}^{(\beta)}_{\beta}(\mathbf{z}_m, t_{m\pm}) = 0, \\ G^{(1,\beta)}(\mathbf{z}_m, t_{m\pm}) &= \mathbf{n}^{\mathrm{T}}_{\partial\Omega_{\alpha\beta}} \cdot D\mathbf{F}^{(\beta)}_{\beta}(\mathbf{z}_m, t_{m\pm}) > 0. \end{aligned}\right\} \tag{3.84}$$

$$\left.\begin{aligned} G^{(0,\alpha)}(\mathbf{z}_m, t_{m\pm}) &= \mathbf{n}^{\mathrm{T}}_{\partial\Omega_{\alpha\beta}} \cdot \mathbf{F}^{(\alpha)}_{\alpha}(\mathbf{z}_m, t_{m\pm}) = 0, \\ G^{(0,\beta)}(\mathbf{z}_m, t_{m-}) &= \mathbf{n}^{\mathrm{T}}_{\partial\Omega_{\alpha\beta}} \cdot \mathbf{F}^{(\beta)}_{\beta}(\mathbf{z}_m, t_{m-}) > 0, \\ G^{(1,\alpha)}(\mathbf{z}_m, t_{m\pm}) &= \mathbf{n}^{\mathrm{T}}_{\partial\Omega_{\alpha\beta}} \cdot D\mathbf{F}^{(\alpha)}_{\alpha}(\mathbf{z}_m, t_{m\pm}) < 0. \end{aligned}\right\} \tag{3.85}$$

On $\widetilde{\partial\Omega}_{\alpha\beta}$ with $\mathbf{n}^{\mathrm{T}}_{\partial\Omega_{\alpha\beta}} \to \Omega_{\alpha}$, a sliding flow vanishing from $\widetilde{\partial\Omega}_{\alpha\beta}$ and going into the domain Ω_{γ} $(\gamma \in \{\alpha, \beta\})$ at $t_m = t_{\mathrm{f}}$ requires:

$$\left.\begin{aligned} G^{(0,\alpha)}(\mathbf{z}_m, t_{m-}) &= \mathbf{n}^{\mathrm{T}}_{\partial\Omega_{\alpha\beta}} \cdot \mathbf{F}^{(\alpha)}_{\alpha}(\mathbf{z}_m, t_{m-}) < 0, \\ G^{(0,\beta)}(\mathbf{z}_m, t_{m\pm}) &= \mathbf{n}^{\mathrm{T}}_{\partial\Omega_{\alpha\beta}} \cdot \mathbf{F}^{(\beta)}_{\beta}(\mathbf{z}_m, t_{m\pm}) = 0, \\ G^{(1,\beta)}(\mathbf{z}_m, t_{m\pm}) &= \mathbf{n}^{\mathrm{T}}_{\partial\Omega_{\alpha\beta}} \cdot D\mathbf{F}^{(\beta)}_{\beta}(\mathbf{z}_m, t_{m\pm}) < 0. \end{aligned}\right\} \tag{3.86}$$

$$\left.\begin{aligned} G^{(0,\beta)}(\mathbf{z}_m, t_{m-}) &= \mathbf{n}^{\mathrm{T}}_{\partial\Omega_{\alpha\beta}} \cdot \mathbf{F}^{(\beta)}_{\beta}(\mathbf{z}_m, t_{m-}) > 0, \\ G^{(0,\alpha)}(\mathbf{z}_m, t_{m\mp}) &= \mathbf{n}^{\mathrm{T}}_{\partial\Omega_{\alpha\beta}} \cdot \mathbf{F}^{(\alpha)}_{\alpha}(\mathbf{z}_m, t_{m\mp}) = 0, \\ G^{(1,\alpha)}(\mathbf{z}_m, t_{m\mp}) &= \mathbf{n}^{\mathrm{T}}_{\partial\Omega_{\alpha\beta}} \cdot D\mathbf{F}^{(\alpha)}_{\alpha}(\mathbf{z}_m, t_{m\mp}) > 0. \end{aligned}\right\} \tag{3.87}$$

From (3.84)–(3.87), the switching conditions for the sliding flows are summarized as

$$\left.\begin{aligned}&\left.\begin{aligned}&F_1(\mathbf{z}_m,\Omega t_{m-})<0 \text{ and } F_2(\mathbf{z}_m,\Omega t_{m\pm})=0,\\&G^{(1,2)}(\mathbf{z}_m,t_{m\pm})>0\end{aligned}\right\}\quad \text{for } \Omega_1\to\widetilde{\partial\Omega}_{12},\\&\left.\begin{aligned}&F_2(\mathbf{z}_m,\Omega t_{m-})>0 \text{ and } F_1(\mathbf{z}_m,\Omega t_{m\pm})=0,\\&G^{(1,1)}(\mathbf{z}_m,t_{m\pm})<0\end{aligned}\right\}\quad \text{for } \Omega_2\to\widetilde{\partial\Omega}_{21}.\end{aligned}\right\}\tag{3.88}$$

For a better understanding of the force characteristic of the sliding flow, the condition for the sliding flow in (3.83) can be re-written as

$$F_1(\mathbf{z}_m,\Omega t_{m-})<0 \text{ and } F_2(\mathbf{z}_m,\Omega t_{m-})>0 \text{ on } \widetilde{\partial\Omega}_{12}.\tag{3.89}$$

With (3.84) and (3.85), the onset (switching) conditions for the sliding flow are,

$$\left.\begin{aligned}&\left.\begin{aligned}&F_1(\mathbf{z}_m,\Omega t_{m-})<0 \text{ and } F_2(\mathbf{z}_m,\Omega t_{m\pm})=0,\\&G^{(1,2)}(\mathbf{z}_m,t_{m\pm})>0\end{aligned}\right\}\quad \text{for } \Omega_1\to\widetilde{\partial\Omega}_{12},\\&\left.\begin{aligned}&F_2(\mathbf{z}_m,\Omega t_{m-})>0 \text{ and } F_1(\mathbf{z}_m,\Omega t_{m\pm})=0,\\&G^{(1,1)}(\mathbf{z}_m,t_{m\pm})<0\end{aligned}\right\}\quad \text{for } \Omega_2\to\widetilde{\partial\Omega}_{21}.\end{aligned}\right\}\tag{3.90}$$

at $t_m = t_c$, and with (3.86)and (3.87), the vanishing condition for the sliding flow are re-written as at $t_m = t_f$

$$\left.\begin{aligned}&\left.\begin{aligned}&F_1(\mathbf{z}_m,\Omega t_{m-})<0 \text{ and } F_2(\mathbf{z}_m,\Omega t_{m\mp})=0,\\&G^{(1,2)}(\mathbf{z}_m,t_{m\mp})<0\end{aligned}\right\}\quad \text{from } \widetilde{\partial\Omega}_{12}\to\Omega_2,\\&\left.\begin{aligned}&F_2(\mathbf{z}_m,\Omega t_{m-})>0 \text{ and } F_1(\mathbf{z}_m,\Omega t_{m\mp})=0,\\&G^{(1,1)}(\mathbf{z}_m,t_{m\mp})>0\end{aligned}\right\}\quad \text{from } \widetilde{\partial\Omega}_{12}\to\Omega_1.\end{aligned}\right\}\tag{3.91}$$

A sketch of the sliding flow on the discontinuous boundary is presented in Figs. 3.16 and 3.17. In Fig. 3.16, the force conditions for the onset and vanishing of sliding flow are sketched in the absolute frame through the vector fields of $\mathbf{F}_1^{(1)} \times (\mathbf{x}, t)$ and $\mathbf{F}_2^{(2)}(\mathbf{x}, t)$. The corresponding force conditions in the relative frame is presented in Fig. 3.17 via the vector fields of $\mathbf{F}_1^{(1)}(\mathbf{z}, t)$ and $\mathbf{F}_2^{(2)}(\mathbf{z}, t)$. The vanishing condition for sliding flow along the velocity boundary is illustrated by $G^{(0,1)}(\mathbf{z}_m, t_{m-}) < 0$ and $G^{(0,2)}(\mathbf{z}_m, t_{m\pm}) = 0$ (or $F_2(\mathbf{z}_m, t_{m\pm}) = 0$) with $G^{(1,2)}(\mathbf{z}_m, t_{m\pm}) < 0$. The vanishing point is labeled by the gray circular symbol. However, the starting points of the sliding flow may not be the switching points from the passable flow to the sliding flow. The switching point satisfies the onset condition of the sliding flow in (3.58). When the flow arrives to the boundary, once (3.90) holds, the sliding flow along the discontinuous boundary will be formed. The onset of the sliding flow from the domain Ω_2 onto the sliding boundary

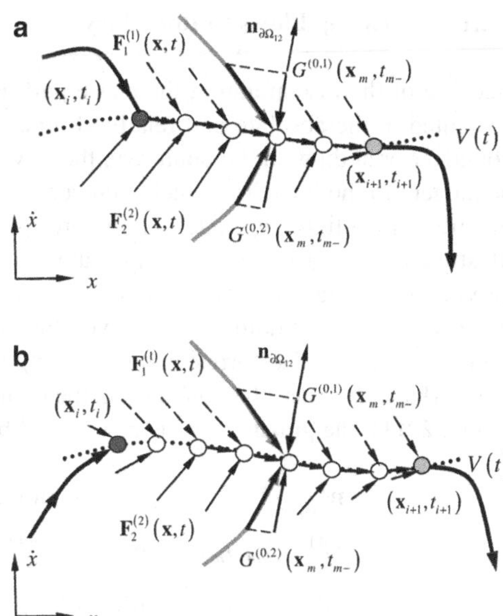

Fig. 3.16 Vector fields in the absolute frame: (**a**) vanishing and (**b**) appearance and vanishing of sliding flow on the time-varying boundary

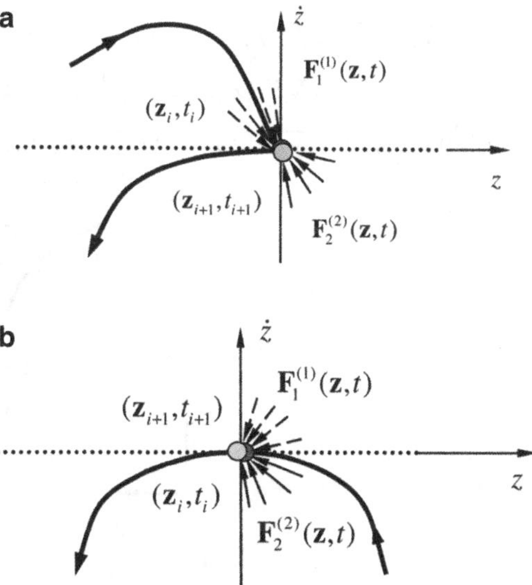

Fig. 3.17 Vector fields in the relative frame: (**a**) vanishing and (**b**) onset and vanishing of sliding flow

$\widetilde{\partial\Omega}_{12}$ is labeled by the dark circular symbol in Fig. 4.8b. From (3.90), $F_1(\mathbf{z}_m, t_{m\pm}) = 0$ and $G^{(1,1)}(\mathbf{z}_m, t_{m\pm}) > 0$ with $F_2(\mathbf{z}_m, t_{m-}) < 0$ should hold for the onset of the sliding flow. Notice that the sliding flow in the relative motion is always at the origin.

3.2.1.4 Grazing Flows to Boundary

Because of the discontinuity, the phase partition for this oscillator with friction is presented in the absolute and relative frames. Except for the flow passable at the boundary or sliding on the boundary, the flow in such a discontinuous system may be tangent to the boundary. Such a tangential flow is often called the grazing flow (or grazing motion). A sketch of grazing motions in domain Ω_α ($\alpha = \{1,2\}$) is illustrated in Fig. 3.18a, b. In the absolute frame, the separation boundary is a curve varying with time. The corresponding vector fields for grazing flows are sketched by the arrows in two domains. However, the discontinuous boundary in the relative frame is constant. The corresponding grazing flow and vector fields are sketched in Fig. 3.19a, b. Because the boundary in the relative frame is a straight line, from Luo (2009, 2011), the grazing flow is guaranteed by

$$\left.\begin{aligned} &\mathbf{n}^{\mathrm{T}}_{\partial\Omega_{\alpha\beta}} \cdot \mathbf{F}^{(\alpha)}_{\alpha}(\mathbf{z}_m, t_{m\pm}) = 0 \quad \text{for } \alpha = 1,2, \\ &G^{(1,1)}(\mathbf{z}_m, t_{m\pm}) = \mathbf{n}^{\mathrm{T}}_{\partial\Omega_{\alpha\beta}} \cdot D\mathbf{F}^{(1)}_{1}(\mathbf{z}_m, t_{m\pm}) > 0, \text{ and} \\ &G^{(2,1)}(\mathbf{z}_m, t_{m\pm}) = \mathbf{n}^{\mathrm{T}}_{\partial\Omega_{\alpha\beta}} \cdot D\mathbf{F}^{(2)}_{2}(\mathbf{z}_m, t_{m\pm}) < 0. \end{aligned}\right\} \tag{3.92}$$

From (3.78) and (3.79), the sufficient and necessary conditions for grazing flows are:

$$\left.\begin{aligned} &G^{(0,\alpha)}(\mathbf{z}_m, t_{m\pm}) = F_\alpha(\mathbf{z}_m, t_{m\pm}) = 0, \quad \alpha \in \{1,2\}, \\ &G^{(1,1)}(\mathbf{z}_m, t_{m\pm}) = \nabla F_1(\mathbf{z}_m, t_{m\pm}) \cdot \mathbf{F}^{(1)}_{1}(\mathbf{z}_m, t_{m\pm}) + \partial_{t_m} F_1(\mathbf{z}_m, t_{m\pm}) > 0, \\ &G^{(1,2)}(\mathbf{z}_m, t_{m\pm}) = \nabla F_2(\mathbf{z}_m, t_{m\pm}) \cdot \mathbf{F}^{(2)}_{2}(\mathbf{z}_m, t_{m\pm}) + \partial_{t_m} F_2(\mathbf{z}_m, t_{m\pm}) < 0. \end{aligned}\right\} \tag{3.93}$$

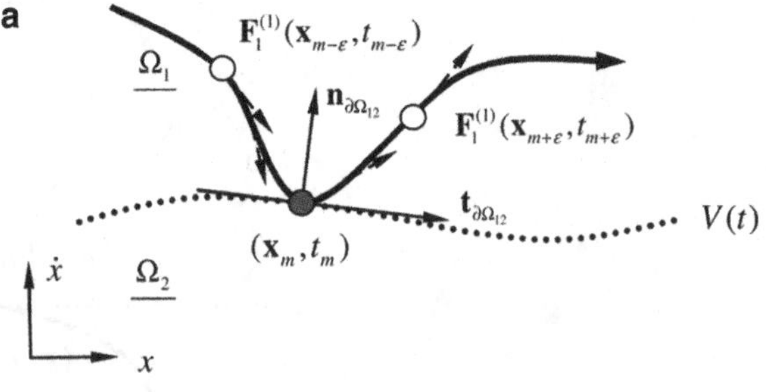

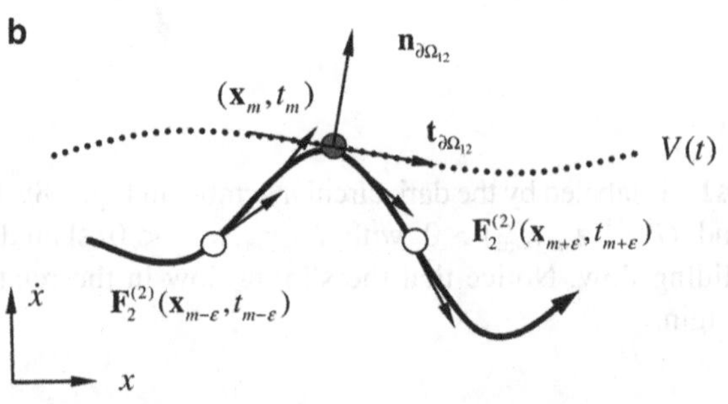

Fig. 3.18 Absolute vector fields of grazing motions in domains (**a**) Ω_1 and (**b**) Ω_2

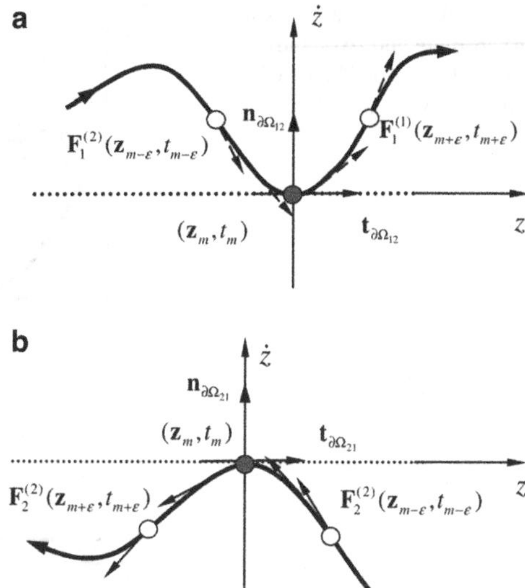

Fig. 3.19 Relative vector fields of grazing motions in domains (**a**) Ω_1 and (**b**) Ω_2

The grazing conditions are also illustrated in Fig. 3.19a, b, and the vector fields in Ω_1 and Ω_2 are expressed by the dashed and solid arrows, respectively. The condition in (3.93) for the grazing motion in Ω_α is presented via the vector fields of $\mathbf{F}_\alpha^{(\alpha)}(t)$. In addition to $F_\alpha(\mathbf{z}_m, t_{m\pm}) = 0$, the sufficient condition requires $F_1(\mathbf{z}, t) < 0$ for time $t \in [t_{m-\varepsilon}, t_m)$ and $F_1(\mathbf{z}, t) > 0$ for time $t \in (t_m, t_{m+\varepsilon}]$ in Ω_1; and $F_2(\mathbf{z}, t) > 0$ for $t \in [t_{m-\varepsilon}, t_m)$ and $F_2(\mathbf{z}, t) < 0$ for $t \in (t_m, t_{m+\varepsilon}]$ in Ω_2. In other words, $G^{(1,1)}(\mathbf{z}_m, t_{m\pm}) > 0$ in Ω_1 and $G^{(1,2)}(\mathbf{z}_m, t_{m\pm}) < 0$ in Ω_2 are required.

3.2.2 *Generic Mappings and Force Product Criteria*

In this section, the basic mappings are introduced. From such mappings, the force product criteria for stick and grazing flows to the boundary are discussed.

3.2.2.1 Generic Mappings

Direct integration of (3.60) with initial condition $(t_i, z_i, 0)$ gives the sliding displacement [i.e., (3.61)]. For a small δ-neighborhood of the sliding flow ($\delta \to 0$), substitution of (3.60) and (3.61) into (3.73) gives the forces in the two domains Ω_α ($\alpha \in \{1, 2\}$). For passable motions, select the initial condition on the velocity boundary (i.e., $\dot{x}_i = V_i$ and $x_i = X_i$) in the absolute frame. Based on (3.66), the basic solutions of the generalized discontinuous linear oscillator for a certain domain in Appendix A are used for the construction of mappings. In absolute phase space, a trajectory in Ω_α starting and ending at the velocity discontinuity (i.e., from $\partial\Omega_{\beta\alpha}$ to $\partial\Omega_{\alpha\beta}$) is sketched in Fig. 3.20. The starting and ending points for

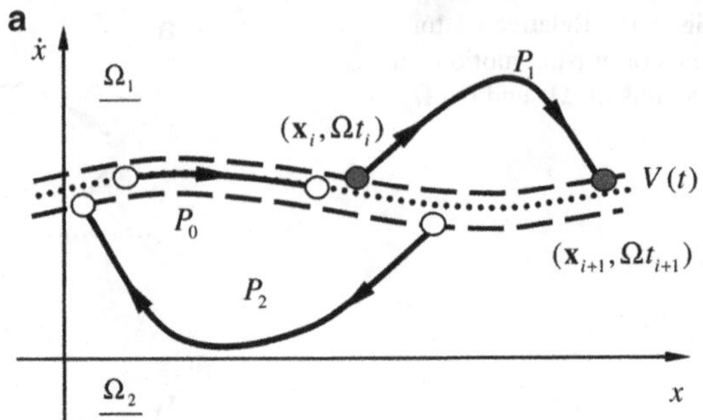

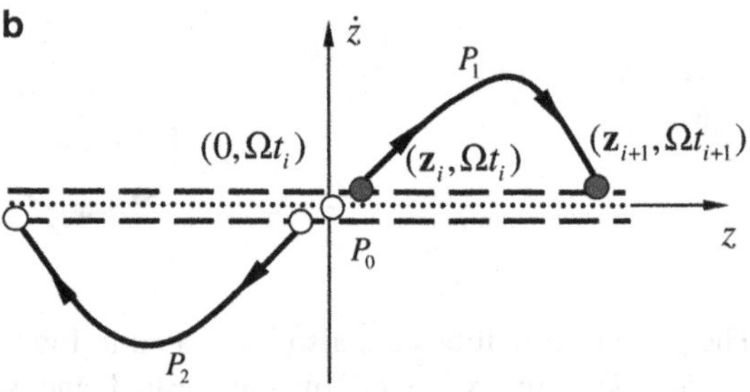

Fig. 3.20 Basic mappings in (**a**) absolute and (**b**) relative frames for the oscillator with a time-varying boundary

mappings P_α in Ω_α are (x_i, V_i, t_i) and $(x_{i+1}, V_{i+1}, t_{i+1})$, respectively. The sliding (or stick) mapping on the boundary is P_0. On the boundary $\partial\Omega_{\alpha\beta}$, letting $z_i = 0$, the switching planes is defined as

$$\left.\begin{aligned} \Xi^0 &= \{(x_i, \Omega t_i) | \dot{x}_i = V_i\}, \\ \Xi^1 &= \{(x_i, \Omega t_i) | \dot{x}_i = V_i^+\}, \\ \Xi^2 &= \{(x_i, \Omega t_i) | \dot{x}_i = V_i^-\}, \end{aligned}\right\} \tag{3.94}$$

where $V_i^- = \lim_{\delta \to 0} (V_i - \delta)$ and $V_i^+ = \lim_{\delta \to 0} (V_i + \delta)$ for arbitrarily small $\delta > 0$. Thus,

$$P_1 : \Xi^1 \to \Xi^1, \quad P_2 : \Xi^2 \to \Xi^2, \quad P_0 : \Xi^0 \to \Xi^0. \tag{3.95}$$

From the preceding two equations, we have

$$\left.\begin{aligned} P_0 &: (x_i, V_i, t_i) \to (x_{i+1}, V_{i+1}, t_{i+1}), \\ P_1 &: (x_i, V_i^+, t_i) \to (x_{i+1}, V_{i+1}^+, t_{i+1}), \\ P_2 &: (x_i, V_i^-, t_i) \to (x_{i+1}, V_{i+1}^-, t_{i+1}). \end{aligned}\right\} \tag{3.96}$$

From (3.61) and $F_\alpha(\mathbf{z}_{i+1}, t_{i+1}) = 0$, with $z_{i+1} = z_i = 0$ and $\dot{z}_{i+1} = \dot{z}_i = 0$, the governing equations for P_0 and $\alpha \in \{1, 2\}$ are

$$\left.\begin{aligned} x_{i+1} - V_0[\sin(t_{i+1} + \beta_i) - \sin(t_i + \beta_i)] - V_1 \times (t_{i+1} - t_i) - x_i = 0, \\ A_0 \cos \Omega t_{i+1} - b_\alpha - 2d_\alpha V_{i+1} - \dot{V}_{i+1} - c_\alpha x_{i+1} = 0. \end{aligned}\right\} \tag{3.97}$$

For this problem, the two domains Ω_α $(\alpha \in \{1,2\})$ are unbounded. From Luo (2005, 2006), the flows of dynamical systems on the corresponding domains should be bounded. For any nonsliding flow, there are three possible stable motions in the two domains Ω_α $(\alpha \in \{1,2\})$, the governing equations of mapping P_α $(\alpha \in \{1,2\})$ are obtained from the displacement and velocity in Appendix A. Thus, the governing equations of mapping P_α $(\alpha \in \{0,1,2\})$ is expressed by

$$\left.\begin{aligned} f_1^{(\alpha)}(x_i, \Omega t_i, x_{i+1}, \Omega t_{i+1}) = 0, \\ f_2^{(\alpha)}(x_i, \Omega t_i, x_{i+1}, \Omega t_{i+1}) = 0. \end{aligned}\right\} \tag{3.98}$$

3.2.2.2 Sliding Flows and Fragmentation

Based on the switching conditions, the corresponding criteria can be developed by the product of normal vector field as in Luo (2006). Two equations with four unknowns in (3.98) cannot give the unique solutions for the sliding flow. Once the initial state is given, the final state of the sliding flow is uniquely determined. Consider switching states $(\Omega t_c, X_c, V(t_c))$ and $(\Omega t_f, X_f, V(t_f))$ as the initial and final conditions of a sliding flow, respectively. For a time interval $[t_i, t_{i+1}] \subseteq [t_c, t_f]$ of any sliding flows, assume $(\Omega t_{i+1}, X_{i+1}, V(t_{i+1})) \equiv (\Omega t_f, X_f, V(t_f))$. Because of $z_{i+1} = z_i = 0$ and $\dot{z}_{i+1} = \dot{z}_i = 0$, we have

$$\begin{aligned} F_\alpha(\mathbf{z}_i, \Omega t_i) &\equiv F_\alpha(x_i, V(t_i), \Omega t_i) \\ &= A_0 \cos \Omega t_i - b_\alpha - 2d_\alpha V(t_i) - \dot{V}(t_i) - c_\alpha x_i \end{aligned} \tag{3.99}$$

From (3.98) and (3.99), the sliding flow on the boundary $\widetilde{\partial\Omega}_{\alpha\beta}$ for all $t_i \in [t_c, t_f)$ satisfies the following force product relation

$$\begin{aligned} L_{12}(t_i) &\equiv L_{12}(x_i, V(t_i), \Omega t_i) \\ &= F_1(x_i, V(t_i), \Omega t_i) \times F_2(x_i, V(t_i), \Omega t_i) < 0. \end{aligned} \tag{3.100}$$

The critical condition in (3.80) or (3.81) gives the *initial force product condition* at the critical time t_c, i.e.,

$$\begin{aligned} L_{12}(t_c) &\equiv L_{12}(x_c, V(t_c), \Omega t_c) \\ &= F_1(x_c, V(t_c), \Omega t_c) \times F_2(x_c, V(t_c), \Omega t_c) = 0, \\ (-1)^\alpha & G^{(1,\alpha)}(\mathbf{z}_c, t_{c\pm}) < 0 \quad \text{for } \alpha \in \{1,2\}. \end{aligned} \tag{3.101}$$

If the initial condition satisfies (3.101), the sliding flow is called the critical sliding flow. The condition in (3.91) for the sliding flow gives the *final force product condition* at the time $t_{i+1} = t_{\mathrm{f}}$, i.e.,

$$\begin{aligned} L_{12}(t_{\mathrm{f}}) &\equiv L_{12}(x_{\mathrm{f}}, V_{\mathrm{f}}, \Omega t_{\mathrm{f}}) \\ &= F_1(x_{\mathrm{f}}, V_{\mathrm{f}}, \Omega t_{\mathrm{f}}) \times F_2(x_{\mathrm{f}}, V_{\mathrm{f}}, \Omega t_{\mathrm{f}}) = 0, \\ (-1)^{\alpha} &G^{(1,\alpha)}(\mathbf{z}_{\mathrm{f}}, t_{\mathrm{f}\pm}) < 0 \quad \text{for } \alpha \in \{1,2\}. \end{aligned} \tag{3.102}$$

Once the force product of the sliding flow changes its sign at $t_m \in (t_i, t_{i+1})$, such a sliding flow will vanish. So the characteristic of the force product for the sliding flow should be further discussed. From Chap. 2, a set of the local maximum of the force product relative to the domains Ω_1 and Ω_2 is defined from Luo (2006) as

$$_{L\max}L_{12}(t_k) = \left\{ L_{12}(t_k) \left| \begin{array}{l} \forall t_k \in (t_i, t_{i+1}), \exists L_{12}(t_k) \leq 0, \\ \frac{\mathrm{d}}{\mathrm{d}t} L_{12}(t)\Big|_{(t_k, \mathbf{x}_k)} = 0 \text{ and } \frac{\mathrm{d}^2}{\mathrm{d}t^2} L_{12}(t)\Big|_{(t_k, \mathbf{x}_k)} < 0 \end{array} \right. \right\} \tag{3.103}$$

and the maximum force product is defined as

$$_{G\max}L_{12}(t_k) = \max_{t_k \in (t_i,\, t_{i+1})} \{L_{12}(t_k) | L_{12}(t_k) \in {}_{L\max}L_{12}(t_k)\}. \tag{3.104}$$

Once $L_{12}(t_k) \in {}_{L\max}L_{12}(t_k)$ is greater than zero, the sliding flow will disappear. Further, the sliding flow will be fragmentized. The corresponding critical condition is

$$_{G\max}L_{12}(t_k) = 0 \quad \text{for } t_k \in (t_i, t_{i+1}). \tag{3.105}$$

The foregoing condition is termed the *sliding fragmentation condition*. After fragmentation, the sliding flow for $t_i < t_{i+1} \leq t_{i+2} < t_{i+3}$ becomes two pieces from $(x_i, V(t_i), \Omega t_i)$ to $(x_{i+1}, V(t_{i+1}), \Omega t_{i+1})$ and $(x_{i+2}, V(t_{i+2}), \Omega t_{i+2})$ to $(x_{i+3}, V \times (t_{i+3}), \Omega t_{i+3})$, respectively. From (3.81), the passable motion between the two fragmentized sliding flows require

$$L_{12}(t_k) > 0 \quad \text{for } t_k \in (t_{i+1}, t_{i+2}). \tag{3.106}$$

The fragmentation is sketched in Fig. 3.21 through the force product characteristics of the sliding flow in phase plane. The sliding, critical sliding, and sliding fragmentation motions are depicted for given parameters and excitation amplitude $A_0^{(1)} < A_0^{(2)} < A_0^{(3)}$. The sliding flow vanishes from the boundary and goes into the domain Ω_1. Two points $\mathbf{x}_i$ and $\mathbf{x}_{i+1}$ represent the starting and vanishing switching displacements of the sliding flow, respectively. The point $\mathbf{x}_k \in (\mathbf{x}_i, \mathbf{x}_{i+1})$ represents a critical point for the fragmentation of the sliding flow. After the fragmentation, the critical point $\mathbf{x}_k$ splits into two new points $\mathbf{x}_{i+1}$ and $\mathbf{x}_{i+2}$, and the index of $\mathbf{x}_{i+1}$ will be shifted as $\mathbf{x}_{i+3}$. Suppose the force product $L_{12}(t_k)$ increases

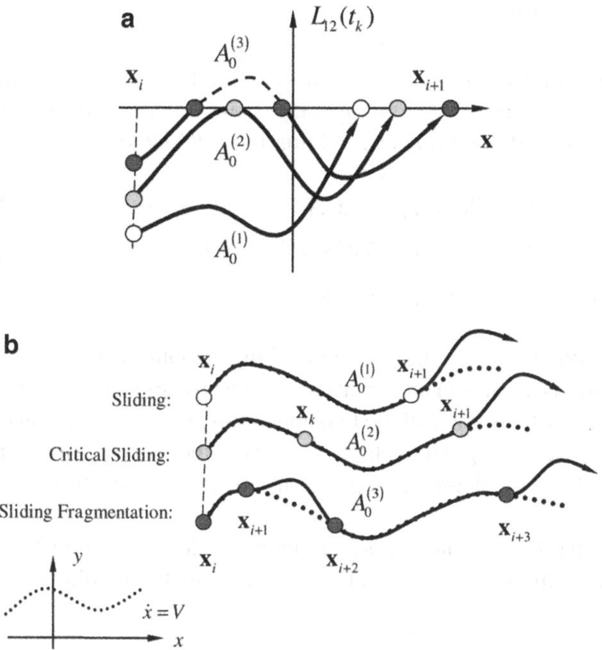

Fig. 3.21 (**a**) A sketch of force product and (**b**) phase plane for sliding, critical sliding, and sliding fragmentation motions under given parameters and $A_0^{(1)}<A_0^{(2)}<A_0^{(3)}$. The ending point of the sliding flow vanishes on the boundary and goes to the domain Ω_1

with increasing excitation amplitude, then the flows from the sliding to the sliding fragmentation are given for increasing A_0 from $A_0^{(1)}$ to $A_0^{(3)}$. The passable portion of the sliding fragmentation needs $L_{12}(t) > 0$ as in (3.106) for $t \in U_n = (t_j^n, t_{j+1}^n) \subset (t_i, t_{i+1})$ where $L_{12}(t_k) = 0$ for $t_k = t_j^n, t_{j+1}^n$. However, for the sliding portion, the force products keep $L_{12}(t_k) < 0$ for $t_k \in (t_i, t_{i+1}) \backslash \cup_n U_n$. Otherwise, suppose the force products at the potential merging points of the two sliding flows decrease with decreasing excitation, then the merging of the two sliding flows can be observed for decreasing A_0 from $A_0^{(3)}$ to $A_0^{(1)}$. Once the merging or fragmentation of a sliding flow occurs, the old sliding flow will be destroyed. For a sliding flow on $\partial\widetilde{\Omega}_{\alpha\beta}$ with starting and ending points $(x_i, V_i, \Omega t_i)$ and $(x_{i+1}, V_{i+1}, \Omega t_{i+1})$, consider $\Delta = \sqrt{\delta_{x_i}^2 + \delta_{t_i}^2 + \delta_{V_i}^2}$ with $\delta_{V_i} = V_{i+1} - V_i$, $\delta_{x_i} = x_{i+1} - x_i$, and $\delta_{t_i} = \Omega t_{i+1} - \Omega t_i \geq 0$. Let us assume that the starting and ending points satisfy the initial and final force product conditions. As $\Delta \to 0$, the following force product condition is termed the *onset* condition for the sliding flow.

$$\begin{aligned} &L_{12}(t_k) = 0 \quad \text{for } k \in \{i, i+1\}, \\ &(x_{i+1}, V_{i+1}, \Omega t_{i+1}) = \lim_{\delta \to 0} (x_i + \delta_{x_i}, V_i + \delta_{V_i}, \Omega t_i + \delta_{t_i}). \end{aligned} \tag{3.107}$$

For detailed discussion, one can refer to Gegg et al. (2008).

3.2.2.3 Grazing Flows

If the grazing for two mappings of passable motions occurs at the final state $(x_{i+1}, V_{i+1}, t_{i+1})$, from (3.81), the grazing conditions based on mappings are obtained. With (3.73), the grazing condition in (3.94) becomes

$$\left.\begin{aligned}&A_0\cos\Omega t_{i+1} - b_\alpha - 2d_\alpha V_{i+1} - c_\alpha x_{i+1} - \dot{V}_{i+1} = 0 \ (\alpha\in\{1,2\}),\\&-2d_\alpha\dot{V}_{i+1} - c_\alpha V_{i+1} - \ddot{V}_{i+1} - A_0\Omega\sin\Omega t_{i+1} > 0 \quad (\alpha = 1).\\&-2d_\alpha\dot{V}_{i+1} - c_\alpha V_{i+1} - \ddot{V}_{i+1} - A_0\Omega\sin\Omega t_{i+1} < 0 \quad (\alpha = 2).\end{aligned}\right\} \tag{3.108}$$

The grazing conditions for the two nonsliding mappings can be illustrated. The grazing conditions in (3.108) are given through the forces and jerks. Hence, both the initial and final switching sets of the two nonsliding mappings will vary with system parameters. Because the grazing characteristics of the two nonsliding mappings are different, illustrations of grazing conditions for the two mappings are presented separately.

To ensure the initial switching sets to be passable, from Luo (2006), the initial switching sets of mapping P_α ($\alpha\in\{1,2\}$) should satisfy the following condition as in Luo and Gegg (2006a, b).

$$L_{12}(t_m) = F_1(\mathbf{z}_m, t_{m\mp}) \times F_2(\mathbf{z}_m, t_{m\pm}) > 0. \tag{3.109}$$

The condition in (3.109) guarantees the initial switching sets of mapping P_α ($\alpha\in\{1,2\}$) which is passable to the discontinuous boundary (i.e., $\dot{x}_i = V_i$). The force product for the initial switching sets is also illustrated to ensure the existence of nonsliding mapping. The force conditions for the final switching sets of mapping P_α ($\alpha\in\{1,2\}$) are presented in (3.108).

The grazing conditions are given through (3.98) and (3.108). Three equations plus an inequality with four unknowns require one unknown be given. For instance, specific values for the initial displacement or phase of mapping P_α ($\alpha\in\{1,2\}$) can be selected from (3.108). Therefore, three equations with three unknowns will give the grazing conditions. Namely, the initial switching phase, the final switching phase, and displacement of mapping P_α ($\alpha\in\{1,2\}$) will be determined by (3.98) and (3.108). From the inequality of (3.108), the critical value for $\mathrm{mod}\,(\Omega t_{i+1}, 2\pi)$ is introduced by

$$\Theta_\alpha^{\mathrm{cr}} = \arcsin\left(-\frac{\gamma_\alpha}{A_0\Omega}\right), \tag{3.110}$$

where $\gamma_\alpha = c_\alpha V_{i+1} + 2d_\alpha\dot{V}_{i+1} + \ddot{V}_{i+1}$ ($\alpha = 1, 2$) and the superscript "cr" represents a critical value relative to grazing and $\alpha\in\{1,2\}$. From the second equation of (3.108), the final switching phase for mapping P_1 has the following six cases:

$$\left.\begin{aligned}&\mathrm{mod}\,(\Omega t_{i+1}, 2\pi) \in (\pi + |\Theta_1^{\mathrm{cr}}|, 2\pi - |\Theta_1^{\mathrm{cr}}|) \subset (\pi, 2\pi), \quad \text{for } 0 < \gamma_1 < A_0\Omega,\\&\mathrm{mod}\,(\Omega t_{i+1}, 2\pi) \in (\pi - \Theta_1^{\mathrm{cr}}, 2\pi] \cup [0, \Theta_1^{\mathrm{cr}}), \quad \text{for } \gamma_1 < 0 \text{ and } A_0\Omega > |\gamma_1|,\\&\mathrm{mod}\,(\Omega t_{i+1}, 2\pi) \in (\pi, 2\pi), \quad \text{for } \gamma_1 = 0,\end{aligned}\right\} \tag{3.111}$$

$$\left.\begin{array}{l}\bmod(\Omega t_{i+1},2\pi)\in\{\varnothing\},\quad\text{for }\gamma_1>0\text{ and }A_0\Omega\le\gamma_1,\\ \bmod(\Omega t_{i+1},2\pi)\in[0,2\pi],\quad\text{for }\gamma_1<0\text{ and }A_0\Omega<|\gamma_1|,\\ \bmod(\Omega t_{i+1},2\pi)\in[0,2\pi]/\{\pi/2\},\quad\text{for }\gamma_1<0\text{ and }A_0\Omega=|\gamma_1|.\end{array}\right\}\tag{3.112}$$

The parameter characteristics of grazing for mapping P_2 are presented as follows. Similarly, from the third equation of (3.108), the six cases of the final switching phase for mapping P_2 are

$$\left.\begin{array}{l}\bmod(\Omega t_{i+1},2\pi)\in[0,\pi+|\Theta_2^{\mathrm{cr}}|)\cup(2\pi-|\Theta_2^{\mathrm{cr}}|,2\pi],\quad\text{for }0<\gamma_2<A_0\Omega,\\ \bmod(\Omega t_{i+1},2\pi)\in(\Theta_2^{\mathrm{cr}},\pi-\Theta_2^{\mathrm{cr}})\subset(0,\pi),\quad\text{for }\gamma_2<0\text{ and }A_0\Omega>|\gamma_2|,\\ \bmod(\Omega t_{i+1},2\pi)\in(0,\pi),\quad\text{for }\gamma_2=0,\end{array}\right\}\tag{3.113}$$

$$\left.\begin{array}{l}\bmod(\Omega t_{i+1},2\pi)\in[0,2\pi],\quad\text{for }\gamma_2>0\text{ and }A_0\Omega<\gamma_2,\\ \bmod(\Omega t_{i+1},2\pi)\in[0,2\pi]/\{3\pi/2\},\quad\text{for }\gamma_2>0\text{ and }A_0\Omega=\gamma_2,\\ \bmod(\Omega t_{i+1},2\pi)\in\{\varnothing\},\quad\text{for }\gamma_2<0\text{ and }A_0\Omega<|\gamma_2|.\end{array}\right\}\tag{3.114}$$

For more details, the reader can refer to Gegg et al. (2009).

3.2.3 *Numerical Simulations*

To illustrate the analytical prediction of sliding motions on the discontinuous boundary, the sliding motion of the friction-induced oscillator will be demonstrated through relative and absolute velocity time-histories and trajectories in phase plane. The grazing and sliding motions strongly depend on the force responses in such a discontinuous dynamical system, and the force responses will be given to illustrate the force criteria. For illustrations, the starting and grazing points of mapping P_α ($\alpha\in\{1,2\}$) are represented by the large, hollow and dark-solid circular symbols, respectively. The oscillating separation boundaries in the relative and absolute frames are represented by the dashed lines and curves. Grazing motions are presented by thick solid curves.

In Fig. 3.22, relative and absolute phase trajectories, force distributions along the relative displacement and velocity, and the relative and absolute velocity time-histories for the grazing motion in domain Ω_1 are illustrated. The parameters ($\Omega=1$, $A_0=100, V_1=1, d_1=1, d_2=0.1, b_1=-b_2=0.5, c_1=c_2=30$) plus the initial conditions $(x_i,y_i)=(-4.,V_i)$ and $\Omega t_i\approx\{3.7297,\ 3.5288,\ 3.3155\}$ corresponding to $V_0=\{0,2.5,5\}$ are used. If $V_0=0$, this friction-induced oscillator moves on the constant translation belt, which was discussed for constant velocity. In relative phase plane, three trajectories are bouncing to the discontinuous boundary ($\dot z=w=0$), as

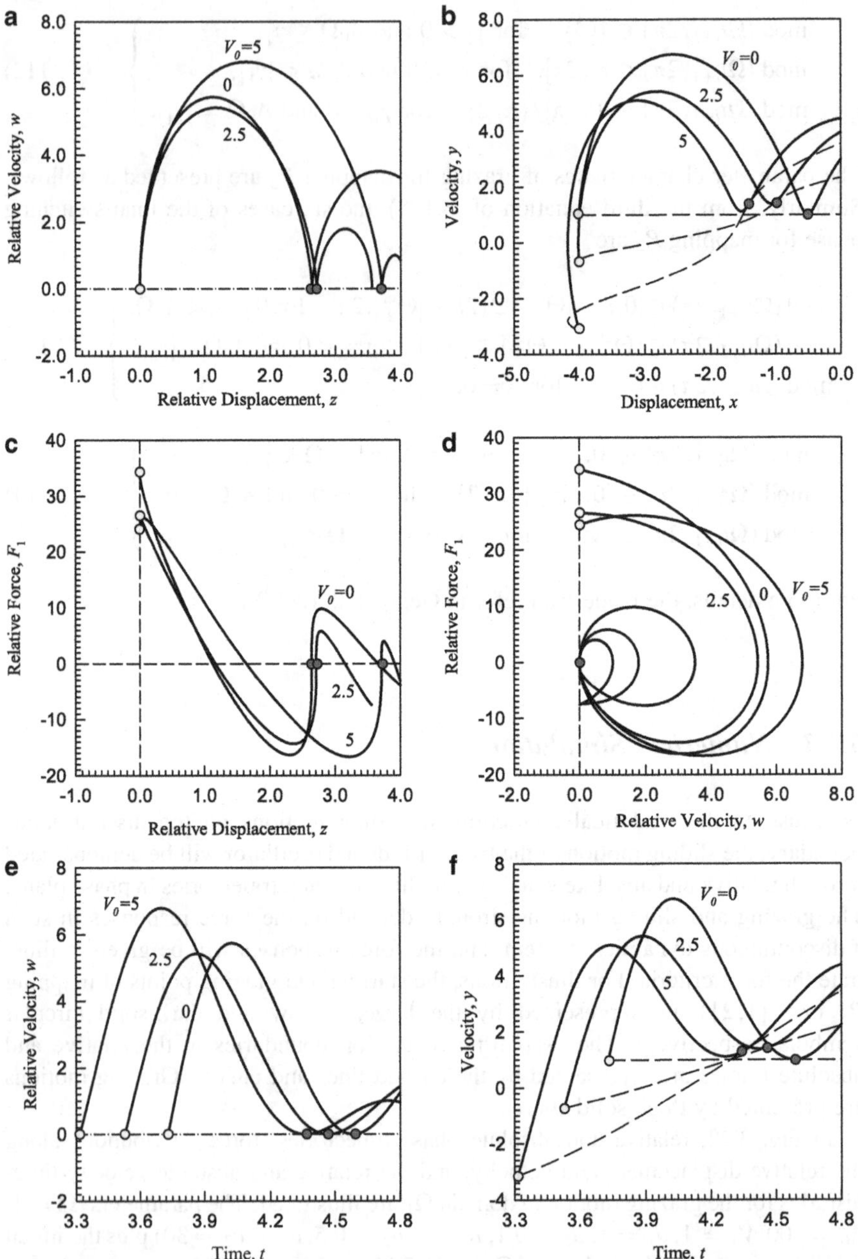

Fig. 3.22 Grazing motion of mapping P_1 in domain Ω_1 for $V_0 = \{0, 2.5, 5\}$: (**a**) relative phase trajectory, (**b**) absolute phase trajectory, (**c**) relative force distribution along relative displacement, (**d**) relative forces distribution on relative velocity, (**e**) relative velocity time-history, and (**f**) absolute velocity time-history ($\Omega = 1$, $A_0 = 100$, $V_1 = 1$, $d_1 = 1$, $d_2 = 0.1$, $b_1 = -b_2 = 0.5$, $c_1 = c_2 = 30$). The initial conditions are $(x_i, y_i) = (-4., V_i)$ and $\Omega t_i \approx \{3.7297, 3.5288, 3.3155\}$ with $V_0 = \{0, 2.5, 5\}$ accordingly

shown in Fig. 3.22a. The grazing points look like a singular points. However, in Fig. 3.22b, the trajectories in the absolute phase plane are tangential to the periodically oscillating boundary. In Fig. 3.22c, the relative forces $F_1(t)$ along the relative displacement in domain Ω_1 are presented. At grazing points, the relative forces $F_1(t)$ are zero. In vicinity of such grazing points, the relative forces $F_1(t)$ have a sign change from negative to positive. In Fig. 3.22d, the relative force distribution along velocity is presented and the curves of relative force $F_1(t)$ at the grazing points is tangential to the relative velocity boundary $w = 0$. This indicates that conditions in (3.92) are satisfied. In Fig. 3.22e, the time-histories of relative velocity responses are plotted, such responses are tangential to the boundary $w = 0$. However, in Fig. 3.22f, the absolute velocity time-history are tangential to the discontinuous boundary $V(t)$.

Consider system parameters (i.e., $V_0 = 0.5$, $V_1 = 1.0$, $\omega = \Omega = 1$, $d_1 = 1$, $d_2 = 0.1$, $b_1 = -b_2 = 0.5$, and $c_1 = c_2 = 30$) with the initial condition ($\Omega t_i = 7\pi/6$, $x_i = -1.5$, and $y_i \approx 0.7077$) for illustration of the sliding motion disappearance and getting into domain Ω_1. For specific excitation amplitudes $A_0 = \{49.75, 50.5, 51.25\}$, the absolute phase trajectories of sliding motions are plotted in Fig. 3.23a. The

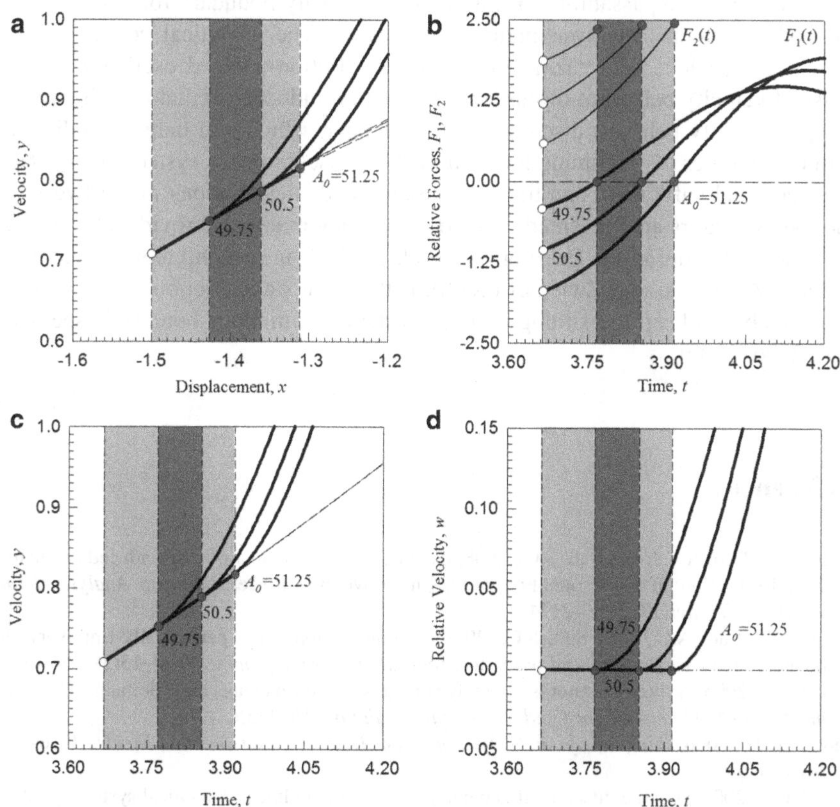

Fig. 3.23 Phase, velocity, and force responses of the sliding motion vanishing on the boundary and going into the domain Ω_1 for $A_0 = (49.75,\ 50.5,\ 51.25)$: (**a**) absolute phase plane, (**b**) relative forces time-history, (**c**) absolute velocity time-history, (**d**) relative velocity time-history ($V_0 = 1$, $V_1 = 0.5$, $\omega = \Omega = 1$, $d_1 = 1$, $d_2 = 0.1$, $b_1 = -b_2 = 0.5$, $c_1 = c_2 = 30$, $x_i = -1.5$, $\Omega t_i = 7\pi/6$)

time-histories of relative forces for the sliding motion are presented in Fig. 3.23b. It is observed that the sliding motions satisfy $F_1 \times F_2 \leq 0$. For $A_0 \in [A_{0\min}, A_{0\max}]$ where $A_{0\min} = 49.6$ and $A_{0\max} = 51.91$, the initial force product is zero. The two values will be the minimum and maximum values for this sliding motion. For $A_{0\min} = 49.6$, $F_1 = 0$; and for $A_{0\max} = 51.91$, $F_2 = 0$. If $A_0 > A_{0\max}$ or $A_0 < A_{0\min}$, the sliding motion will not exist on the boundary and the passable motion with $F_1 \times F_2 > 0$ exists. Therefore, under the above given parameters, the sliding motion starting at $(x_i, y_i, \Omega t_i) = (-1.5, 0.7077, 7\pi/6)$ will exist for $A_{0\min} \leq A_0 \leq A_{0\max}$. Further, the absolute and relative velocity time-histories are also presented in Fig. 3.23c, d. The sliding motion along the velocity boundary is clearly observed. The relative force product conditions are clearly illustrated.

3.3 Concluding Remarks

In this chapter, the passability of a flow to the velocity boundary for two friction-induced oscillators was systematically addressed. The analytical conditions for the passable, stick, and grazing motion in the friction-induced oscillator with a constant velocity belt were discussed. The friction-induced oscillator with a time-varying velocity belt was discussed in this chapter, which will help us to discuss cutting dynamics of machining tools through the discontinuous system theory. The necessary and sufficient conditions for grazing and stick motions in the two dry-friction oscillators are presented. The corresponding materials provide a technique for how to determine the grazing and stick motions in discontinuous dynamical systems. As in Luo and Zwiegart (2008), if the large static friction is considered, it is possible to keep the sliding motion on the discontinuous boundary (see also Luo, 2006, 2009).

References

Gegg, B.C., Luo, A.C.J. and Suh, S.C., 2008, Grazing phenomena in a friction-induced oscillator with dry-friction on a sinusoidally time-varying, traveling surface, *Nonlinear Analysis: Real Word Applications*, **9**, 2156–2174.

Gegg, B.C., Luo, A.C.J. and Suh, S.C., 2009, Sliding motions on a periodically time-varying boundary for a friction-induced oscillator, *Journal of Vibration and Control*, **15**(5), 671–703.

Luo, A.C.J., 2005, A theory for non-smooth dynamical systems on connectable domains, *Communication in Nonlinear Science and Numerical Simulation*, **10**, 1–55.

Luo, A.C.J., 2006, *Singularity and Dynamics on Discontinuous Vector Fields*, Elsevier: Amsterdam.

Luo, A.C.J., 2008a, On the differential geometry of flows in nonlinear dynamical systems, *ASME Journal of Computational and Nonlinear Dynamics*, **3**, 021104 (1–10).

Luo, A.C.J., 2008b, A theory for flow switchability in discontinuous dynamical systems, *Nonlinear Analysis: Hybrid Systems*, **2**(4), 1030–1061.

Luo, A.C.J., 2009, *Discontinuous Dynamical Systems on Time-varying Domains*, HEP-Springer: Heidelberg.

Luo, A.C.J., 2011, *Discontinuous Dynamical Systems*, HEP- Springer: Heidelberg.

Luo, A.C.J. and Gegg, B.C., 2006a, On the mechanism of stick and non-stick, periodic motions in a forced linear oscillator including dry friction, *ASME Journal of Vibration and Acoustics*, **128**, 97–105.

Luo, A.C.J. and Gegg, B.C., 2006b, Stick and non-stick, periodic motions of a periodically forced, linear oscillator with dry friction, *Journal of Sound and Vibration*, **291**, 132–168.

Luo, A.C.J. and Gegg, B.C., 2006c, Grazing phenomena in a periodically forced, friction-induced, linear oscillator, *Communications in Nonlinear Science and Numerical Simulation*, **11**, 777–802.

Luo, A.C.J. and Gegg, B.C., 2006d, Periodic motions in a periodically forced oscillator moving on an oscillating belt with dry friction, *ASME Journal of Computational and Nonlinear Dynamics*, **1**, 212–220.

Luo, A.C.J. and Gegg, B.C., 2006e, Dynamics of a periodically excited oscillator with dry friction on a sinusoidally time-varying, traveling surface, *International Journal of Bifurcation and Chaos*, **16**, 3539–3566.

Luo, A.C.J. and Gegg, B.C., 2007, An analytical prediction of sliding motions along discontinuous boundary in non-smooth dynamical systems, *Nonlinear Dynamics*, **49**, 401–424.

Luo, A.C.J. and Zwiegart Jr., P., 2008, Existence and analytical prediction of periodic motions in a periodically forced, nonlinear friction oscillator, *Journal of Sound and Vibration*, **309**, 129–149.

Chapter 4
Cutting Dynamics Mechanism

In this chapter, a mechanical model is developed for cutting dynamics caused by coupled interactions among the tool-piece, work-piece, and chip during the complex cutting operation. Four regions are used to describe four distinct cutting processes. The switching from one motion to another in the adjacent region is to characterize the specific tool–workpiece interaction at the boundary, and such a switching is a key to understand cutting mechanism. The domains and boundaries for cutting dynamical systems are discussed using the discontinuous system theory in Chap. 2. The passable motions and nonpassable flow to the boundaries are discussed, and the corresponding criteria are given. In addition, the switching bifurcations from passable motions (nonpassable motion) to nonpassable motion (passable motion) are developed for cutting process.

4.1 Machine-Tool Dynamics

The cutting process in manufacturing can be modeled by a two-degree-of-freedom tool-piece model with external force effects due to contact and cutting of a work-piece (e.g., Moon and Kalmar-Nagy, 2001). As stated in Shaw (2005), the turning process is a common machining practice consisting of "a single point tool that removes unwanted material to produce a surface of revolution," as shown in Fig. 4.1. In addition to orthogonal cutting by lathing, cutting actions include sawing, planning, and broaching (e.g., Shaw 2005). Orthogonal cutting is a flow of removed material across the tool surface at a perpendicular angle to the cutting edge. The evolution of the mechanical model describing orthogonal cutting begins where the tool- and work-pieces are described by the two masses m and m_{eq} in Fig. 4.2a. The tool-piece is governed by viscous damping of coefficients d_x and d_y, and linear equivalent spring forces of stiffness k_x and k_y, respectively. In Fig. 4.2a, the large arrows mark two paths or phases a machine-tool could experience. Phase one is the contact of the tool- and work-piece without cutting, which is sketched in

B.C. Gegg et al., *Machine Tool Vibrations and Cutting Dynamics*,
DOI 10.1007/978-1-4419-9801-9_4,

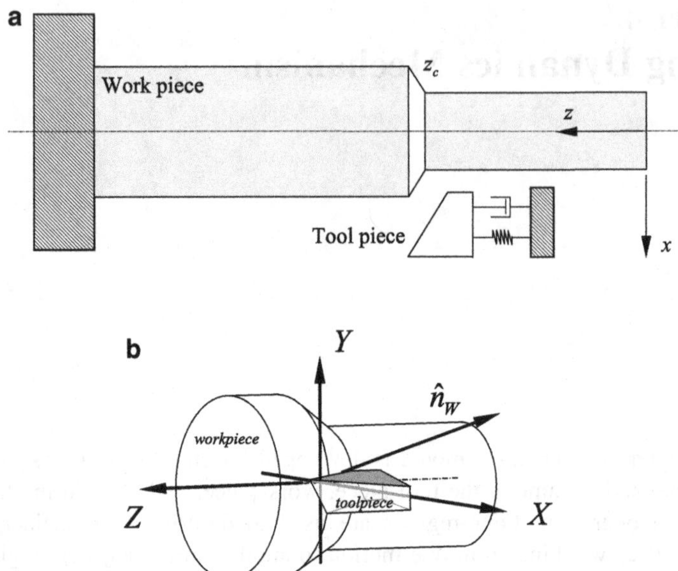

Fig. 4.1 (**a**) Tool-piece and work-piece configuration and (**b**) engagement at point of cutting

Fig. 4.2b. The contact forces transmitted by the work-piece are resolved into the normal direction with respect to the contact surface. The equivalent response of the work-piece is modeled by viscous damping and linear stiffness forces of coefficients d_1 and k_1, respectively, which is presented in Fig. 4.2c. In this case, the contact surface is the bottom face (or flank surface). The noncutting phase can be resolved to the model given in Fig. 4.2e, where no forces act on the tool-piece rake face. The tool-rake surface, cutting edge, and flank surface are presented in Fig. 4.2a. Phase two is the case where the tool contacts the work-piece and the cutting conditions are satisfied. A chip will form on the top-left face (or rake surface) of the tool-piece, as shown in Fig. 4.2b.

Since the typical mass of a chip m_{ch} is quite small, the chip simply transmits forces and has no significant acceleration independent of the tool- or work-piece in Dassanayake and Suh (2007). The forces transmitted by a chip to the tool-piece are governed by the shearing action of the cutting process (cutting depth, width, and shearing angle) and the motion of the work-piece. In Fig. 4.2d, the cutting process is modeled by viscous damping and linear stiffness forces of coefficients d_2 and k_2, respectively. The exertion of the chip on the rake surface creates a normal force and a frictional force. Since the frictional force is typically dependent on the relative velocity of the chip and rake surfaces, the chip can be modeled by a traveling belt with a dynamical normal force, as shown in Fig. 4.2f (e.g., Wiercigroch 1997 and Warminski et al. 2003). The forces transmitted to the tool-piece by the chip and work-pieces are resolved into their vector forms in Fig. 4.2e. The model presented in Fig. 4.3b will be used for machine-tool to investigate the cutting process.

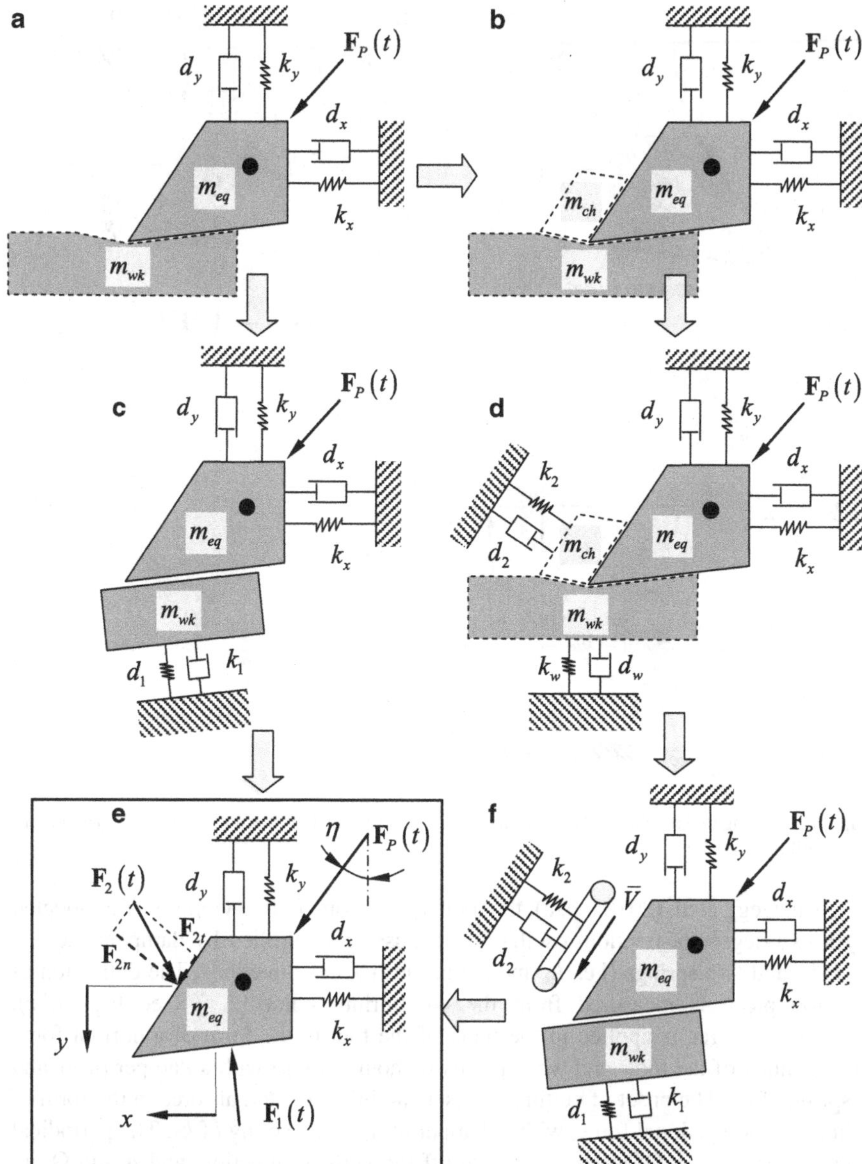

Fig. 4.2 Cutting tool mechanical model: (**a**) tool- and work-pieces in contact (no cutting), (**b**) tool-piece, work-piece, and chip in contact (cutting), (**c**) tool-piece and work-piece equivalent forces, (**d**) tool-piece, work-piece, and chip dynamic system, (**e**) equivalent machine-tool model, and (**f**) tool-piece, equivalent work-piece, and chip dynamics with frictional surface

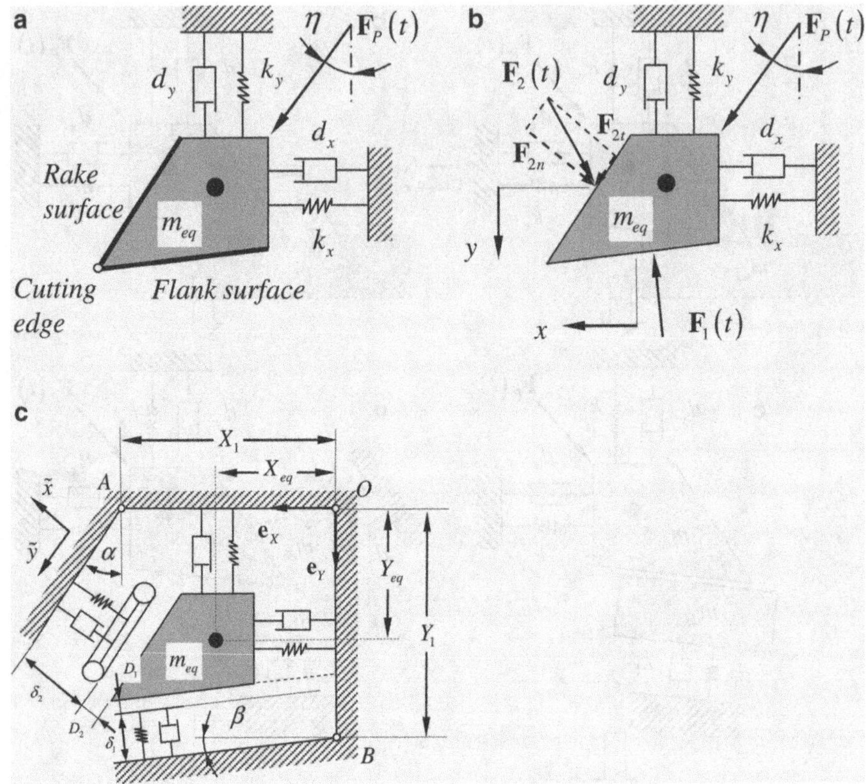

Fig. 4.3 Cutting tool mechanical model: (**a**) surface description, (**b**) external forces, and (**c**) geometry and equilibrium

As in Gegg et al. (2010a, b, c), the tool and supports in free vibration are modeled by a two-degree-of-freedom oscillator, of mass m, controlled by dampers (i.e., d_x and d_y) and two springs (i.e., k_x and k_y) in the $(\mathbf{e}_X, \mathbf{e}_Y)$ directions. The deflection of the tool-piece is measured from the equilibrium point $(\bar{x}, \bar{y})$ (see Fig. 4.3c). An external force is applied to the flank of the tool in the form of a normal force (i.e., contact of the tool- and work-pieces without cutting) with a damper of d_1 and a spring of k_1. The onset of cutting exerts an additional external force in the form of a normal and frictional force, with a damper of d_2 and a spring of k_2. The periodical force $F_p(\bar{t}) = A\cos\Omega\bar{t}$ at an angle of η off the vertical direction, and A and Ω are excitation amplitude and frequency, respectively. Such a mechanical model is shown in Fig. 4.3c. The flank surface has an angle of β from the horizontal surface.

The rake surface has an angle of α from the vertical surface, and the distance between the origin and point A (i.e., $\overline{OA}$) is X_1. The distances from the *flank* and *rake* surfaces to the equilibrium of the unstretched springs are δ_1 and δ_2, respectively. For the flank and rake surfaces, two distances are defined as

$$D_1 = (X_{eq} + \bar{x})\sin\beta + (Y_1 - Y_{eq} - \bar{y})\cos\beta - \delta_1. \tag{4.1}$$

$$D_2 = (Y_{eq} + \bar{y}) \sin\alpha + (X_1 - X_{eq} - \bar{x}) \cos\alpha - \delta_2. \tag{4.2}$$

For D_1>0 and D_2>0, the machine tool is free running (no external forces from the work-piece in any form). The equations of motion for the tool-piece are

$$\begin{aligned}&\begin{bmatrix}1 & 0\\ 0 & 1\end{bmatrix}\begin{Bmatrix}\ddot{\bar{x}}\\ \ddot{\bar{y}}\end{Bmatrix} + \frac{1}{m}\begin{bmatrix}d_x & 0\\ 0 & d_y\end{bmatrix}\begin{Bmatrix}\dot{\bar{x}}\\ \dot{\bar{y}}\end{Bmatrix} + \frac{1}{m}\begin{bmatrix}k_x & 0\\ 0 & k_y\end{bmatrix}\begin{Bmatrix}\bar{x}\\ \bar{y}\end{Bmatrix}\\ &= A_F \cos(\Omega \bar{t})\begin{Bmatrix}\sin(\eta)\\ \cos(\eta)\end{Bmatrix},\end{aligned} \tag{4.3}$$

where $A_F = A_e/m$. This case is considered as the *first region* in Fig. 4.4a.

If $D_1 \leq 0$, the special case may occur. From (4.1), the initial contact for the flank surface and work-piece ($D_1 = 0$) gives

$$(X_{eq} + \bar{x}) \sin\beta + (Y_1 - Y_{eq} - \bar{y}) \cos\beta = \delta_1. \tag{4.4}$$

The force per unit mass at the contact point with mass m on the flank surface is

$$\mathbf{F}_1(\bar{\mathbf{x}}, \bar{\mathbf{y}}) = \begin{Bmatrix}F_{1x}\\ F_{1y}\end{Bmatrix} = F_{n1}(\bar{\mathbf{x}}, \bar{\mathbf{y}})\begin{Bmatrix}-\sin\beta\\ \cos\beta\end{Bmatrix}, \tag{4.5}$$

where $\bar{\mathbf{x}} = (\bar{x}, \dot{\bar{x}})^{\mathrm{T}}$ and $\bar{\mathbf{y}} = (\bar{y}, \dot{\bar{y}})^{\mathrm{T}}$ with

$$F_{n1}(\bar{\mathbf{x}}, \bar{\mathbf{y}}) = d_1\dot{\bar{x}} \sin\beta - d_1\dot{\bar{y}} \cos\beta + k_1(\bar{x} - \bar{x}_1^*) \sin\beta - k_1(\bar{y} - \bar{y}_1^*) \cos\beta, \tag{4.6}$$

which is considered as the *second region*, as sketched in Fig. 4.4a, b. The forces in (4.6) express a continuous stiffness force at the point of switching. This situation is similar to the results in Luo and O'Connor (2009a, b, c) or Luo (2009). Without contact, this force will disappear. Thus, the forces are C^0-continuous but C^1 discontinuous at $D_1 = 0$ because of $F_{n1}(\bar{\mathbf{x}}, \bar{\mathbf{y}}) = 0$. This is a directly physical observation, which can be analytically derived from discontinuous dynamical system theory. At this point, equation (4.4) is satisfied and the contact of the work-piece is defined to exert an impact force through the conservation of momentum, in the direction of normal contact, by restitution coefficient e_{r}. In fact, the impacting chatter exists before the flank surface and work-piece completely contact each other without any relative velocity. In this book, such an issue will not be discussed. Herein, the steady-state cutting process on the friction boundary is focused. If readers are interested in this issue, the reader can refer to Luo and O'Connor (2009a, b, c) and Luo (2009).

For $D_1 \leq 0$ and $D_2 \geq 0$, equations (4.1) and (4.2) give

$$\begin{aligned}&(X_{eq} + \bar{x}) \sin\beta + (Y_1 - Y_{eq} - \bar{y}) \cos\beta = \delta_1,\\ &(Y_{eq} + \bar{y}) \sin\alpha + (X_1 - X_{eq} - \bar{x}) \cos\alpha \geq \delta_2.\end{aligned} \tag{4.7}$$

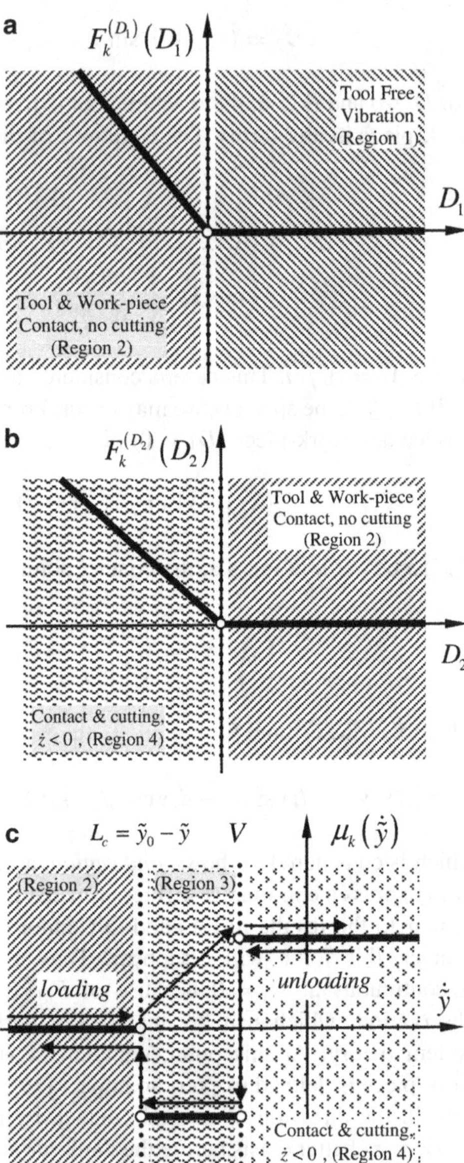

Fig. 4.4 Force definitions for this machine-tool system: (**a**) region one to region two force condition, (**b**) region two to region four force condition, and (**c**) loading and unloading paths

In this case, the machine tool will not cut the work-piece, but just slide across the flank surface as shown in Fig. 4.2a, c. The equations of motion for this case are

$$
\begin{aligned}
&\begin{bmatrix} 1 & 0 \\ 0 & 1 \end{bmatrix} \begin{Bmatrix} \ddot{\bar{x}} \\ \ddot{\bar{y}} \end{Bmatrix} + \begin{bmatrix} \bar{D}_{11} & \bar{D}_{12} \\ \bar{D}_{21} & \bar{D}_{22} \end{bmatrix} \begin{Bmatrix} \dot{\bar{x}} \\ \dot{\bar{y}} \end{Bmatrix} + \begin{bmatrix} \bar{K}_{11} & \bar{K}_{12} \\ \bar{K}_{21} & \bar{K}_{22} \end{bmatrix} \begin{Bmatrix} \bar{x} \\ \bar{y} \end{Bmatrix} \\
&\quad = A_F \cos(\Omega \bar{t}) \begin{Bmatrix} \sin(\eta) \\ \cos(\eta) \end{Bmatrix} + \begin{Bmatrix} F_{1x} \\ F_{1y} \end{Bmatrix}.
\end{aligned}
\tag{4.8}
$$

For $D_1 < D_2 \leq 0$, two external forces $\mathbf{F}_i(\bar{\mathbf{x}}, \bar{\mathbf{y}})$ $(i = 1, 2)$ will act on the tool-piece, as in Fig. 4.2b, d, and f. The force $\mathbf{F}_1(\bar{\mathbf{x}}, \bar{\mathbf{y}})$ is given in (4.5). The supporting normal force and friction force per unit mass on the left inclined surface (rake surface) acting on mass m provides the total force $\mathbf{F}_2(\bar{\mathbf{x}}, \bar{\mathbf{y}})$ as

$$\mathbf{F}_2(\bar{\mathbf{x}}, \bar{\mathbf{y}}) = \begin{Bmatrix} F_{2x} \\ F_{2y} \end{Bmatrix} = F_{n2}(\bar{\mathbf{x}}, \bar{\mathbf{y}}) \begin{Bmatrix} M(\dot{\bar{x}}, \dot{\bar{y}}) \\ N(\dot{\bar{x}}, \dot{\bar{y}}) \end{Bmatrix}, \tag{4.9}$$

where

$$F_{n2}(\bar{\mathbf{x}}, \bar{\mathbf{y}}) = d_2 \dot{\bar{x}} \sin\alpha - d_2 \dot{\bar{y}} \cos\alpha + k_2(\bar{x} - \bar{x}_2^*) \sin\alpha - k_2(\bar{y} - \bar{y}_2^*) \cos\alpha, \tag{4.10}$$

and

$$\left.\begin{aligned} M(\dot{\bar{x}}, \dot{\bar{y}}) &= -[\cos\alpha + \mu \operatorname{sgn}(\dot{\bar{z}}) \sin\alpha], \\ N(\dot{\bar{x}}, \dot{\bar{y}}) &= \sin\alpha - \mu \operatorname{sgn}(\dot{\bar{z}}) \cos\alpha, \\ \dot{\bar{z}} &= \dot{\bar{x}} \sin\alpha + \dot{\bar{y}} \cos\alpha - \bar{V}. \end{aligned}\right\} \tag{4.11}$$

The forces in (4.9) express a continuous stiffness force at the point of switching in Fig. 4.4b. The forces are C^0 discontinuous. Due to the nature of friction, if $\dot{\bar{z}}<0$, the region of motion is considered as the *third region*. If $\dot{\bar{z}}>0$, the region of motion is called the *fourth region* in Fig. 4.4c.

The corresponding equation of motion in region three is

$$\begin{aligned} &\begin{bmatrix} 1 & 0 \\ 0 & 1 \end{bmatrix} \begin{Bmatrix} \ddot{\bar{x}} \\ \ddot{\bar{y}} \end{Bmatrix} + \begin{bmatrix} \bar{D}_{11} & \bar{D}_{12} \\ \bar{D}_{21} & \bar{D}_{22} \end{bmatrix} \begin{Bmatrix} \dot{\bar{x}} \\ \dot{\bar{y}} \end{Bmatrix} + \begin{bmatrix} \bar{K}_{11} & \bar{K}_{12} \\ \bar{K}_{21} & \bar{K}_{22} \end{bmatrix} \begin{Bmatrix} \bar{x} \\ \bar{y} \end{Bmatrix} \\ &= A_F \cos(\Omega \bar{t}) \begin{Bmatrix} \sin(\eta) \\ \cos(\eta) \end{Bmatrix} + \begin{Bmatrix} F_{1x} \\ F_{1y} \end{Bmatrix} + \begin{Bmatrix} F_{2x}^{(i)} \\ F_{2y}^{(i)} \end{Bmatrix}. \end{aligned} \tag{4.12}$$

with

$$\begin{Bmatrix} F_{2x}^{(i)} \\ F_{2y}^{(i)} \end{Bmatrix} = F_{n2}(\bar{\mathbf{x}}, \bar{\mathbf{y}}) \begin{Bmatrix} M^{(i)}(\dot{\bar{x}}, \dot{\bar{y}}) \\ N^{(i)}(\dot{\bar{x}}, \dot{\bar{y}}) \end{Bmatrix}, \tag{4.13}$$

where for region three

$$\left.\begin{aligned} M^{(3)}(\dot{\bar{x}}, \dot{\bar{y}}) &= -(\cos\alpha + \mu \sin\alpha), \\ N^{(3)}(\dot{\bar{x}}, \dot{\bar{y}}) &= \sin\alpha - \mu \cos\alpha, \\ \text{if } \dot{\bar{z}} &= \dot{\bar{x}} \sin\alpha + \dot{\bar{y}} \cos\alpha - \bar{V} > 0. \end{aligned}\right\} \tag{4.14}$$

and for region four

$$\left.\begin{aligned} M^{(4)}(\dot{\bar{x}}, \dot{\bar{y}}) &= -(\cos\alpha - \mu \sin\alpha), \\ N^{(4)}(\dot{\bar{x}}, \dot{\bar{y}}) &= \sin\alpha + \mu \cos\alpha, \\ \text{if } \dot{\bar{z}} &= \dot{\bar{x}} \sin\alpha + \dot{\bar{y}} \cos\alpha - \bar{V} < 0. \end{aligned}\right\} \tag{4.15}$$

If $\dot{z} = \dot{x}\sin\alpha + \dot{y}\cos\alpha - \bar{V} = 0$, the following relation exists

$$\begin{aligned}\begin{Bmatrix} F_{2nfx} \\ F_{2nfy} \end{Bmatrix} &= \begin{bmatrix} 1 & 0 \\ 0 & 1 \end{bmatrix}\begin{Bmatrix} \ddot{\bar{x}} \\ \ddot{\bar{y}} \end{Bmatrix} + \begin{bmatrix} \bar{D}_{11} & \bar{D}_{12} \\ \bar{D}_{21} & \bar{D}_{22} \end{bmatrix}\begin{Bmatrix} \dot{\bar{x}} \\ \dot{\bar{y}} \end{Bmatrix} + \begin{bmatrix} \bar{K}_{11} & \bar{K}_{12} \\ \bar{K}_{21} & \bar{K}_{22} \end{bmatrix}\begin{Bmatrix} \bar{x} \\ \bar{y} \end{Bmatrix} \\ &- A_F\cos(\Omega\bar{t})\begin{Bmatrix} \sin(\eta) \\ \cos(\eta) \end{Bmatrix} - \begin{Bmatrix} F_{1x} \\ F_{1y} \end{Bmatrix} - F_{n2}(\bar{\mathbf{x}}, \bar{\mathbf{y}})\begin{Bmatrix} -\cos\alpha \\ \sin\alpha \end{Bmatrix}.\end{aligned} \tag{4.16}$$

$$\begin{Bmatrix} F_{2nfx} \\ F_{2nfy} \end{Bmatrix} = F_{2nf}(\bar{\mathbf{x}}, \bar{\mathbf{y}})\begin{Bmatrix} -\sin\alpha \\ \cos\alpha \end{Bmatrix}. \tag{4.17}$$

If $|F_{2nf}(\bar{\mathbf{x}}, \bar{\mathbf{y}})| \le \mu F_{n2}(\bar{\mathbf{x}}, \bar{\mathbf{y}})$,

$$\begin{aligned}\dot{\bar{z}} &= \dot{\bar{x}}\sin\alpha + \dot{\bar{y}}\cos\alpha - \bar{V} = 0, \\ \ddot{\bar{z}} &= \ddot{\bar{x}}\sin\alpha + \ddot{\bar{y}}\cos\alpha = 0.\end{aligned} \tag{4.18}$$

In different regions, the governing equations are different. The nondimensional time $t = \Omega\bar{t}$ is introduced, and

$$\bar{\mathbf{r}} = (\bar{x}, \bar{y})^{\mathrm{T}} = (x, y)^{\mathrm{T}}, \tag{4.19}$$

$$\frac{\mathrm{d}\bar{\mathbf{r}}}{\mathrm{d}\bar{t}} = \left(\frac{\mathrm{d}\bar{x}(t)}{\mathrm{d}\bar{t}}, \frac{\mathrm{d}\bar{y}(t)}{\mathrm{d}\bar{t}}\right)^{\mathrm{T}} = \Omega\left(\frac{\mathrm{d}x(t)}{\mathrm{d}t}, \frac{\mathrm{d}y(t)}{\mathrm{d}t}\right)^{\mathrm{T}}, \tag{4.20}$$

$$\frac{\mathrm{d}^2\bar{\mathbf{r}}}{\mathrm{d}\bar{t}^2} = \left(\frac{\mathrm{d}^2\bar{x}(t)}{\mathrm{d}\bar{t}^2}, \frac{\mathrm{d}^2\bar{y}(t)}{\mathrm{d}\bar{t}^2}\right)^{\mathrm{T}} = \Omega^2\left(\frac{\mathrm{d}^2x(t)}{\mathrm{d}t^2}, \frac{\mathrm{d}^2y(t)}{\mathrm{d}t^2}\right)^{\mathrm{T}}. \tag{4.21}$$

Therefore, the equations of motion with the forces of $\mathbf{F}_1(\bar{\mathbf{x}}, \bar{\mathbf{y}})$ and $\mathbf{F}_2(\bar{\mathbf{x}}, \bar{\mathbf{y}})$ for nonstick motion (pure cutting, no chip seizure) are

$$\begin{aligned}&\begin{bmatrix} 1 & 0 \\ 0 & 1 \end{bmatrix}\begin{Bmatrix} \ddot{x} \\ \ddot{y} \end{Bmatrix} + \begin{bmatrix} D_{11}^{(i)} & D_{12}^{(i)} \\ D_{21}^{(i)} & D_{22}^{(i)} \end{bmatrix}\begin{Bmatrix} \dot{x} \\ \dot{y} \end{Bmatrix} + \begin{bmatrix} K_{11}^{(i)} & K_{12}^{(i)} \\ K_{21}^{(i)} & K_{22}^{(i)} \end{bmatrix}\begin{Bmatrix} x \\ y \end{Bmatrix} \\ &= \cos(t)\begin{Bmatrix} A_x^{(i)} \\ A_y^{(i)} \end{Bmatrix} + \begin{Bmatrix} C_x^{(i)} \\ C_y^{(i)} \end{Bmatrix},\end{aligned} \tag{4.22}$$

for $i = 1, 2, 3, 4$, where i is the region index for four regions.

During the cutting process, if the chip material adheres to the tool-piece, the relative velocity on the cutting surface (rake surface) should be zero, i.e., a velocity boundary is defined for the frictional force exerted by chip material on the tool-rake surface,

$$\dot{x}\sin\alpha + \dot{y}\cos\alpha = V, \tag{4.23}$$

$$D_3 = \dot{z} = \dot{x}\sin\alpha + \dot{y}\cos\alpha - V, \tag{4.24}$$

where $V = -\bar{V}/\Omega$ ($\bar{V}$ is the chip velocity in the $\tilde{y}$-direction or in tool-rake surface direction). In order to discuss this phenomenon, a new coordinate system $(\tilde{x}, \tilde{y})$ is introduced. The transformation for the two coordinates (x,y) and $(\tilde{x}, \tilde{y})$ is

$$\begin{Bmatrix} x \\ y \end{Bmatrix} = \begin{bmatrix} \cos\alpha & \sin\alpha \\ -\sin\alpha & \cos\alpha \end{bmatrix} \begin{Bmatrix} \tilde{x} \\ \tilde{y} \end{Bmatrix} = \Lambda \begin{Bmatrix} \tilde{x} \\ \tilde{y} \end{Bmatrix}. \tag{4.25}$$

The corresponding velocity can be given by

$$\begin{Bmatrix} \dot{x} \\ \dot{y} \end{Bmatrix} = \Lambda \begin{Bmatrix} \dot{\tilde{x}} \\ \dot{\tilde{y}} \end{Bmatrix}, \begin{Bmatrix} \dot{\tilde{x}} \\ \dot{\tilde{y}} \end{Bmatrix} = \Lambda^{-1} \begin{Bmatrix} \dot{x} \\ \dot{y} \end{Bmatrix}. \tag{4.26}$$

From (4.25) and (4.26), equation (4.22) becomes

$$\begin{aligned} &\begin{bmatrix} 1 & 0 \\ 0 & 1 \end{bmatrix} \begin{Bmatrix} \ddot{\tilde{x}} \\ \ddot{\tilde{y}} \end{Bmatrix} + \Lambda^{-1} \begin{bmatrix} D_{11}^{(i)} & D_{12}^{(i)} \\ D_{21}^{(i)} & D_{22}^{(i)} \end{bmatrix} \Lambda \begin{Bmatrix} \dot{\tilde{x}} \\ \dot{\tilde{y}} \end{Bmatrix} + \Lambda^{-1} \begin{bmatrix} K_{11}^{(i)} & K_{12}^{(i)} \\ K_{21}^{(i)} & K_{22}^{(i)} \end{bmatrix} \Lambda \begin{Bmatrix} \tilde{x} \\ \tilde{y} \end{Bmatrix} \\ &= \cos(t)\Lambda^{-1} \begin{Bmatrix} A_x^{(i)} \\ A_y^{(i)} \end{Bmatrix} + \Lambda^{-1} \begin{Bmatrix} C_x^{(i)} \\ C_y^{(i)} \end{Bmatrix}. \end{aligned} \tag{4.27}$$

For stick motion (chip material adheres to the tool-piece, $\dot{z} \equiv 0$) in the $\tilde{y}$-direction, the following equations hold

$$\tilde{y} = \tilde{y}_0 - V(t - t_0), \quad \dot{\tilde{y}} = V, \quad \ddot{\tilde{y}} = 0. \tag{4.28}$$

The governing equation for stick motion in the $\tilde{x}$-direction is

$$\ddot{\tilde{x}} + 2d\dot{\tilde{x}} + \omega^2 \tilde{x} = A_0 \cos(t) + B_0 t + C_0. \tag{4.29}$$

The undefined parameters in (4.27) and (4.29) are given in Appendix A.3. There are four boundaries including three displacement boundaries and one velocity boundary. Two of the displacement boundaries are due to contact with the corresponding vector fields $\mathbf{F}_1(\bar{\mathbf{x}}, \bar{\mathbf{y}})$ and $\mathbf{F}_2(\bar{\mathbf{x}}, \bar{\mathbf{y}})$ (or tool- and work-piece contact and onset of cutting).

The third displacement boundary is for nonstick motion (cutting) where $D_1 < D_2 \leq 0$ and $\dot{z} < 0$. If the effective contact between the chip and tool-rake face maintains, the forces of the shearing action and work-piece motion will be transmitted through the tool-rake surface, as sketched in Fig. 4.5a. If the loss of effective chip–tool-rake surface contact occurs, the corresponding route is presented in Fig. 4.5b. The vanishing of effective cutting force contact induces a transition to region two, as shown in Fig. 4.5c.

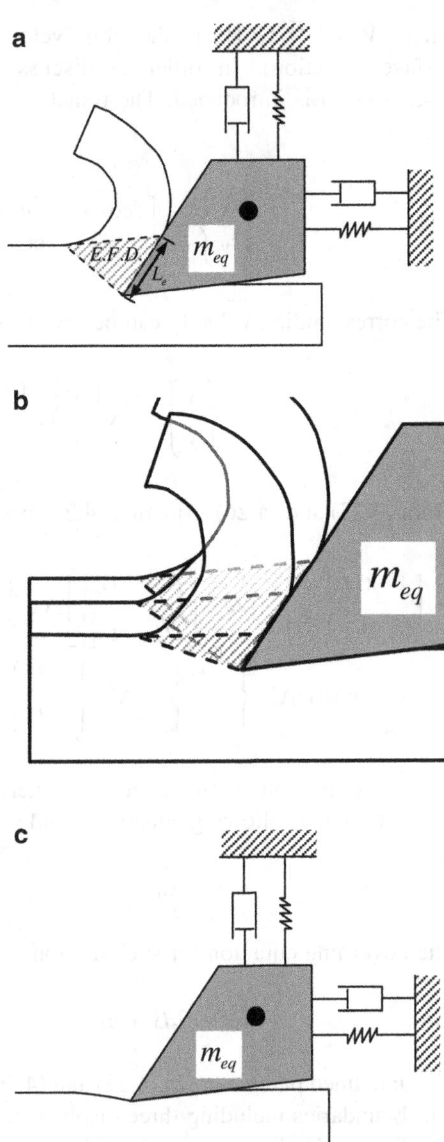

Fig. 4.5 Chip and tool-piece (**a**) effective cutting force contact (region four) and (**b**) route to loss of effective cutting force contact (region three), and (**c**) loss of effective cutting force contact (region two)

The effective force contact length is L_c. Hence a displacement boundary is defined as

$$L_c = \tilde{y}_0 - \tilde{y} \tag{4.30}$$

and

$$D_4 = L_c - (\tilde{y}_0 - \tilde{y}). \tag{4.31}$$

The force conditions at the point of switching from one region to another are shown in Fig. 4.4. The total force is given by $F_k^{(D_\gamma)}$ ($\gamma = 1, 2, 3$). For $\gamma = 3$, the friction coefficient distribution is presented. The motion switches from *region one* to *region two* and *region two* to *region four* are shown in Fig. 4.4a, b, respectively. The kinetic friction coefficient distribution switching from *region two* to *region four* jumps over *region three* on a loading path, which is given in Fig. 4.4c. The unloading path begins in *region four, and* moves through *region three* and ends at *region two.*

4.2 Domains and Boundaries

To apply the discontinuous system theory, the dynamical systems in domains relative to each boundary need to be described. The mechanical model of the machining system possesses four boundaries. The first boundary is the contact of the tool- and work-pieces. The phase planes are partitioned to identify the discontinuities in the machine-tool system, which are given in Figs. 4.6 and 4.7.

(A) *Contact Boundary.* The contact boundary exists between the tool- and work-pieces. If D_1 is positive ($D_1>0$), the tool- and work-pieces are not in contact. The contact exists initially when $D_1 = 0$ and continues for $D_1<0$. The state variable vector and vector fields relative to the contact boundary are defined as

$$\mathbf{D}_1 \triangleq (D_1, \dot{D}_1)^{\mathrm{T}} \text{ and } \mathbf{F}_1 \triangleq (\dot{D}_1, F_{D_1}^{(i)}(\mathbf{D}_1, t))^{\mathrm{T}}, \tag{4.32}$$

where

$$\begin{aligned} F_{D_1}^{(i)}(\mathbf{D}_1, t) = \ddot{D}_1 &= \Omega^2\left(\ddot{x}^{(i)} \sin\beta + \ddot{y}^{(i)} \cos\beta\right) \\ &= \Omega^2\left[F_x^{(i)}(\mathbf{x}, \mathbf{y}, t) \sin\beta + F_y^{(i)}(\mathbf{x}, \mathbf{y}, t) \cos\beta\right], \end{aligned} \tag{4.33}$$

where $i = 1, 2, 4$. As in Figs. 4.6a and 4.7a, domains for (4.32) are

$$\Omega_1 = \{(x, y, \dot{x}, \dot{y}) | D_1(x, y) \in (0, \infty)\}, \tag{4.34}$$

$$\Omega_2 = \{(x, y, \dot{x}, \dot{y}) | D_1(x, y) \in (-\infty, 0)\}, \tag{4.35}$$

$$\Omega_3 = \{(x, y, \dot{x}, \dot{y}) | D_1(x, y) \in (-\infty, 0)\}, \tag{4.36}$$

$$\Omega_4 = \{(x, y, \dot{x}, \dot{y}) | D_1(x, y) \in (-\infty, 0)\}. \tag{4.37}$$

The equation of motion in such domains relative to the contact boundary is

$$\dot{\mathbf{D}}_1^{(i)} = \mathbf{F}_1^{(i)} \triangleq (\dot{D}_1, F_{D_1}^{(i)}(\mathbf{D}_1, t))^{\mathrm{T}} \quad (i = 1, 2, 3, 4) \tag{4.38}$$

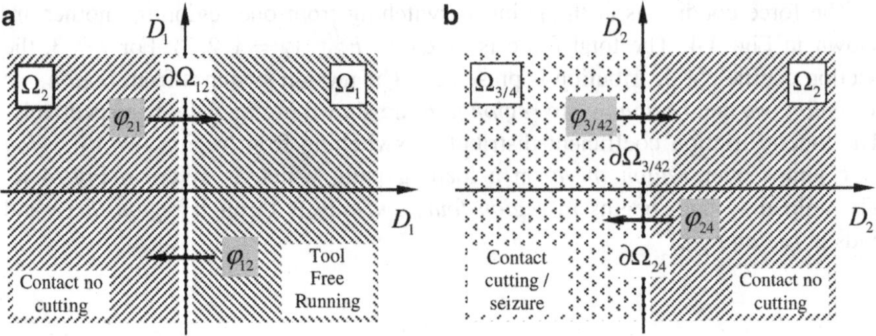

Fig. 4.6 Partitions in phase space for the displacement and velocity discontinuities of this machine-tool system: (**a**) D_1 phase plane and (**b**) D_2 phase plane

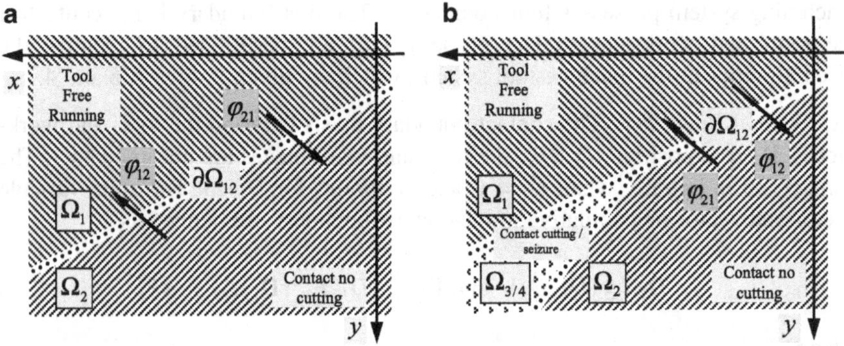

Fig. 4.7 Partitions in (x, y) and $(\dot{x}, \dot{y})$ space for the displacement and velocity discontinuities of this machine-tool system; boundaries (**a**) $\partial\Omega_{12}$and (**b**) φ_{12}

(B) *Cutting Boundary*. Cutting boundary is the cutting condition. In this case, the measure is a displacement constraint, where cutting will not occur if $D_2 > 0$. The cutting will initiate if $D_2 = 0$ and continue such that $D_2 \leq 0$. The state variable vector and vector fields are

$$\mathbf{D}_2 \triangleq (D_2, \dot{D}_2)^{\mathrm{T}} \text{ and } \mathbf{F}_2 \triangleq (\dot{D}_2, F_{D_2}^{(i)}(\mathbf{D}_2, t))^{\mathrm{T}}, \tag{4.39}$$

where

$$\begin{aligned} F_{D_2}^{(i)}(\mathbf{D}_2, t) &= \ddot{D}_2 = \Omega^2\left(\ddot{x}^{(i)} \sin\alpha + \ddot{y}^{(i)} \cos\alpha\right) \\ &= \Omega^2\left[F_x^{(i)}(\mathbf{x}, \mathbf{y}, t) \sin\alpha + F_y^{(i)}(\mathbf{x}, \mathbf{y}, t) \cos\alpha\right], \end{aligned} \tag{4.40}$$

with $i = 1, 4$ for Figs. 4.6b and 4.7b.

The domains with respect to (4.39) are defined as

$$\Omega_2 = \{(D_2, \dot{D}_2) | D_2 \in (0, \infty)\}, \tag{4.41}$$

$$\Omega_3 = \{(D_2, \dot{D}_2) | D_2 \in (-\infty, 0)\}, \tag{4.42}$$

$$\Omega_4 = \{(D_2, \dot{D}_2) | D_2 \in (-\infty, 0)\}. \tag{4.43}$$

The equation of motion becomes

$$\dot{\mathbf{D}}_2^{(i)} = \mathbf{F}_2^{(i)} \triangleq \left(\dot{D}_2, F_{D_2}^{(i)}(\mathbf{D}_2, t)\right)^{\mathrm{T}} (i = 2, 3, 4). \tag{4.44}$$

(C) *Friction Boundary*. Frictional boundary is for the frictional discontinuity. For specific values $\dot{\tilde{y}}$, there are different frictional forces. This is because the friction force varies with the relative velocity. If $\dot{\tilde{y}} > V$, then the friction force acts in the negative direction of $\tilde{y}$. Also, if $\dot{\tilde{y}} < V$, then the friction force acts in the positive direction of $\tilde{y}$. The state variable vector and vector fields for the frictional boundary are

$$\begin{aligned} &\mathbf{D}_3 = \tilde{\mathbf{y}} = (\tilde{y}, \dot{\tilde{y}})^{\mathrm{T}}, \\ &\mathbf{F}_{\tilde{y}}^{(\kappa)}(\tilde{\mathbf{x}}, \tilde{\mathbf{y}}, t) \triangleq \left(\dot{\tilde{y}}, F_{\tilde{y}}^{(\kappa)}(\tilde{\mathbf{x}}, \tilde{\mathbf{y}}, t)\right)^{\mathrm{T}} \quad (\kappa \in \{0, 3, 4\}), \end{aligned} \tag{4.45}$$

where

$$\begin{aligned} F_{\tilde{y}}^{(i)}(\tilde{\mathbf{x}}, \tilde{\mathbf{y}}, t) &= F_{D_3}^{(i)}(\tilde{\mathbf{x}}, \tilde{\mathbf{y}}, t) = F_{D_4}^{(i)}(\tilde{\mathbf{x}}, \tilde{\mathbf{y}}, t), \\ &= \ddot{\tilde{y}}^{(i)}(t) = \ddot{x}^{(i)} \sin\alpha + \ddot{y}^{(i)} \cos\alpha, \\ &= F_x^{(i)}(\mathbf{x}, \mathbf{y}, t) \sin\alpha + F_y^{(i)}(\mathbf{x}, \mathbf{y}, t) \cos\alpha \end{aligned} \tag{4.46}$$

and $\mathbf{x} = (x, \dot{x})^{\mathrm{T}}, \mathbf{y} = (y, \dot{y})^{\mathrm{T}}, \tilde{\mathbf{x}} = (\tilde{x}, \dot{\tilde{x}})^{\mathrm{T}}, \tilde{\mathbf{y}} = (\tilde{y}, \dot{\tilde{y}})^{\mathrm{T}}$ with $i = 3, 4$, as shown in Figs. 4.8a, c and 4.9b. The domains with respect to (4.39) and (4.45) are defined as

$$\Omega_2 = \{(D_2, \dot{D}_2) | D_2 \in (0, \infty) \text{ and } (\tilde{y}, \dot{\tilde{y}}) | \dot{\tilde{y}} \in (-\infty, V)\}, \tag{4.47}$$

$$\Omega_3 = \{(D_2, \dot{D}_2) | D_2 \in (-\infty, 0) \text{ and } (\tilde{y}, \dot{\tilde{y}}) | \dot{\tilde{y}} \in (-\infty, V)\}, \tag{4.48}$$

$$\Omega_4 = \{(D_2, \dot{D}_2) | D_2 \in (-\infty, 0) \text{ and } (\tilde{y}, \dot{\tilde{y}}) | \dot{\tilde{y}} \in (V, \infty)\}, \tag{4.49}$$

where

$$\left.\begin{aligned} &\tilde{y} = (x - x_0) \sin\alpha + (y - y_0) \cos\alpha, \\ &\dot{\tilde{y}} = \dot{x} \sin(\alpha) + \dot{y} \cos(\alpha). \end{aligned}\right\} \tag{4.50}$$

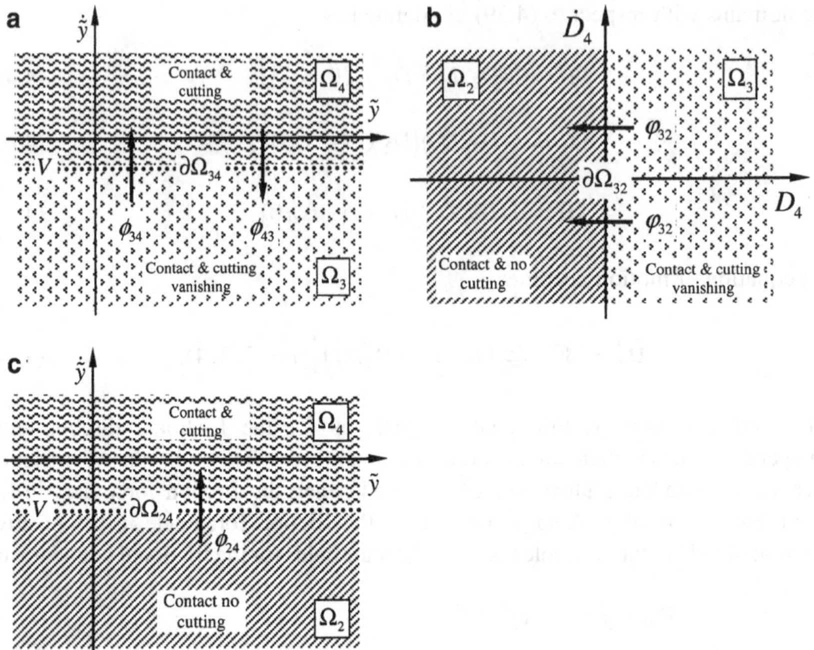

Fig. 4.8 Partitions in phase space for the displacement and velocity discontinuities of this machine-tool system; (**a**) $\tilde{y}$ phase plane, (**b**) D_4 phase plane, (**c**) $\partial\Omega_{24}$ boundary in phase plane $(\tilde{y}, \dot{\tilde{y}})$

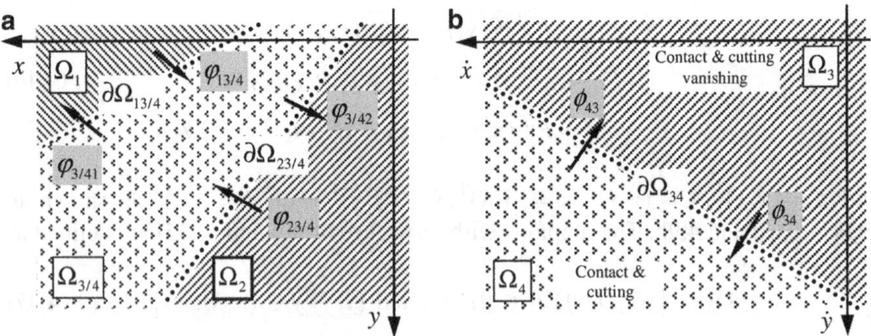

Fig. 4.9 Partitions in (x, y) and $(\dot{x}, \dot{y})$ space for the displacement and velocity discontinuities of this machine-tool system; boundaries (**a**) $\partial\Omega_{23/4}$, $\partial\Omega_{13/4}$ and (**b**) $\partial\Omega_{34}$

The equation of motion is

$$\dot{\mathbf{D}}_{D_3}^{(\kappa)} = \dot{\tilde{\mathbf{y}}}^{(\kappa)} = \mathbf{F}_{D_3}^{(\kappa)}(\tilde{\mathbf{x}}, \tilde{\mathbf{y}}, t) \quad (\kappa = 0, 3, 4). \tag{4.51}$$

(D) *Chip Contact Boundary*. Chip contact boundary is for the chip contact measure. Such a boundary determines whether the tool will continue to maintain effective force and contact with the chip while $\dot{\tilde{y}}<V$. Consider the initial point $\dot{\tilde{y}}=V$ at $\tilde{y}_0$. If the tool-piece continues to maintain $\dot{\tilde{y}}<V$ to a point where the difference of $\tilde{y}$ and $\tilde{y}_0$ is equal to L_c, the effective force contact is lost and the cutting terminates. For more information about the similar phenomena, the interested reader can refer Luo and Thapa (2009) or Luo (2009). The state variable vector and vector fields for the chip contact boundary are defined as

$$\begin{aligned}&\mathbf{D}_4=(D_4,\dot{D}_4)^{\mathrm{T}} \text{ and}\\&\mathbf{F}_{D_4}^{(i)}(\tilde{\mathbf{x}},\tilde{\mathbf{y}},t)\triangleq(\dot{D}_4,F_{D_4}^{(i)}(\tilde{\mathbf{x}},\tilde{\mathbf{y}},t))^{\mathrm{T}}\quad(i\in\{2,3\}),\end{aligned}\tag{4.52}$$

where

$$\begin{aligned}F_{D_4}^{(i)}(\tilde{\mathbf{x}},\tilde{\mathbf{y}},t)&=F_{\tilde{y}}^{(i)}(\tilde{\mathbf{x}},\tilde{\mathbf{y}},t)=F_{D_3}^{(i)}(\tilde{\mathbf{x}},\tilde{\mathbf{y}},t)\\&=\ddot{\tilde{y}}^{(i)}(t)=\ddot{x}^{(i)}\sin\alpha+\ddot{y}^{(i)}\cos\alpha\\&=F_x^{(i)}(\mathbf{x},\mathbf{y},t)\sin\alpha+F_y^{(i)}(\mathbf{x},\mathbf{y},t)\cos\alpha.\end{aligned}\tag{4.53}$$

For $i=2$, the corresponding description is presented in Fig. 4.8b. The domains with respect to (4.45) and (4.53) are defined as

$$\Omega_2=\{(D_4,\dot{D}_4)|D_4\in(-\infty,0)\text{ and }(\tilde{y},\dot{\tilde{y}})|\dot{\tilde{y}}\in(-\infty,V)\},\tag{4.54}$$

$$\Omega_3=\{(D_4,\dot{D}_4)|D_4\in(0,L_c)\text{ and }(\tilde{y},\dot{\tilde{y}})|\dot{\tilde{y}}\in(-\infty,V)\},\tag{4.55}$$

where

$$D_4=L_c-(\tilde{y}_0-\tilde{y}),\quad \dot{D}_4=\dot{z}=\dot{\tilde{y}}-V.\tag{4.56}$$

The equation of motion in domain relative to the chip/contact boundary is

$$\dot{\mathbf{D}}_4^{(i)}=\mathbf{F}_{D_4}^{(i)}(\tilde{\mathbf{x}},\tilde{\mathbf{y}},t)\triangleq(\dot{D}_4,F_{D_4}^{(i)}(\tilde{\mathbf{x}},\tilde{\mathbf{y}},t))^{\mathrm{T}}\quad(i=2,3),\tag{4.57}$$

(E) *Summary of Domains and Boundaries*: The four domains overlap each other. Thus, from the previous description, domain one is for the vibration of the tool-piece without contacting the work-piece, i.e.,

$$\Omega_1=\{(x,y,\dot{x},\dot{y})|D_1(x,y)\in(0,\infty)\}\tag{4.58}$$

or

$$\Omega_1=\{(x,y,\dot{x},\dot{y})|y\in((X_{eq}+x)\tan\beta+\delta_1/\cos\beta+Y_1-Y_{eq},\infty)\}.\tag{4.59}$$

Domain two is the contact of the tool- and work-piece without cutting,

$$\Omega_2 = \left\{ (x,y,\dot{x},\dot{y}) \middle| \begin{array}{l} D_1(x,y)<0,\ D_2(x,y)>0 \\ D_4(x,y)<0 \text{ and } \tilde{w}(\dot{x},\dot{y})<V \end{array} \right\} \tag{4.60}$$

or

$$\Omega_2 = \left\{ (x,y,\dot{x},\dot{y}) \middle| \begin{array}{l} x>(Y_{eq}+y)\tan\alpha - \delta_2/\cos\alpha + X_1 - X_{eq}, \\ y<(X_{eq}+x)\tan\beta + \delta_1/\cos\beta + Y_1 - Y_{eq}, \\ y<y_0 - (x-x_0)\tan\alpha + L_c/\cos\alpha, \\ \dot{y}<V/\cos\alpha - \dot{x}\tan\alpha. \end{array} \right\} \tag{4.61}$$

Domain three exists purely during route to chip vanishing ($\dot{\tilde{y}}<V$),

$$\Omega_3 = \left\{ (x,y,\dot{x},\dot{y}) \middle| \begin{array}{ll} D_1(x,y)<0, & D_2(x,y)<0 \\ D_4(x,y) \in (0,L_c), & \dot{\tilde{y}}(\dot{x},\dot{y})<V \end{array} \right\} \tag{4.62}$$

or

$$\Omega_3 = \left\{ (x,y,\dot{x},\dot{y}) \middle| \begin{array}{l} y<(X_{eq}+x)\tan\beta + \delta_1/\cos\beta + Y_1 - Y_{eq} \\ x<(Y_{eq}+y)\tan\alpha - \delta_2/\cos\alpha + X_1 - X_{eq} \\ y \in (y_0 - (x-x_0)\tan\alpha + L_c/\cos\alpha,\ y_0 - (x-x_0)\tan\alpha) \\ \dot{y}<V/\cos\alpha - \dot{x}\tan\alpha \end{array} \right\} \tag{4.63}$$

and domain four is defined by normal cutting ($\dot{\tilde{y}}>V$),

$$\Omega_4 = \left\{ (x,y,\dot{x},\dot{y}) \middle| \begin{array}{ll} D_1(x,y)<0, & D_2(x,y)<0 \\ \dot{\tilde{y}}(\dot{x},\dot{y})>V & \end{array} \right\} \tag{4.64}$$

or

$$\Omega_4 = \left\{ (x,y,\dot{x},\dot{y}) \middle| \begin{array}{l} y<(X_{eq}+x)\tan\beta + \delta_1/\cos\beta + Y_1 - Y_{eq}, \\ x<(Y_{eq}+y)\tan\alpha - \delta_2/\cos\alpha + X_1 - X_{eq}, \\ \dot{y}>V/\cos\alpha - \dot{x}\tan\alpha. \end{array} \right\} \tag{4.65}$$

The boundary between Ω_1 and Ω_2 in (4.58) and (4.60) is

$$\partial\Omega_{12} = \{(x,y,\dot{x},\dot{y}) | \varphi_{12}(x,y) = \varphi_{21}(x,y) = D_1(x,y) = 0\}, \tag{4.66}$$

which is boundary one (tool- and work-piece contact/impact boundary).

The boundaries between Ω_1 and Ω_2 in (4.61) and (4.62) are

$$\partial\Omega_{24} = \begin{cases} \{(x,y,\dot{x},\dot{y})|\varphi_{24}(x,y) = \varphi_{42}(x,y) = D_2(x,y) = 0,\ \dot{\tilde{y}}(\dot{x},\dot{y})>V\} \\ \{(x,y,\dot{x},\dot{y})|\varphi_{24}(\dot{x},\dot{y}) = \dot{\tilde{y}}(\dot{x},\dot{y}) - V = 0,\ D_2(x,y)<0\} \end{cases} \tag{4.67}$$

which is boundary two (onset of cutting boundary) and boundary three (chip/tool friction boundary). The boundaries between domain Ω_2 and Ω_3 in (4.58) and (4.60) are

$$\partial\Omega_{32} = \begin{cases} \{(x,y,\dot{x},\dot{y})|\varphi_{32} = D_4(x,y) = 0, D_2(x,y)<0\}, \\ \{(x,y,\dot{x},\dot{y})|\varphi_{32} = D_2(x,y) = 0, \dot{\tilde{y}}(\dot{x},\dot{y})<V\}. \end{cases} \tag{4.68}$$

which is for the chip vanishing boundary and cutting disappearance boundary, as shown in Fig. 4.9a. The boundaries between domains Ω_3 and Ω_4 in (4.62) and (4.64) are

$$\partial\Omega_{34} = \{(x,y,\dot{x},\dot{y})|\varphi_{34}(\dot{x},\dot{y}) = \dot{\tilde{y}}(\dot{x},\dot{y}) - V = 0, D_2(x,y)<0\}, \tag{4.69}$$

which is presented in Fig. 4.9b.

4.3 Motion Passability

The criteria for motion passability at the boundary in the machining dynamical system are presented in this section.

(A) *Passable Motion*: Consider the starting time t_s and ending time t_e on the boundary. From Chap. 2, the passable motion is guaranteed for $t_n \subset (t_s, t_e)$ by

$$\left[\mathbf{n}^{\mathrm{T}}_{\partial\Omega_{ij}} \cdot \mathbf{F}^{(i)}_{D_i}(\mathbf{D}_i, t_{n-})\right] \times \left[\mathbf{n}^{\mathrm{T}}_{\partial\Omega_{ij}} \cdot \mathbf{F}^{(j)}_{D_i}(\mathbf{D}_i, t_{n+})\right] > 0, \tag{4.70}$$

$i \neq j$ and $i,j = 1,2,3,4$. The normal vectors for the boundaries in (4.66)–(4.69) are

$$\mathbf{n}_{\partial\Omega_{12}} = \nabla\varphi_{12} = \left(\frac{\partial\varphi_{12}}{\partial D_1}, \frac{\partial\varphi_{12}}{\partial \dot{D}_1}\right)^{\mathrm{T}}_{(D_{1m},\dot{D}_{1m})} = (1,0)^{\mathrm{T}}, \tag{4.71}$$

$$\begin{aligned} \mathbf{n}_{\partial\Omega_{24}} &= \nabla\varphi_{24} = \left(\frac{\partial\varphi_{24}}{\partial D_2}, \frac{\partial\varphi_{24}}{\partial \dot{D}_2}\right)^{\mathrm{T}}_{(D_{2m},\dot{D}_{2m})} = (1,0)^{\mathrm{T}} \quad \text{if } \dot{\tilde{y}}>V, \\ \mathbf{n}_{\partial\Omega_{24}} &= \nabla\phi_{24} = \left(\frac{\partial\phi_{24}}{\partial \tilde{y}}, \frac{\partial\phi_{24}}{\partial \dot{\tilde{y}}}\right)^{\mathrm{T}}_{(\tilde{y}_m,\dot{\tilde{y}}_m)} = (0,1)^{\mathrm{T}} \quad \text{if } D_2<0, \end{aligned} \tag{4.72}$$

$$
\begin{aligned}
\mathbf{n}_{\partial\Omega_{32}} &= \nabla\varphi_{32} = \left(\frac{\partial\varphi_{32}}{\partial D_4}, \frac{\partial\varphi_{32}}{\partial \dot{D}_4}\right)^{\mathrm{T}}_{(D_{4m},\dot{D}_{4m})} = (1,0)^{\mathrm{T}} \quad \text{if } D_2<0,\\
\mathbf{n}_{\partial\Omega_{32}} &= \nabla\varphi_{32} = \left(\frac{\partial\varphi_{32}}{\partial D_2}, \frac{\partial\varphi_{32}}{\partial \dot{D}_2}\right)^{\mathrm{T}}_{(D_{2m},\dot{D}_{2m})} = (1,0)^{\mathrm{T}} \quad \text{if } \dot{\tilde{y}}<V,
\end{aligned}
\tag{4.73}
$$

$$
\mathbf{n}_{\partial\Omega_{34}} = \nabla\varphi_{34} = \left(\frac{\partial\varphi_{34}}{\partial \tilde{y}}, \frac{\partial\varphi_{34}}{\partial \dot{\tilde{y}}}\right)^{\mathrm{T}}_{(\tilde{y}_m,\dot{\tilde{y}}_m)} = (0,1)^{\mathrm{T}}, \tag{4.74}
$$

respectively. With (4.71)–(4.74), one can obtain

$$
\mathbf{n}^{\mathrm{T}}_{\partial\Omega_{12}} \cdot \mathbf{F}^{(j)}_{D_1}(\mathbf{D}_1, t_{n-}) = \dot{D}_1, \quad \text{for } j = 1,2, \tag{4.75}
$$

$$
\left.
\begin{aligned}
&\mathbf{n}^{\mathrm{T}}_{\partial\Omega_{24}} \cdot \mathbf{F}^{(j)}_{D_2}(\mathbf{D}_2, t_{n-}) = \dot{D}_2 \text{ if } \dot{\tilde{y}}>V,\\
&\mathbf{n}^{T}_{\partial\Omega_{24}} \cdot \mathbf{F}^{(j)}_{D_3}(\mathbf{D}_3, t_{n-}) = F^{(j)}_{\tilde{y}}(\tilde{\mathbf{x}}, \tilde{\mathbf{y}}, , t) = \ddot{\tilde{y}}^{(j)}(t) \quad \text{if } D_2<0
\end{aligned}
\right\} \text{ for } j = 2,4, \tag{4.76}
$$

$$
\left.
\begin{aligned}
&\mathbf{n}^{\mathrm{T}}_{\partial\Omega_{32}} \cdot \mathbf{F}^{(j)}_{D_4}(\mathbf{D}_4, t_{n-}) = \dot{D}_4 = \dot{\tilde{y}} - V \quad \text{if } \dot{\tilde{y}}<V,\\
&\mathbf{n}^{\mathrm{T}}_{\partial\Omega_{32}} \cdot \mathbf{F}^{(j)}_{D_2}(\mathbf{D}_2, t_{n-}) = \dot{D}_2 \quad \text{if } D_2<0
\end{aligned}
\right\} \text{ for } j = 2,3, \tag{4.77}
$$

$$
\mathbf{n}^{\mathrm{T}}_{\partial\Omega_{34}} \cdot \mathbf{F}^{(j)}_{D_3}(\mathbf{D}_3, t_{n-}) = F^{(j)}_{\tilde{y}}(\tilde{\mathbf{x}}, \tilde{\mathbf{y}}, t) = \ddot{\tilde{y}}^{(j)}(t), \quad \text{for } j = 0,3,4, \tag{4.78}
$$

respectively. Substitution of (4.71) into (4.70) gives

$$
\dot{D}^{(i)}_1 \dot{D}^{(j)}_1 > 0 \text{ on } \partial\Omega_{12}, \quad \text{for } i \neq j; i,j = 1,2. \tag{4.79}
$$

This indicates that a passable motion exists for the tool- and work-piece contact boundary from domain Ω_1 to domain Ω_2 and vice versa. Substitution of (4.76) into (4.70) gives

$$
\left.
\begin{aligned}
&\dot{D}^{(i)}_2 \dot{D}^{(j)}_2 > 0 \text{ if } \dot{\tilde{y}}>V,\\
&F^{(i)}_{\tilde{y}}(\tilde{\mathbf{x}}, \tilde{\mathbf{y}}, t) \times F^{(j)}_{\tilde{y}}(\tilde{\mathbf{x}}, \tilde{\mathbf{y}}, t) > 0 \quad \text{if } D_2<0
\end{aligned}
\right\} \text{ on } \partial\Omega_{ij}, \tag{4.80}
$$

for $i \neq j; i,j = 2,4$. This tells that a passable motion exists for the chip vanishing boundary and the chip/tool friction boundary from domain Ω_2 to domain Ω_4, and vice versa. There are two conditions in (4.80) since there are two entrance and exit planes for domain Ω_2 and domain Ω_4 for the cutting boundary and the frictional boundary. Substitution of (4.77) into (4.70) gives

$$
\left.
\begin{aligned}
&\dot{D}^{(3)}_4 \dot{D}^{(2)}_4 > 0 \quad \text{if } \dot{\tilde{y}}<V\\
&\dot{D}^{(3)}_2 \dot{D}^{(2)}_2 > 0 \quad \text{if } D_2<0
\end{aligned}
\right\} \text{ on } \partial\Omega_{32}. \tag{4.81}
$$

This implies passable motion for the chip vanishing boundary and the chip/tool friction boundary, from domain Ω_3 to domain Ω_2 only. There are two conditions in (4.81) for the cutting boundary and chip contact boundary. Substitution of (4.78) into (4.70) gives

$$F_{\tilde{y}}^{(i)}(\tilde{\mathbf{x}},\tilde{\mathbf{y}},t) \times F_{\tilde{y}}^{(j)}(\tilde{\mathbf{x}},\tilde{\mathbf{y}},t) > 0 \quad \text{on } \partial\Omega_{ij} \tag{4.82}$$

for $i \neq j; i,j = 3,4$. This implies that passable motion exits through the chip/tool friction boundary from domain Ω_3 to domain Ω_4, and vice versa. In Figs. 4.10 and 4.11, the vectors fields and the actual motion flows are presented by short line arrows and long arrows, respectively.

(B) *Nonpassable Motion.* The nonpassable motion and vanishing of the nonpassable motion is guaranteed for $t_m \in [t_k, t_{k+1})$ by

$$\left[\mathbf{n}_{\partial\Omega_{ij}}^{\mathrm{T}} \cdot \mathbf{F}_{D_s}^{(i)}(\mathbf{D}_s, t_{m-})\right] \times \left[\mathbf{n}_{\partial\Omega_{ij}}^{\mathrm{T}} \cdot \mathbf{F}_{D_s}^{(j)}(\mathbf{D}_s, t_{m-})\right] < 0, \tag{4.83}$$

$$\left.\begin{aligned} &\mathbf{n}_{\partial\Omega_{ij}}^{\mathrm{T}} \cdot \mathbf{F}_{D_s}^{(\kappa)}(\mathbf{D}_s, t_{m-}) = 0 \text{ and } \mathbf{n}_{\partial\Omega_{ij}}^{\mathrm{T}} \cdot \mathbf{F}_{D_s}^{(\lambda)}(\mathbf{D}_s, t_{m-}) \neq 0;\\ &\mathbf{n}_{\partial\Omega_{ij}}^{\mathrm{T}} \cdot D\mathbf{F}_{D_s}^{(\kappa)}(\mathbf{D}_s, t_{m\mp}) < 0, \quad \text{if } \mathbf{n}_{\partial\Omega_{ij}} \to \Omega_\lambda\\ &\mathbf{n}_{\partial\Omega_{ij}}^{\mathrm{T}} \cdot D\mathbf{F}_{D_s}^{(\kappa)}(\mathbf{D}_s, t_{m\mp}) > 0, \quad \text{if } \mathbf{n}_{\partial\Omega_{ij}} \to \Omega_\kappa \end{aligned}\right\} \kappa,\lambda \in \{i,j\} \text{ and } \lambda \neq \kappa. \tag{4.84}$$

for $i \neq j$ and $i,j = 1,2,3,4$ with $s = 1,2,3,4$, respectively.

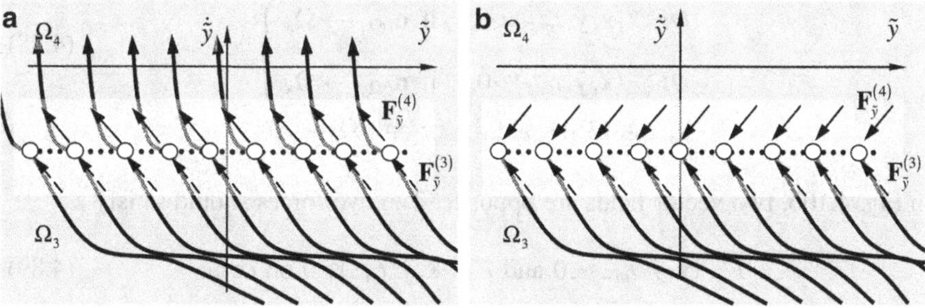

Fig. 4.10 Vector fields for: (**a**) passable and (**b**) nonpassable motion

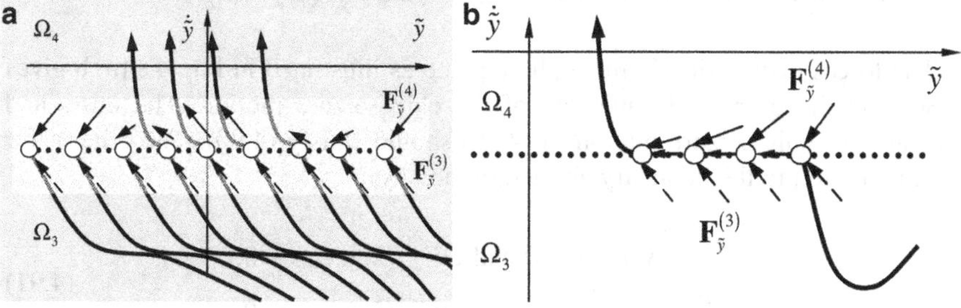

Fig. 4.11 Vector fields for: (**a**) passable and nonpassable with appearance and vanishing points, and (**b**) specific example of nonpassable motion and vanishing point

The only velocity boundaries have the possibility of a sink boundary. Substitution of (4.77) into (4.83) and (4.84) gives

$$F_{\tilde{y}}^{(2)}(\tilde{\mathbf{x}},\tilde{\mathbf{y}},t_{m-}) \times F_{\tilde{y}}^{(4)}(\tilde{\mathbf{x}},\tilde{\mathbf{y}},t_{m-})<0 \quad \text{if } D_2<0, \text{ on } \partial\Omega_{24}. \tag{4.85}$$

$$\begin{aligned}
&F_{\tilde{y}}^{(\kappa)}(\tilde{\mathbf{x}},\tilde{\mathbf{y}},t_{m-}) = 0, \quad F_{\tilde{y}}^{(\lambda)}(\tilde{\mathbf{x}},\tilde{\mathbf{y}},t_{m-}) \neq 0;\\
&\left.\begin{aligned}
&\mathbf{n}_{\partial\Omega_{ij}}^{\mathrm{T}} \cdot DF_{\tilde{y}}^{(\kappa)}(\tilde{\mathbf{x}},\tilde{\mathbf{y}},t_{m\mp})<0, \quad \text{if } \mathbf{n}_{\partial\Omega_{ij}} \to \Omega_\lambda\\
&\mathbf{n}_{\partial\Omega_{ij}}^{\mathrm{T}} \cdot DF_{\tilde{y}}^{(\kappa)}(\tilde{\mathbf{x}},\tilde{\mathbf{y}},,t_{m\mp})>0, \quad \text{if } \mathbf{n}_{\partial\Omega_{ij}} \to \Omega_\kappa
\end{aligned}\right\}\\
&\kappa,\lambda \in \{2,4\} \text{ and } \lambda \neq \kappa \quad \text{if } D_2<0, \text{ on } \partial\Omega_{24}.
\end{aligned} \tag{4.86}$$

Provided (4.85) is satisfied, the boundary is indeed nonpassable. When the tool returns to cutting in the positive $\tilde{y}$-direction, the friction boundary is crossed at which point the contact conditions are reset to ensure a continuous stiffness force.

Since the system equilibrium point is relocated to accommodate the continuous stiffness force across the frictional boundary, the boundary is permanently passable. The boundary dynamics are negated due to local exceptions. Substitution of (4.78) into (4.83) and (4.84) gives

$$F_{\tilde{y}}^{(3)}(\tilde{\mathbf{x}},\tilde{\mathbf{y}},t) \times F_{\tilde{y}}^{(4)}(\tilde{\mathbf{x}},\tilde{\mathbf{y}},t)<0 \quad \text{on } \partial\Omega_{34}. \tag{4.87}$$

$$\begin{aligned}
&F_{\tilde{y}}^{(\kappa)}(\tilde{\mathbf{x}},\tilde{\mathbf{y}},t_{m-}) = 0, \; F_{\tilde{y}}^{(\lambda)}(\tilde{\mathbf{x}},\tilde{\mathbf{y}},t_{m-}) \neq 0;\\
&\left.\begin{aligned}
&DF_{\tilde{y}}^{(\kappa)}(\tilde{\mathbf{x}},\tilde{\mathbf{y}},t_{m\mp})<0, \quad \text{if } \mathbf{n}_{\partial\Omega_{ij}} \to \Omega_\lambda\\
&DF_{\tilde{y}}^{(\kappa)}(\tilde{\mathbf{x}},\tilde{\mathbf{y}},t_{m\mp})>0, \quad \text{if } \mathbf{n}_{\partial\Omega_{ij}} \to \Omega_\kappa
\end{aligned}\right\}\\
&\kappa,\lambda \in \{3,4\} \text{ and } \lambda \neq \kappa \text{ on } \partial\Omega_{34}.
\end{aligned} \tag{4.88}$$

In Fig. 4.10b, two vector fields are opposite. The two forces should satisfy

$$F_{\tilde{y}}^{(4)}(\tilde{\mathbf{x}},\tilde{\mathbf{y}},t_{m-})<0 \text{ and } F_{\tilde{y}}^{(3)}(\tilde{\mathbf{x}},\tilde{\mathbf{y}},t_{m-})>0 \text{ on } \partial\Omega_{ij}, \tag{4.89}$$

$$\begin{aligned}
&F_{\tilde{y}}^{(3)}(\tilde{\mathbf{x}},\tilde{\mathbf{y}},t_{m-}) = 0, \; DF_{\tilde{y}}^{(3)}(\tilde{\mathbf{x}},\tilde{\mathbf{y}},t_{m\mp})<0 \text{ and } F_{\tilde{y}}^{(4)}(\tilde{\mathbf{x}},\tilde{\mathbf{y}},t_{m-})<0;\\
&F_{\tilde{y}}^{(4)}(\tilde{\mathbf{x}},\tilde{\mathbf{y}},t_{m-}) = 0, \; DF_{\tilde{y}}^{(4)}(\tilde{\mathbf{x}},\tilde{\mathbf{y}},t_{m\mp})>0 \text{ and } F_{\tilde{y}}^{(3)}(\tilde{\mathbf{x}},\tilde{\mathbf{y}},t_{m-})>0.
\end{aligned} \tag{4.90}$$

The forces distribution in the $\tilde{y}$ phase plane as illustrated in Fig. 4.11a, b gives (4.86) and the onset and vanishing of the chip seizure motion. The onset and vanishing of chip seizure motion in (4.89) should satisfy (4.90). The criteria for grazing motion to the boundary are determined by

$$\begin{aligned}
&F_{\tilde{y}}^{(3)}(\tilde{\mathbf{x}},\tilde{\mathbf{y}},t_{m-}) = 0 \text{ and } DF_{\tilde{y}}^{(3)}(\tilde{\mathbf{x}},\tilde{\mathbf{y}},t_{m\mp})<0,\\
&F_{\tilde{y}}^{(4)}(\tilde{\mathbf{x}},\tilde{\mathbf{y}},t_{m-}) = 0 \text{ and } DF_{\tilde{y}}^{(4)}(\tilde{\mathbf{x}},\tilde{\mathbf{y}},t_{m\mp})>0.
\end{aligned} \tag{4.91}$$

For $i, j = 3, 4$ and $j \neq i$, such criteria tells that a possible grazing flow to the friction boundary. This phenomenon has been presented in Chap. 2 and the theory derived there will be directly applied for this machine-tool.

References

Dassanayake, A.V. and Suh, C.S., 2007, Machining dynamics involving whirling part I: model development and validation, *Journal of Vibration and Control*, **13**(5), pp. 475–506.

Gegg, B.C., Suh, S.C.S. and Luo, A.C.J., 2010a, Analytical prediction of interrupted cutting periodic motion in a machine toll with a friction boundary, Nonlinear Science and Complexity (Eds, J.A. Tenreiro Machado, A.C.J. Luo, R.S. Barbosa, M.F. Silva, and L. B. Figueiredo, Springer: Dordrecht), 26–36.

Gegg, B.C., Suh, S.C.S. and Luo, A.C.J., 2010b, Modeling and theory of intermittent motions in a machine toll with a friction boundary, *ASME Journal of Manufacturing Science and Engineering*, **132**, 041001(1–9)

Gegg, B.C., Suh, S.C.S. and Luo, A.C.J., 2010c, A parameter study of the eccentricity frequency and amplitude and chip length effects on a machine tool with multiple boundaries, *ASME Journal of Manufacturing Science and Engineering*, **15**, 2575–2602.

Luo, A.C.J., 2009, *Discontinuous Dynamical Systems on Time-varying Domains*, HEP-Springer: Heidelberg.

Luo, A.C.J. and O'Connor, D., 2009a, Periodic motions with impacting chatter and stick in a gear transmission system, *ASME Journal of Vibration and Acoustics*, **131**.

Luo, A.C.J. and O'Connor, D., 2009b, Mechanism of impacting chatter with stick in a gear transmission system, *International Journal of Bifurcation and Chaos*, **19**, 2093–2105.

Luo, A.C.J. and O'Connor, D., 2009c, Periodic motions and chaos with impacting chatter and stick in a gear transmission system, *International Journal of Bifurcation and Chaos*, **19**, 1975–1944.

Luo, A.C.J. and Thapa, S., 2009, periodic motions in a simplified brake system with a periodic excitation, *Communications in Nonlinear Science and Numerical Simulation*, **14**, 2389–2414.

Moon, F.C. and Kalmar-Nagy, T., 2001, Nonlinear models for complex dynamics in cutting materials, *Philosophical Transactions of the Royal Society of London A*, **359**, 695–711.

Shaw, M., 2005, *Metal Cutting Principles*, Oxford University Press: Oxford.

Warminski, J., Litak, G., Cartmell, M.P., Khanin, R., and Wiercigroch, M., 2003, Approximate analytical solution for primary chatter in the non-linear metal cutting model, *Journal of Sound and Vibration*, **259**(4), 917–933.

Wiercigroch, M., 1997, Chaotic vibration of a simple model of the machine tool-cutting process system, *Transactions of the ASME: Journal of Vibration and Acoustics*, **119**, 468–475.

Chapter 5
Complex Cutting Motions

In this chapter, complex motions for cutting process in manufacturing are discussed. Based on the boundaries and domain, the switching planes and generic mappings are introduced to analyze complex motions. From generic mappings, complex cutting motions varying with system parameters are presented. The motion complexity is measured through mapping structures. Periodic cutting motions in manufacturing are predicted analytically. The motion switching is determined by force product criteria. Numerical illustrations of cutting motions near the chip seizure for the machine-tool system are given for a better understanding of complex cutting process in manufacturing.

5.1 Switching Planes and Mappings

Introduce a notation for mapping

$$P_{(ij)\alpha} \quad \text{for } i = 1,2,3,4;\ j = 1,2,3,4;\ \text{and } \alpha = 0,1,2,3,4; \tag{5.1}$$

where i is for the initial boundary, j is for the final boundary, and α is for domain Ω_α, as shown in Figs. 5.1 and 5.2. The mappings describe the following cases: vibration of the tool with no contact of the work-piece ($i = 1$); the tool in contact with the work-piece but no cutting ($i = 2$); the tool in contact with the work-piece with cutting where $\dot{z} < 0$ ($i = 3$); the tool in contact with the work-piece with cutting where $\dot{z} > 0$ ($i = 4$); and contact with the tool in special case where the chip/tool rake face seizure occurs, $\dot{z} \equiv 0$ ($i = 0$). The action of one mapping given a set of initial conditions yields a set of final conditions for this machine-tool system. The mappings can be combined to describe the trajectory of periodic orbit in phase plane. For instance, consider the mappings $P_{(11)1}$ and $P_{(11)2}$ in series; which can be simplified using the notation $P_{21} = P_{(11)2} \circ P_{(11)1}$, where the boundary definitions

B.C. Gegg et al., *Machine Tool Vibrations and Cutting Dynamics*,
DOI 10.1007/978-1-4419-9801-9_5,

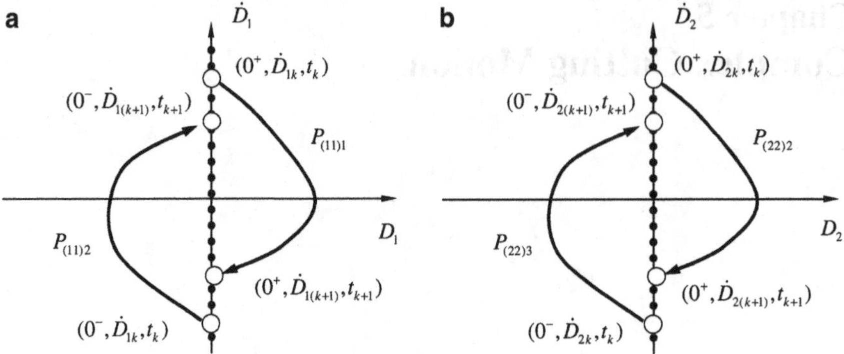

Fig. 5.1 Mappings in phase planes in (**a**) D_1 and (**b**) D_2

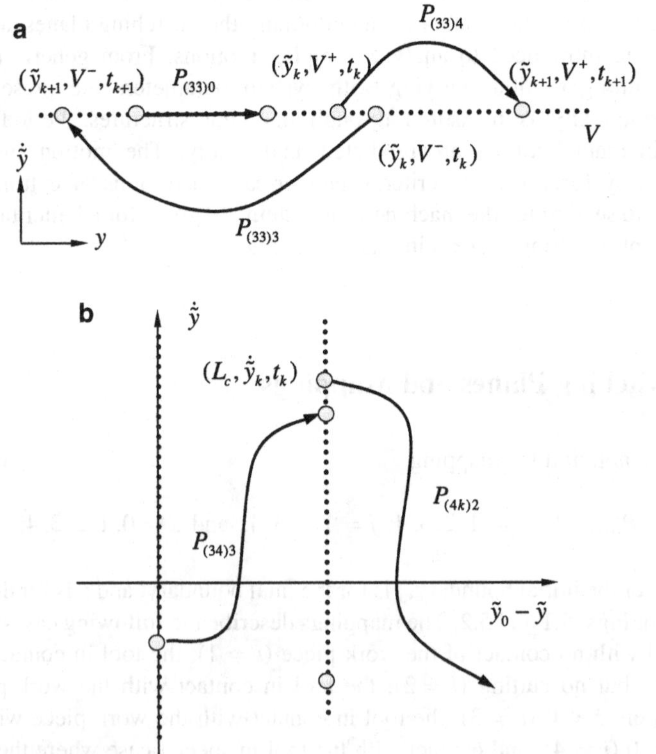

Fig. 5.2 Mappings in phase planes in (**a**) D_3 and (**b**) D_4

are discarded and only the domains traveled remain in the notation. Through this notation, varying a system parameter gives a description of how the orbits change. The following section will develop the switching sets with regard to boundary one (tool- and work-piece contact/impact boundary).

In this section, the switching planes and corresponding mappings are introduced in order to investigate periodic and chaotic motions in cutting process.

(A) *Contact Boundary*: The motion of the machine tool can be tracked through mappings in phase plane, as shown in Figs. 5.1 and 5.2. The initial and final states of these mappings are on switching planes. The switching planes for the tool- and work-piece contact/impact boundary (the first boundary) are

$$\begin{aligned}\Xi^1_{D_1} &= \{(x_k, y_k, \dot{x}_k, \dot{y}_k)|D_1(x_k, y_k) = 0^+\},\\ \Xi^2_{D_1} &= \{(x_k, y_k, \dot{x}_k, \dot{y}_k)|D_1(x_k, y_k) = 0^-\},\end{aligned} \quad (5.2)$$

$$\Xi^3_{D_1} = \{(x_k, y_k, \dot{x}_k, \dot{y}_k)|D_1(x_k, y_k) = 0^-, D_2 \le 0\}, \quad (5.3)$$

$$\Xi^4_{D_1} = \{(x_k, y_k, \dot{x}_k, \dot{y}_k)|D_1(x_k, y_k) = 0^-, D_2 \le 0\}, \quad (5.4)$$

$$\Xi^0_{D_1} = \left\{(x_k, y_k, \dot{x}_k, \dot{y}_k)\middle|D_1(x_k, y_k) = 0^-, D_2 \le 0 \text{ and } \dot{\tilde{y}} \equiv V\right\}. \quad (5.5)$$

Note that $0^- = \lim_{\delta\to 0}(0 - \delta)$ and $0^+ = \lim_{\delta\to 0}(0 + \delta)$. The mapping starting on switching plane $\Xi^0_{D_3}$ and ending at switching plane $\Xi^0_{D_1}$ on the boundary $\partial\Omega_{34}$ is defined as

$$P_{(31)0} : \Xi^0_{D_3} \to \Xi^0_{D_1}. \quad (5.6)$$

This mapping shows the motion from the three-dimensional boundary to the two-dimensional edge. The foregoing mapping describes cutting motion with an initial condition on the chip/tool friction boundary (switching set) sliding on the tool- and work-piece contact boundary (or chip seizure). The definition of $\Xi^0_{D_3}$ is given later. The mapping beginning and ending at switching plane $\Xi^1_{D_1}$ through domain Ω_1 is defined as

$$P_{(11)1} : \Xi^1_{D_1} \to \Xi^1_{D_1}. \quad (5.7)$$

The above mapping represents motion beginning on the tool- and work-piece contact boundary, tool-piece free running in domain Ω_1, and ending on the tool- and work-piece contact/impacting boundary, as shown in Fig. 5.1a. The mapping beginning and ending at switching plane $\Xi^2_{D_1}$ via domain Ω_2 is defined as

$$P_{(11)2} : \Xi^2_{D_1} \to \Xi^2_{D_1}. \quad (5.8)$$

The proceeding mapping describes motion through domain Ω_2 (i.e., tool and work-piece contact but not cutting) and beginning and ending at boundary $\partial\Omega_{12}$, as shown in Fig. 5.1b. The mapping beginning and ending at switching plane $\Xi^3_{D_1}$ through domain Ω_3 is defined as

$$P_{(11)3} : \Xi^3_{D_1} \to \Xi^3_{D_1}. \quad (5.9)$$

The foregoing mapping describes motion beginning and ending on boundary $\partial\Omega_{12}$ through domain Ω_3 (i.e., tool- and work-piece contact, and cutting with $\dot{y}(\dot{x},\dot{y}) \in (-\infty, V)$). The mapping beginning and ending at switching plane $\Xi^4_{D_1}$ through domain Ω_4 is defined as

$$P_{(11)4} : \Xi^4_{D_1} \to \Xi^4_{D_1}. \tag{5.10}$$

The foregoing mapping describes motion beginning and ending on boundary $\partial\Omega_{12}$ through domain Ω_4 (i.e., tool- and work-piece contact, and cutting with $\dot{y}(\dot{x},\dot{y}) \in (V, \infty)$). In the next section, the switching sets with regard to the cutting boundary are developed.

(B) *Cutting Boundary*: The switching planes for the onset of cutting boundary (boundary two) are

$$\Xi^1_{D_2} = \{(x_k, y_k, \dot{x}_k, \dot{y}_k)|D_2(x_k, y_k) = 0^+ \text{ and } D_1 \equiv 0\}, \tag{5.11}$$

$$\Xi^2_{D_2} = \{(x_k, y_k, \dot{x}_k, \dot{y}_k)|D_2(x_k, y_k) = 0^+\}, \tag{5.12}$$

$$\Xi^3_{D_2} = \left\{(x_k, y_k, \dot{x}_k, \dot{y}_k)|D_2(x_k, y_k) = 0^- \text{ and } \dot{y} < V\right\}, \tag{5.13}$$

$$\Xi^4_{D_2} = \left\{(x_k, y_k, \dot{x}_k, \dot{y}_k)|D_2(x_k, y_k) = 0^- \text{ and } \dot{y} > V\right\}, \tag{5.14}$$

and

$$\Xi^0_{D_2} = \left\{(x_k, y_k, \dot{x}_k, \dot{y}_k)|D_2(x_k, y_k) = 0^- \text{ and } \dot{y} \equiv V\right\}. \tag{5.15}$$

As in Luo and Mao (2010), the switching planes in (5.11) and (5.15) are the two-dimensional edges, and the rest of switching planes in (5.12)–(5.13) is on the three-dimensional surface. Readers who are interested in edge dynamics can refer to Luo (2011).

The mapping beginning and ending at switching plane $\Xi^0_{D_2}$ sliding on the edge is expressed as

$$P_{(32)0} : \Xi^0_{D_3} \to \Xi^0_{D_2}. \tag{5.16}$$

The foregoing mapping describes motion beginning on the frictional boundary (i.e., chip/tool friction boundary), sliding on the boundary (i.e., chip/tool seizure), and ending on cutting boundary (onset/vanishing of cutting boundary). The switching plane relative to the frictional boundary is presented later. The mapping beginning and ending at switching plane $\Xi^1_{D_2}$ in domain Ω_1 is defined as

$$P_{(22)1} : \Xi^1_{D_2} \to \Xi^1_{D_2}. \tag{5.17}$$

The foregoing mapping describes motion beginning and ending on the cutting boundary while the motion is in domain Ω_1 (i.e., tool-free running).

Although the switching planes in (5.7) and (5.17) are different, the motion is in domain Ω_1. The mapping beginning and ending at switching plane $\Xi^2_{D_2}$ in domain Ω_2 is defined as

$$P_{(22)2} : \Xi^2_{D_2} \to \Xi^2_{D_2}, \tag{5.18}$$

The foregoing mapping describes motion in domain Ω_2 beginning and ending on the cutting boundary. In other words, tool- and work-piece contact with no cutting will be described. Although the switching planes in (5.8) and (5.18) are different, the domain for motion to travel is domain Ω_2 as shown in Fig. 5.1b. The mapping beginning and ending at switching plane $\Xi^3_{D_2}$ in domain Ω_3 is defined as

$$P_{(22)3} : \Xi^3_{D_2} \to \Xi^3_{D_2}, \tag{5.19}$$

The foregoing mapping describes motion beginning and ending on the cutting boundary in domain Ω_3. The motion implies that the tool- and work-piece will contact and the cutting with $\dot{\tilde{y}}(\dot{x}, \dot{y}) \in (-\infty, V)$ will be completed, as shown in Fig. 5.1b. The mapping beginning and ending at switching plane $\Xi^4_{D_2}$ in domain Ω_4 is defined as

$$P_{(22)4} : \Xi^4_{D_2} \to \Xi^4_{D_2}. \tag{5.20}$$

The above mapping describes motion beginning and ending on boundary two while traversing domain Ω_4. Like mapping $P_{(22)3}$, the tool- and work-piece will contact, and the cutting with $\dot{\tilde{y}}(\dot{x}, \dot{y}) \in (V, \infty)$ will be completed. In the next section, the switching sets with regard to the frictional boundary (i.e., chip/tool friction boundary) are developed.

(C) *Friction Boundary*: The switching planes for chip/tool friction boundary (boundary three) are

$$\Xi^1_{D_3} = \{(x, y, \dot{x}, \dot{y}) | \dot{\tilde{y}}(\dot{x}, \dot{y}) = V, D_1 = 0 \text{ and } D_2 < 0\}, \tag{5.21}$$

$$\Xi^2_{D_3} = \{(x, y, \dot{x}, \dot{y}) | \dot{\tilde{y}}(\dot{x}, \dot{y}) = V, D_2 = 0\}, \tag{5.22}$$

$$\Xi^3_{D_3} = \{(x, y, \dot{x}, \dot{y}) | \dot{\tilde{y}}(\dot{x}, \dot{y}) = V^-, D_1 < D_2 < 0\}, \tag{5.23}$$

$$\Xi^4_{D_3} = \{(x, y, \dot{x}, \dot{y}) | \dot{\tilde{y}}(\dot{x}, \dot{y}) = V^+, D_1 < D_2 < 0\}, \tag{5.24}$$

$$\Xi^0_{D_3} = \{(x, y, \dot{x}, \dot{y}) | \dot{\tilde{y}}(\dot{x}, \dot{y}) = V, D_1 < D_2 < 0\}. \tag{5.25}$$

The mapping beginning and ending at $\Xi^0_{D_3}$ along the friction boundary is defined as

$$P_{(33)0} : \Xi^0_{D_3} \to \Xi^0_{D_3}. \tag{5.26}$$

The foregoing mapping gives the sliding motion beginning and ending on the friction boundary for chip/tool seizure, as shown in Fig. 5.2a. The mapping beginning and ending at switching plane $\Xi^1_{D_3}$ through domain Ω_1 is

$$P_{(33)1} : \Xi^1_{D_3} \to \Xi^1_{D_3}. \tag{5.27}$$

The mapping describes the motion of tool-free running. The mapping beginning and ending at this switching plane $\Xi^2_{D_3}$ in domain Ω_2 is

$$P_{(33)2} : \Xi^2_{D_3} \to \Xi^2_{D_3}. \tag{5.28}$$

The mapping describes the motion of tool- and work-piece contact with no cutting. The mapping beginning and ending at switching plane $\Xi^3_{D_3}$ in domain Ω_3 is

$$P_{(33)3} : \Xi^3_{D_3} \to \Xi^3_{D_3}. \tag{5.29}$$

The mapping describes the motion of tool- and work-piece contact and cutting with $\dot{\tilde{y}}(\dot{x},\dot{y}) \in (-\infty, V)$, as shown in Fig. 5.2a. The mapping beginning and ending at switching plane $\Xi^4_{\dot{y}}$ in domain Ω_4 is

$$P_{(33)4} : \Xi^4_{D_3} \to \Xi^4_{D_3}. \tag{5.30}$$

The mapping gives the motion of tool- and work-piece contact and cutting with $\dot{\tilde{y}}(\dot{x},\dot{y}) \in (V, \infty)$ as shown in Fig. 5.2b. In the next section, the switching sets relative to the chip vanishing boundary are discussed.

(D) *Chip Vanishing Boundary*: The switching planes for the chip vanishing boundary are

$$\Xi^2_{D_4} = \{(x,y,\dot{x},\dot{y}) | D_4(x,y) = 0^- \text{ and } D_1 < D_2 < 0\}, \tag{5.31}$$

$$\Xi^3_{D_4} = \{(x,y,\dot{x},\dot{y}) | D_4(x,y) = 0^+ \text{ and } D_1 < D_2 < 0\}. \tag{5.32}$$

The mapping beginning at switching plane $\Xi^3_{D_3}$ and ending at switching plane $\Xi^3_{D_4}$, in domain Ω_3 is defined as

$$P_{(34)3} : \Xi^3_{D_3} \to \Xi^3_{D_4}. \tag{5.33}$$

The mapping describes the motion of tool- and work-piece contact, and cutting with $\dot{\tilde{y}}(\dot{x},\dot{y}) \in (-\infty, V)$ as shown in Fig. 5.2b. The mapping beginning at $\Xi^2_{D_4}$ and ending at $\Xi^2_{D_k}$ $(k = 1, 2, 3,)$ in domain Ω_4 is defined as

$$P_{(4k)2} : \Xi^2_{D_4} \to \Xi^2_{D_k}. \tag{5.34}$$

The mapping describes the motion for the chip vanishing boundary to the other boundary in domain Ω_4 for tool and work-piece contact and cutting with $\dot{\tilde{y}}(\dot{x},\dot{y}) \in (V, \infty)$.

Table 5.1 Possible switching set combinations and mappings

Initial boundary	Final boundary	Domain Ω_α	Mappings	j, k
j	k	0	$P_{(jk)0} : \Xi^0_{D_j} \to \Xi^0_{D_k}$	$j = 1,2,3; k = 1,2,3$
j	k	1	$P_{(jk)1} : \Xi^1_{D_j} \to \Xi^1_{D_k}$	$j = 1; k = 1$
j	k	2	$P_{(jk)2} : \Xi^2_{D_j} \to \Xi^2_{D_k}$	$j = 1,2,4; k = 1,2,3$
j	k	3	$P_{(jk)3} : \Xi^3_{D_j} \to \Xi^3_{D_k}$	$j = 1,2,3; k = 1,2,3,4$
j	k	4	$P_{(jk)4} : \Xi^4_{D_j} \to \Xi^4_{D_k}$	$j = 1,2,3; k = 1,2,3$
1	j	α	$P_{(1j)\alpha} : \Xi^\alpha_{D_1} \to \Xi^\alpha_{D_j}$	$j = 1,2,3; \alpha = 0,1,2,4$
2	j	α	$P_{(2j)\alpha} : \Xi^\alpha_{D_2} \to \Xi^\alpha_{D_j}$	$j = 1,2,3; \alpha = 0,2,4$
3	j	α	$P_{(3j)k} : \Xi^\alpha_{D_3} \to \Xi^\alpha_{D_j}$	$j = 1,2,3; \alpha = 0,3,4$
4	j	α	$P_{(4j)\alpha} : \Xi^\alpha_{D_4} \to \Xi^\alpha_{D_j}$	$j = 1,2,3; \alpha = 2$

For $j = 3$, $\alpha \neq 1$; for $j = 1,2,3$, $\alpha = 0,1,2,4$; for $j = 4$, $\alpha = 3$

(E) *Possible Combinations*: The physical combinations of initial and final boundaries are summarized in Table 5.1. The possible combinations for motion include four possible initial boundaries and three possible final boundaries through five possible domains. Certain configurations of these boundaries and domains are not physically possible. Thus, the potential motions are shown in Table 5.1. Consider the jth and kth initial and final switching planes of a mapping, which is defined as

$$P_{(jk)0} : \Xi^0_{D_j} \to \Xi^0_{D_k} \tag{5.35}$$

along the boundary for chip/tool seizure ($j, k = 1,2,3$),

$$P_{(jk)1} : \Xi^1_{D_j} \to \Xi^1_{D_k} \tag{5.36}$$

in domain Ω_1 for tool-free running ($j = 1$ and $k = 1$),

$$P_{(jk)2} : \Xi^2_{D_j} \to \Xi^2_{D_k} \tag{5.37}$$

in domain Ω_2 for tool- and work-piece contact but no cutting ($j = 1,2,4$ and $k = 1,2,3$),

$$P_{(jk)3} : \Xi^3_{D_j} \to \Xi^3_{D_k} \tag{5.38}$$

in domain Ω_3 for tool- and work-piece contact with cutting and $\dot{y}(\dot{x}, \dot{y}) \in (-\infty, V)$ ($j = 1,2,3$ and $k = 1,2,3,4$),

$$P_{(jk)4} : \Xi^4_{D_j} \to \Xi^4_{D_k} \tag{5.39}$$

in domain Ω_4 for tool- and work-piece contact, with cutting and $\dot{y}(\dot{x}, \dot{y}) \in (V, \infty)$ ($j = 1,2,3$ and $k = 1,2,3$). The two-dimensional and one-dimensional edges will not be discussed herein. The following section discusses combinations of the switching planes with respect to initial boundaries.

Consider physical motion with an initial condition on the tool- and work-piece contact/impact boundary where the final boundary is the jth boundary through domain Ω_α. The possible mappings can be defined as

$$\Xi^{\alpha}_{D_1} \to \Xi^{\alpha}_{D_j} : P_{(1j)\alpha} \quad \text{for } j = 1,2,3 \text{ and } \alpha = 0,1,2,3 \tag{5.40}$$

but for $j = 3$ then $\alpha \neq 1$.

Similarly, for physical motion with an initial condition on the boundary of tool- and work-piece contact and no cutting, the possible mappings are

$$\Xi^{\alpha}_{D_2} \to \Xi^{\alpha}_{D_j} : P_{(2j)\alpha} \quad \text{for } j = 1,2,3 \text{ and } \alpha = 0,2,4 \tag{5.41}$$

but for $j = 3$ then $\alpha \neq 2$.

For physical motions with the initial conditions on the boundary of tool and work-piece contact, and cutting with $\dot{\tilde{y}}(\dot{x},\dot{y}) \in (-\infty, V)$, the possible mappings are

$$P_{(3j)\alpha} : \Xi^{\alpha}_{D_3} \to \Xi^{\alpha}_{D_j} \quad \text{for } j = 1,2,3,4 \text{ and } \alpha = 0,3,4 \tag{5.42}$$

but for $j = 3$ then $\alpha \neq 0, 4$.

For physical motion with initial conditions on the boundary of tool and work-piece contact and cutting with $\dot{\tilde{y}}(\dot{x},\dot{y}) \in (V, \infty)$, the possible mappings are

$$P_{(4j)\alpha} : \Xi^{\alpha}_{D_3} \to \Xi^{\alpha}_{D_j} \quad \text{for } j = 1,2,3 \text{ and } \alpha = 2. \tag{5.43}$$

The summary of the above combinations is shown in Table 5.1. The next section employs the switching planes and mapping to develop mapping structure for periodic motions for predicting and labeling cutting motions.

5.2 Local Mapping Structures

The switching sets and mappings developed in the previous sections are used for periodic motions. Local mapping structures for periodic motions near the boundaries are presented in this section.

(A) *Contact Boundary*. Consider local periodic motion near the tool- and work-piece contact/impact boundary. A simple periodic motion near such a boundary is discussed first. Select the initial condition on switching plane $\Xi^1_{D_1}$; and under mapping $P_{(11)1}$, the initial condition will map to the final condition on switching plane $\Xi^1_{D_1}$. In fact, the switching planes $\Xi^1_{D_1}$ and $\Xi^2_{D_1}$ are the same. Thus, the final condition in $\Xi^1_{D_1}$ can be used as the initial condition in switching plane $\Xi^2_{D_1}$.

Under the mapping $P_{(11)2}$, such an initial condition will bring to the final condition on switching plane $\Xi^2_{D_1}$. Thus,

$$P_{(11)1} : \Xi^1_{D_1} \to \Xi^1_{D_1} \text{ and } P_{(11)2} : \Xi^2_{D_1} \to \Xi^2_{D_1} \tag{5.44}$$

The combination of two mappings can form a resultant mapping. For simplicity, the boundary index will be dropped. The combination action is represented by the action symbol ($\circ$), which can read "composite" in mathematics. Therefore, the resultant mapping is given by

$$P_{21} = P_{(11)2} \circ P_{(11)1} : \Xi^1_{D_1} \to \Xi^1_{D_1}. \tag{5.45}$$

If the final condition on switching plane $\Xi^2_{D_1}$ is the same as the initial condition on $\Xi^1_{D_1}$ during a certain time period, then the above mapping will describe a periodic motion near the contact boundary.

Consider three times of the above mapping cycle during a certain period as

$$\begin{aligned} &P_{(11)1} : \Xi^1_{D_1} \to \Xi^1_{D_1} \text{ and } P_{(11)2} : \Xi^2_{D_1} \to \Xi^2_{D_1}, \\ &P_{(11)1} : \Xi^1_{D_1} \to \Xi^1_{D_1} \text{ and } P_{(11)2} : \Xi^2_{D_1} \to \Xi^2_{D_1}, \\ &P_{(11)1} : \Xi^1_{D_1} \to \Xi^1_{D_1} \text{ and } P_{(11)2} : \Xi^2_{D_1} \to \Xi^2_{D_1}. \end{aligned} \tag{5.46}$$

The corresponding resultant mapping can be simplified as

$$P_{212121} = P_{(11)2} \circ P_{(11)1} \circ P_{(11)2} \circ P_{(11)1} \circ P_{(11)2} \circ P_{(11)1}. \tag{5.47}$$

As in Luo (2006), the sample notation will be adopted as

$$P_{(21)^3} = P_{212121}. \tag{5.48}$$

In general, n times of such a resultant mapping P_{21} during a certain period yields the resultant mapping, i.e.,

$$P_{(21)^n} = P_{\underbrace{21\ldots21}_{n \text{ pairs}}}. \tag{5.49}$$

In addition, the resultant motion near the contract boundary through the domain Ω_1 and Ω_4 with $\dot{y} > V$ is discussed herein, which includes short cutting action. In a similar fashion, the resultant mapping can be defined as

$$P_{41} = P_{(11)4} \circ P_{(11)1}. \tag{5.50}$$

If the chip–tool seizure ($\dot{y} \equiv V$) with short cutting ($\dot{y} < V$) exists, the resultant mapping is defined as

$$P_{041} = P_{(31)0} \circ P_{(13)4} \circ P_{(11)1}. \tag{5.51}$$

Without the chip/tool seizure ($\dot{y} \equiv V$), the cutting motion between cutting ($\dot{y} < V$) and without cutting ($\dot{y} > V$) exists, the resultant mapping is defined as

$$P_{341} = P_{(31)3} \circ P_{(13)4} \circ P_{(11)1}. \tag{5.52}$$

(B) *Cutting Boundary*: As before, the mapping structure of local periodic motions near the cutting boundary is discussed. A simple periodic motion near such a boundary is discussed first. Select the initial condition on switching plane $\Xi^2_{D_2}$; and under mapping $P_{(22)2}$, the initial condition will map to the final condition on switching plane $\Xi^2_{D_2}$. Thus, the final condition in $\Xi^4_{D_2}$ can be used as the initial condition in switching plane $\Xi^4_{D_2}$. Under the mapping $P_{(22)4}$, such an initial condition will bring to the final condition on switching plane $\Xi^4_{D_2}$.

$$P_{(22)2} : \Xi^2_{D_2} \to \Xi^2_{D_2} \text{ and } P_{(22)4}{:}\Xi^4_{D_2} \to \Xi^4_{D_2}. \tag{5.53}$$

The resultant mapping is given by

$$P_{42} = P_{(22)4} \circ P_{(22)2}. \tag{5.54}$$

The motion given by the foregoing resultant mapping corresponds to the tool-piece and work-piece in contact without cutting directly followed by the tool-piece and work-piece in contact with cutting.

The resultant mapping for a chip vanishing is

$$P_{342} = P_{(32)3} \circ P_{(23)4} \circ P_{(22)2}, \tag{5.55}$$

with an inclusion of a mapping in domain Ω_3 ($\dot{y} < V$) or

$$P_{042} = P_{(32)0} \circ P_{(23)4} \circ P_{(22)2}, \tag{5.56}$$

with an inclusion of a mapping for a motion along the boundary (i.e., chip/tool seizure, $\dot{y} \equiv V$).

(C) *Friction Boundary*. The local structures of periodic motions near the frictional boundary are discussed. A simple periodic motion near such a boundary is discussed first. Select the initial condition on switching plane $\Xi^3_{D_3}$; and under mapping $P_{(33)3}$, the initial condition will map to the final condition on switching plane $\Xi^3_{D_3}$. Thus, the final condition in $\Xi^4_{D_3}$ can be used as the initial condition in switching plane $\Xi^4_{D_3}$. Under the mapping $P_{(33)4}$, such an initial condition will map to the final condition on switching plane $\Xi^4_{D_2}$.

$$P_{(33)3} : \Xi^3_{D_3} \to \Xi^3_{D_3} \text{ and } P_{(33)4}{:}\Xi^4_{D_3} \to \Xi^4_{D_3}. \tag{5.57}$$

The resultant mapping is given by

$$P_{34} = P_{(33)3} \circ P_{(33)4}. \tag{5.58}$$

The resultant mapping gives the tool-piece and work-piece in contact with cutting ($\dot{\bar{y}} > V$) directly followed by the reduction in chip length ($\dot{\bar{y}} < V$).

The resultant mapping for a chip vanishing is

$$P_{04} = P_{(33)0} \circ P_{(33)4}, \tag{5.59}$$

with an inclusion of motion along the boundary of chip/tool seizure ($\dot{\bar{y}} \equiv V$),

$$P_{034} = P_{(33)0} \circ P_{(33)3} \circ P_{(33)4}, \tag{5.60}$$

with an inclusion of a chip vanishing motion and motion in domain Ω_3 ($\dot{\bar{y}} < V$).

(D) *Chip Vanishing Boundary*: The local structures of periodic motions near the chip vanishing boundary are discussed. The individual mappings are

$$P_{(43)2} : \Xi^2_{D_4} \rightarrow \Xi^2_{D_4}; \quad P_{(33)4} : \Xi^4_{D_3} \rightarrow \Xi^4_{D_3}; \quad P_{(34)3} : \Xi^3_{D_3} \rightarrow \Xi^3_{D_4}. \tag{5.61}$$

The resultant mapping is given by

$$P_{342} = P_{(34)3} \circ P_{(33)4} \circ P_{(43)2}. \tag{5.62}$$

This resultant mapping gives the motion corresponding to the tool-piece and work-piece in contact without cutting directly, followed by cutting ($\dot{\bar{y}} > V$) which then crosses the frictional boundary where the chip reduces in length to zero from domain Ω_3 ($\dot{\bar{y}} < V$) to domain Ω_4 ($\dot{\bar{y}} > V$).

The resultant mapping for the chip vanishing boundary is

$$P_{3042} = P_{(34)3} \circ P_{(33)0} \circ P_{(33)4} \circ P_{(43)2}, \tag{5.63}$$

with an inclusion of motion along the boundary (i.e., chip seizure)

$$P_{1234} = P_{(11)1} \circ P_{(41)2} \circ P_{(34)3} \circ P_{(13)4}. \tag{5.64}$$

with an inclusion of motion in domain Ω_1. The machine tool will have a loss of contact (i.e., tool-free running). The global structure is discussed later.

5.3 Periodic Interrupted Cutting Motions

As in Chaps. 2 and 3, consider one of the simplest mappings as

$$P = P_3 \circ P_4. \tag{5.65}$$

With initial and final conditions for each mapping, the corresponding relations are

$$\left.\begin{aligned} P_4 &: (\tilde{x}_k, \dot{\tilde{x}}_k, \tilde{y}_k, V^+, t_k) \rightarrow (\tilde{x}_{k+1}, \dot{\tilde{x}}_{k+1}, \tilde{y}_{k+1}, V^+, t_{k+1}), \\ P_3 &: (\tilde{x}_{k+1}, \dot{\tilde{x}}_{k+1}, \tilde{y}_{i+1}, V^-, t_{i+1}) \rightarrow (\tilde{x}_{k+2}, \dot{\tilde{x}}_{k+2}, \tilde{y}_{i+2}, V^-, t_{i+2}), \end{aligned}\right\} \tag{5.66}$$

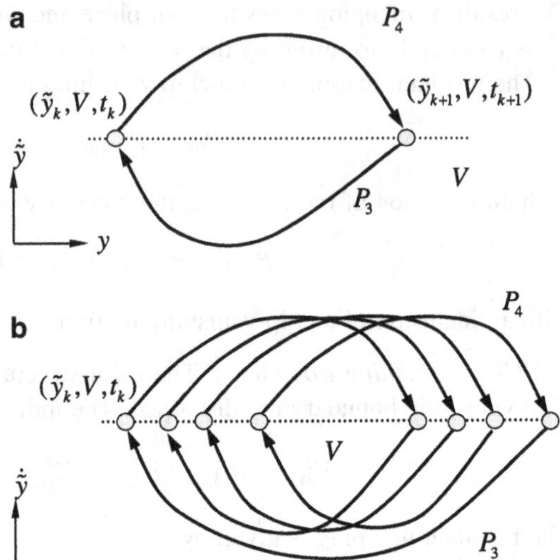

Fig. 5.3 Nonstick periodic motion in the absolute phase plane: (**a**) period one motion P_{34} and (**b**) period four motion $P_{(34)^4}$

without chip seizure ($\dot{\tilde{y}} \equiv V$) motion, $V = V^+ = V^- = -\bar{V}/\Omega$. The composition of mappings is to form periodic motion as shown in Fig. 5.3.

For the periodic motion of $\tilde{\mathbf{Y}}_{i+2} = P\tilde{\mathbf{Y}}_i$ where $\tilde{\mathbf{Y}}_k = (\tilde{x}_k, \dot{\tilde{x}}_k, \tilde{y}_k, t_k)^{\mathrm{T}}$ during N excitation periods, the periodicity conditions are

$$\left.\begin{aligned} \tilde{x}_{k+2} &= \tilde{x}_k, \quad \dot{\tilde{x}}_{k+2} = \dot{\tilde{x}}_k, \\ \tilde{y}_{k+2} &= \tilde{y}_k, \quad \Omega t_{k+2} = \Omega t_k + 2N\pi. \end{aligned}\right\} \tag{5.67}$$

Using the notation in Gegg et al. (2008), a generalized mapping for (5.65) is

$$P = P_{(34)^m} \triangleq \underbrace{(P_3 \circ P_4) \circ \cdots \circ (P_3 \circ P_4)}_{m-\text{pairs}}. \tag{5.68}$$

The periodic motion for the foregoing mapping requires

$$\tilde{\mathbf{Y}}_{k+2m} = P\tilde{\mathbf{Y}}_k. \tag{5.69}$$

The periodicity conditions of periodic motion for the foregoing mapping are

$$\tilde{x}_{k+2m} = \tilde{x}_k, \quad \dot{\tilde{x}}_{k+2m} = \dot{\tilde{x}}_k, \tag{5.70}$$

$$\tilde{y}_{k+2m} = \tilde{y}_k, \quad \Omega t_{k+2m} = \Omega t_k + 2N\pi. \tag{5.71}$$

Such periodicity relations will be used to develop the solution structure for the prediction of periodic motion. The governing equations and an appropriate solving method (e.g., Newton–Raphson Method) are employed to obtain the solution set. Because the machine tool is modeled by a two-degree-of-freedom oscillator, the governing equations of each mapping P_κ ($\kappa \in \{3,4\}$) can be expressed by

$$\left.\begin{aligned} f_1^{(\kappa)}(\tilde{\mathbf{x}}_k, \tilde{\mathbf{y}}_k, t_k, \tilde{\mathbf{x}}_{k+1}, \tilde{\mathbf{y}}_{k+1}, t_{k+1}) = 0, \\ f_2^{(\kappa)}(\tilde{\mathbf{x}}_k, \tilde{\mathbf{y}}_k, t_k, \tilde{\mathbf{x}}_{k+1}, \tilde{\mathbf{y}}_{k+1}, t_{k+1}) = 0, \\ f_3^{(\kappa)}(\tilde{\mathbf{x}}_k, \tilde{\mathbf{y}}_k, t_k, \tilde{\mathbf{x}}_{k+1}, \tilde{\mathbf{y}}_{k+1}, t_{k+1}) = 0, \\ f_4^{(\kappa)}(\tilde{\mathbf{x}}_k, \tilde{\mathbf{y}}_k, t_k, \tilde{\mathbf{x}}_{k+1}, \tilde{\mathbf{y}}_{k+1}, t_{k+1}) = 0. \end{aligned}\right\} \tag{5.72}$$

(A) *Cutting Motion with Stick*: Consider the mapping structure for periodic motion with chip seizure ($\dot{\tilde{y}} \equiv V$) as

$$P_{034} = P_0 \circ P_3 \circ P_4. \tag{5.73}$$

The mapping P_0 describes the starting and ending of the stick motion. The disappearance of stick motion requires $F_{\tilde{y}}^{(i)}(\tilde{\mathbf{x}}_{k+1}, \tilde{\mathbf{y}}_{k+1}, t_{k+1}) = 0$ and $\dot{\tilde{y}} \equiv V$ in (4.76). The general mapping structure for periodic orbit with stick motion is

$$P = \underbrace{(P_0^{(j_{m0})} \circ P_3^{(j_{m3})} \circ P_4^{(j_{m4})}) \circ \cdots \circ (P_0^{(j_{10})} \circ P_3^{(j_{13})} \circ P_4^{(j_{14})})}_{m-\text{terms}}, \tag{5.74}$$

where $j_{lk} \in \{0,1\}$ with $l \in \{1,2,\cdots,m\}$ and $k = 0,3,4$. $P_k^{(0)} = 1$ and $P_k^{(n)} = P_k \circ P_k^{(n-1)}$. In domain Ω_α ($\alpha \in \{3,4\}$), there are three possible stable motions. The governing equations of mapping P_i ($i \in \{3,4\}$) are obtained from the solutions in Appendix A.4. The governing equations of mapping P_κ ($\kappa \in \{0,3,4\}$) can be expressed as in (5.72).

(B) *Cutting Motions Without Stick*: Consider a mapping structure as

$$P = P_2 \circ P_3 \circ P_4. \tag{5.75}$$

The resultant mapping structure is for the cutting motion interrupted by the frictional boundary and loss of chip contact as shown in Fig. 5.4a. The corresponding mapping relations are

$$\left.\begin{aligned} &P_4 : (\tilde{x}_k, \dot{\tilde{x}}_k, \tilde{y}_k, V^+, t_k) \to (\tilde{x}_{k+1}, \dot{\tilde{x}}_{k+1}, \tilde{y}_{k+1}, V^+, t_{k+1}), \\ &P_3 : (\tilde{x}_{k+1}, \dot{\tilde{x}}_{k+1}, \tilde{y}_{k+1}, V^-, t_{k+1}) \to (\tilde{x}_{k+2}, \dot{\tilde{x}}_{k+2}, \tilde{y}_{k+1} - L_c, \dot{\tilde{y}}_{k+2}, t_{k+2}), \\ &P_2 : (\tilde{x}_{k+2}, \dot{\tilde{x}}_{k+2}, \tilde{y}_{k+1} - L_c, \dot{\tilde{y}}_{k+2}, t_{k+2}) \to (\tilde{x}_{k+3}, \dot{\tilde{x}}_{k+3}, \tilde{y}_{k+3}, V^-, t_{k+3}). \end{aligned}\right\} \tag{5.76}$$

without chip adhesion (stick), $V^+ = V^- = \bar{V}/\Omega$ exists.

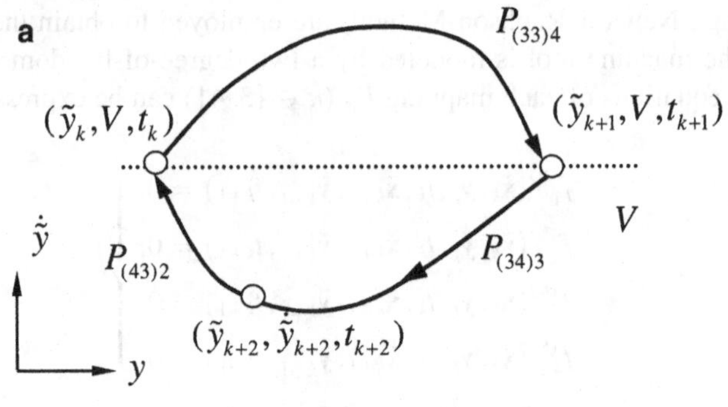

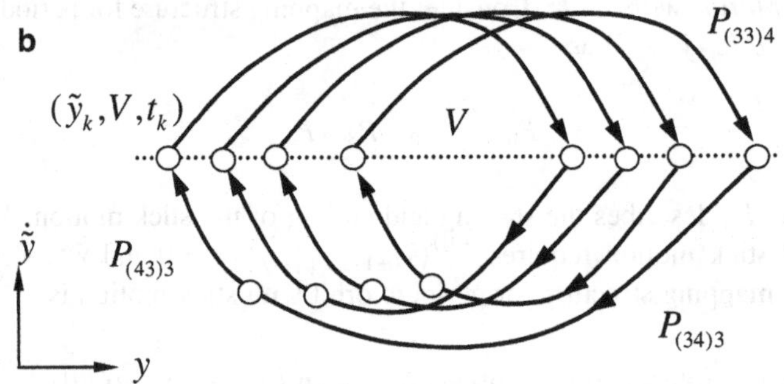

Fig. 5.4 Nonstick periodic motion in the absolute phase plane: (**a**) period one motion P_{234} and (**b**) period four motion $P_{(234)^4}$

For a periodic motion of $\tilde{\mathbf{Y}}_{k+3} = P\tilde{\mathbf{Y}}_k$ where $\tilde{\mathbf{Y}}_k = (\tilde{x}_k, \dot{\tilde{x}}_k, \tilde{y}_k, \dot{\tilde{y}}_k, t_k)^{\mathrm{T}}$ during N excitation periods, the periodicity conditions are

$$\tilde{x}_{k+3} = \tilde{x}_k, \dot{\tilde{x}}_{k+3} = \dot{\tilde{x}}_k; \quad \tilde{y}_{k+3} = \tilde{y}_k, \dot{\tilde{y}}_{k+3} = \dot{\tilde{y}}_k; \quad t_{k+3} = t_k + 2N\pi. \tag{5.77}$$

Similarly, a generalized mapping structure is

$$P = \underbrace{(P_2^{(j_{m2})} \circ P_3^{(j_{m3})} \circ P_4^{(j_{m4})}) \circ \cdots \circ (P_2^{(j_{12})} \circ P_3^{(j_{13})} \circ P_4^{(j_{14})})}_{m-\text{terms}} \tag{5.78}$$

as sketched in Fig. 5.4b. The periodic motion for the foregoing mapping requires

$$\tilde{\mathbf{Y}}_{k+3m} = P\tilde{\mathbf{Y}}_k. \tag{5.79}$$

The periodicity of periodic motion for the foregoing mapping is

$$\tilde{x}_{k+3m} = \tilde{x}_k, \dot{\tilde{x}}_{k+3m} = \dot{\tilde{x}}_k; \quad \tilde{y}_{k+3m} = \tilde{y}_k, \dot{\tilde{y}}_{k+3m} = \dot{\tilde{y}}_k; \quad t_{k+3m} = t_k + 2N\pi. \tag{5.80}$$

Similarly, the solution for $\mathbf{y}_k = (\tilde{y}_k, \dot{\tilde{y}}_k, t_k)^{\mathrm{T}}$ can be determined. Consider the mapping structure for periodic orbit with stick motion is

$$P = P_{2340} \triangleq P_2 \circ P_3 \circ P_4 \circ P_0. \tag{5.81}$$

The general mapping structure for periodic motion with stick motion is

$$P = \underbrace{(P_2^{(j_{m2})} \circ P_3^{(j_{m3})} \circ P_4^{(j_{m4})} \circ P_0^{(j_{m0})}) \circ \cdots \circ (P_2^{(j_{12})} \circ P_3^{(j_{13})} \circ P_4^{(j_{14})} \circ P_0^{(j_{10})})}_{m-\text{terms}}, \tag{5.82}$$

where $j_{lk} \in \{0,1\}$ with $l \in \{1,2,\ldots,m\}$ and $k = 0,2,3,4$. $P_k^{(n)} = P_k \circ P_k^{(n-1)}$ and $P_k^{(0)} = 1$. For nonstick motion, there are three possible stable motions in the two domains Ω_k ($k \in \{2,3,4\}$). The governing equations of mapping P_κ ($\kappa \in \{0,2,3,4\}$) are obtained from the exact solutions presented in Appendix A. So the governing equations of mapping P_κ ($\kappa \in \{0,2,3,4\}$) can be expressed in (5.72).

5.4 Cutting Motions and Chip Adhesion

Numerical predictions of the periodic chip seizure ($\dot{\tilde{y}} \equiv V$) and cutting motions for the machine-tool system subject to an eccentricity force are considered. The dynamical system parameters are

$$\begin{aligned} &m_{\mathrm{eq}}/m_{\mathrm{e}} = 10^3, d_x = 740\mathrm{Ns/mm}, d_y = 630\mathrm{Ns/mm}, d_1 = d_2 = 0;\\ &k_x = k_y = 560\ \mathrm{kN/mm}, k_1 = 10\ \mathrm{kN/mm}, k_2 = 100\ \mathrm{kN/mm} \end{aligned} \tag{5.83}$$

and the external force and geometry parameters are

$$\begin{aligned} &\Omega V = 20\ \mathrm{mm/s}, \delta_1 = \delta_2 = 10^{-3}\ \mathrm{m}, \mu = 0.7, \alpha = \pi/4\ \mathrm{rad},\\ &\beta = 0.1\ \mathrm{rad}, \eta = \pi/4\ \mathrm{rad}, A = em_{\mathrm{e}}\Omega^2, L_{\mathrm{c}} = 10^{-3}\ \mathrm{m},\\ &X_1 = Y_1 = 10^{-3}\ \mathrm{m}, X_{\mathrm{eq}} = Y_{\mathrm{eq}} = 5 \times 10^{-3}\ \mathrm{m}. \end{aligned} \tag{5.84}$$

The corresponding initial contact conditions are

$$\begin{aligned} &x_1^* = 0.3941\ \mathrm{mm}, y_1^* = -4.4638\ \mathrm{mm},\\ &x_2^* = 0.3710\ \mathrm{mm}, y_2^* = -3.2244\ \mathrm{mm}. \end{aligned} \tag{5.85}$$

The motions for the following parameter ranges are summarized in Table 5.2. The term bifurcation will be used throughout the description of the following results.

The definition of bifurcation according to Devaney (2003) is "a division in two, a splitting apart, or a change." The definition of bifurcation used here is any change in the mapping structure; hence a change in the state along the respective boundary. The two definitions noted are similar, but as the numerical results will show the

Table 5.2 Summary of numerical predictions

Parameter	Boundary	Pure cutting	Interrupted cutting	Figures
$e@\Omega = 200$	3	$e < 0.0803$ mm	$[0.0803, 0.4]$ mm[a]	5.5 and 5.6
$\Omega@e = 0.1$ mm	3	$\Omega < 191.6$ rad/s	$[191.6, 1k]$ rad/s[b]	5.7 and 5.8
$\Omega V@\Omega = 200$ rad/s	3	$V > 30.1$ mm/s	$[0, 30.1]$ mm/s[c]	5.9
$\mu@\Omega = 200$ rad/s	3	None	$[0, 3]$[d]	5.10 and 5.11
$k_2@\Omega = 200$ rad/s	3	None	$[0, 567.6]$ kN[e]	5.12 and 5.13

[a] $P_{(0(34)^m)^n} : (0.0803, 0.0870]$ mm, $P_{0(34)^2} : (0.0870, 0.1398]$ mm,
$P_{(034)^3 34} : (0.1398, 0.1406]$ mm, $P_{(034)^2} : (0.1406, 0.1586]$ mm, $P_{34} : (0.1586, 0.4000]$ mm

[b] $P_{(0(34)^m)^n} : (0.1912, 0.1916] \times 10^3$ rad/s, $P_{(034)^2} : (0.1916, 0.1940] \times 10^3$rad/s,
$P_{(0(34)^m)^n} : (0.1941, 0.1969] \times 10^3$rad/s, $P_{0(34)^2} : (0.1970, 0.2028] \times 10^3$rad/s,
$P_{(0(34)^m)^n} : (0.2028, 0.2256] \times 10^3$rad/s, $P_{34} : (0.2256, 1.0] \times 10^3$rad/s

[c] $P_{(0(34)^2)^2} : (16.66, 17.08]$ mm/s, $P_{0(34)^2} : (17.08, 24.26]$ mm/s, $P_{(034)^2} : (24.26, 29.14]$ mm/s,
$P_{034} : (29.14, 32.28]$ mm/s, $P_{(0(34)^m)^n} : (32.28, 32.82]$ mm/s

[d] $P_{34} : (0.0000, 0.2100]$, $P_{0(34)^3} : (0.2100, 0.3140]$, $P_{(0(34)^m)^n} : (0.3140, 0.4380]$
$P_{0(34)^2} : (0.4380, 0.8840]$, $P_{(0(34)^m)^n} : (0.8840, 0.300]$

[e] $P_{34} : (0, 29.40]$ kN, $P_{0(34)^3} : (29.40, 46.80]$ kN, $P_{(0(34)^m)^n} : (46.80, 74.00]$ kN,
$P_{0(34)^2} : (74.00, 120.60]$ kN, $P_{(034)^2} : (120.60, 133.20]$ kN, $P_{034} : (133.20, 227.60]$ kN,
$P_{(0(34)^m)^n} : (227.60, 243.80]$ kN, $P_{04} : (243.80, 386.20]$ kN,
$P_{(04)^2} : (386.20, 405.00]$ kN, $P_{(04)^m} : (405.00, 415.40]$ kN,
$P_{(04)^5} : (415.40, 430.00]$ kN, $P_{(04)^6} : (430.00, 431.20]$ kN,
$P_{(0(34)^m)^n} : (431.20, 486.60]$ kN, $P_{(340)^2 4} : (486.60, 523.80]$ kN,
$P_{(4034043)^2} : (523.80, 533.40]$ kN, $P_{0(40)^2 34} : (533.40, 526.00]$ kN,
$P_{(0(34)^m)^n} : (561.00, 567.60]$ kN

term "bifurcation" defines not only a splitting of the motion (or a change) but also a combining of solution paths (or a disappearance/appearance of a new solution branch). From now on, all the illustrations are based on the coordinates of $(\tilde{x}, \tilde{y})$. Due to labeling problems in plots, the tildes will be dropped.

(A) *Eccentricity Excitation Amplitude*: The numerical predictions of the periodic chip seizure ($\dot{\tilde{y}} \equiv V$) and cutting motions are presented over the range of eccentricity amplitude $e \in [0.0803, 0.4]$ mm, see Fig. 5.5. The eccentricity frequency for this parameter range is held constant at $\Omega = 200$ rad/s. The switching phase $\mathrm{mod}(\varphi_k, 2\pi)$, switching displacement y_k, switching displacement x_k, and switching velocity dy_k/dt versus eccentricity amplitude e are illustrated in Fig. 5.5a–d, respectively. In Fig. 5.6a, b, the switching forces $F^{(3)}_{\tilde{y}_k}$ and $F^{(4)}_{\tilde{y}_k}$ and switching force products $(F^{(3)}_{\tilde{y}_k} \times F^{(4)}_{\tilde{y}_k})$ versus eccentricity amplitude e are shown. Note that $\varphi_k = \Omega \bar{t}_k$. The periodic motions observed through a range of eccentricity amplitude e are

$$
\begin{aligned}
&P_{(0(34)^m)^n} : (0.0803, 0.0870]\ \text{mm},\ P_{0(34)^2} : (0.0870, 0.1398]\ \text{mm},\\
&P_{(034)^3 34} : (0.1398, 0.1406]\ \text{mm},\ P_{(034)^2} : (0.1406, 0.1586]\ \text{mm},\\
&P_{34} : (0.1586, 0.4000]\ \text{mm}.
\end{aligned}
\tag{5.86}
$$

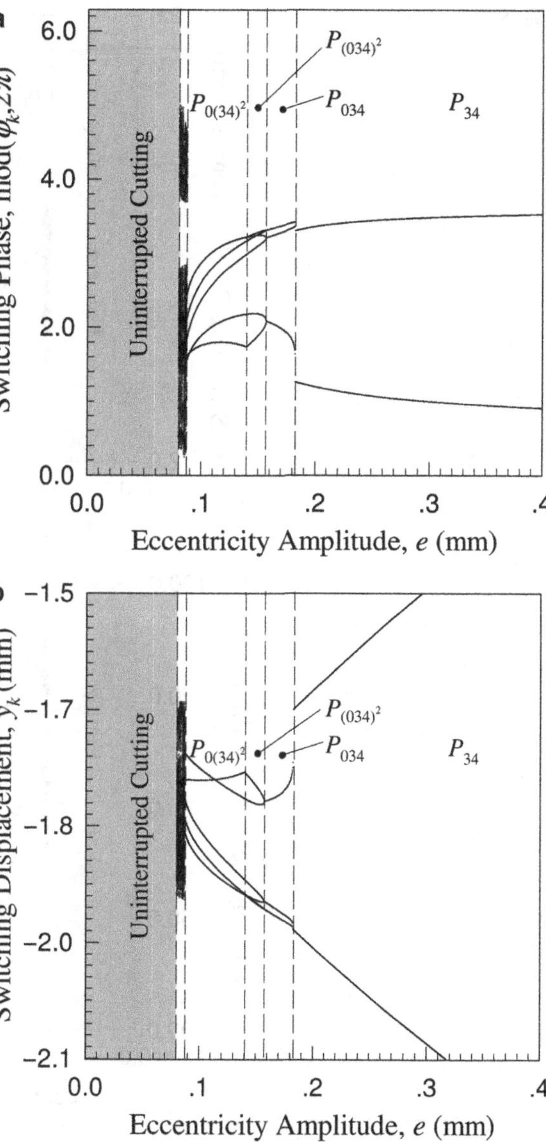

Fig. 5.5 Numerical prediction of (**a**) switching phase $\mathrm{mod}(\Omega t_k, 2\pi)$, (**b**) switching displacement ($y_k = \tilde{y}_k$), (**c**) switching displacement ($x_k = \tilde{x}_k$), and (**d**) switching velocity ($\dot{x}_k = \dot{\tilde{x}}_k$) over a range of eccentricity amplitude e; $L_c = 1.0$ mm; and $\Omega = 200$ rad/s

As observed, the periodic motion becomes simplified as eccentricity amplitude increases. The stick–slip combination can be forced to purely slip (nonstick) motion by excitation amplitude in Gegg et al. (2007). For $e < 0.0803$ mm, no motion intersects the discontinuity (or pure cutting occurs; no interruptions). The lower boundary of the eccentricity amplitude e exhibits complex motions, pseudo-periodic/chaotic motion. The switching phase is observed to have a dense area of switching points which originate from the onset of *chip adhesion.* This *qualitative* observation of onset or route to unstable motion has also been observed in theory and experiment

Fig. 5.5 (continued)

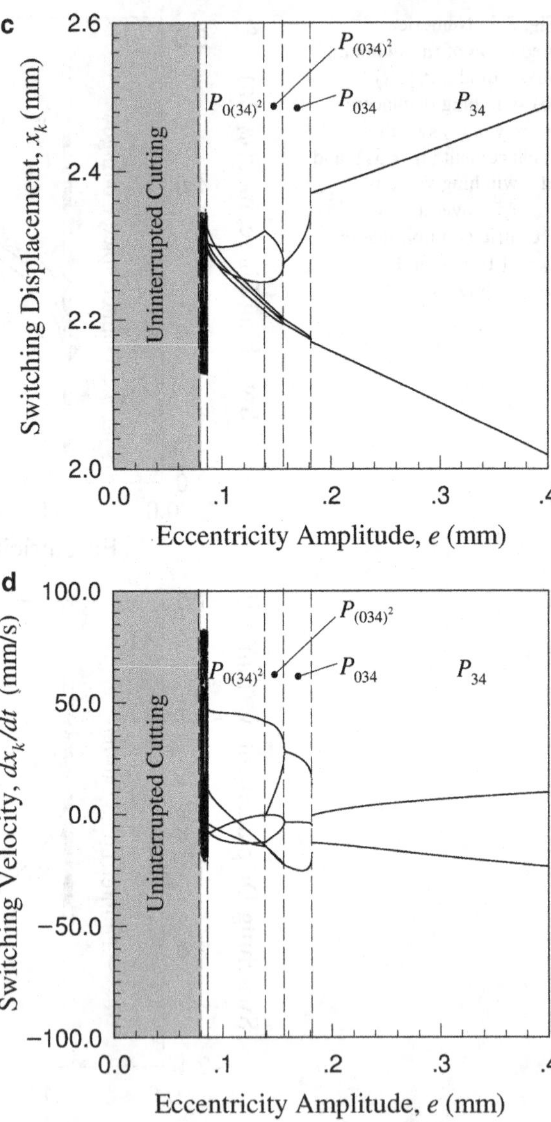

by Astakhov et al. (1997). In such a study, the stability of a chip structure is attributed to the seizure of the work-piece chip to the tool-piece rake surface. Two chip structures contributed to this phenomenon are the continuous and fragmentary hump-backed chip. Such chip adhesion (stick or seizure) is validated to occur at $e \approx 0.1822$ mm where added complexity in motion structure appears as the eccentricity amplitude e decreases, as shown in Fig. 5.6a, b. The forces and force product distributions verify the onset of chip seizure ($\dot{\tilde{y}} \equiv V$) with $F_{\tilde{y}_k}^{(3)} \times F_{\tilde{y}_k}^{(4)} = 0$. The addition of the chip seizure dynamics induces complex motions which are inherently

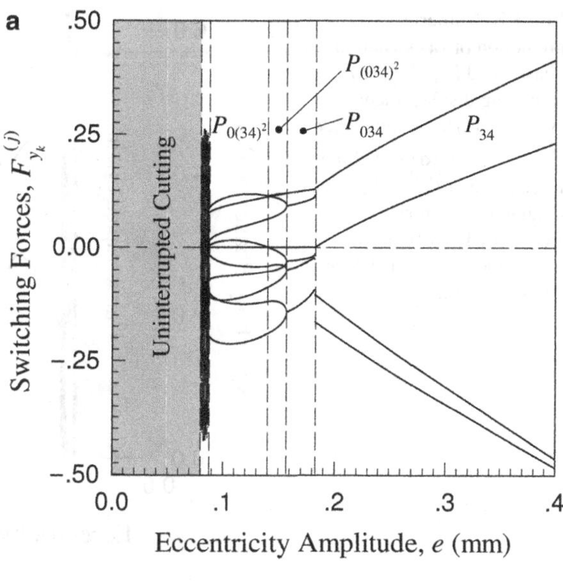

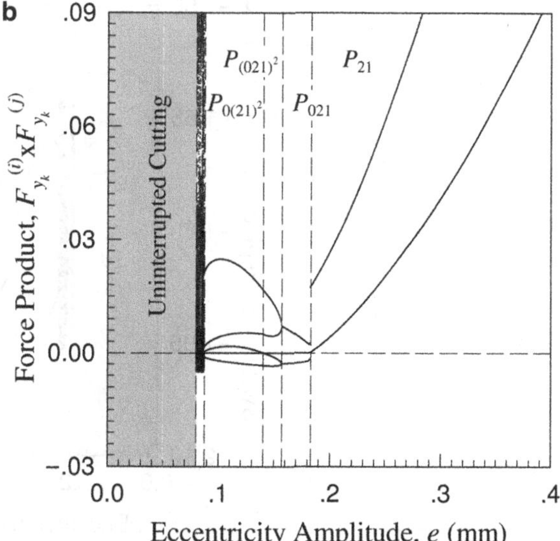

Fig. 5.6 Numerical prediction of (**a**) switching forces $F_{y_k}^{(3)}, F_{y_k}^{(4)}$ and (**b**) switching force product $F_{y_k}^{(3)} \times F_{y_k}^{(4)}$ for chip seizure and cutting periodic motions over a range of eccentricity amplitude e; $L_c = 1.0$ mm; and $\Omega = 200$ rad/s

detrimental to the surface finish of the work-piece and wear of the tool-piece. From now on, the switching points relative to $(x_k, \dot{x}_k)$ will not be presented because cutting motion is relative to the cutting boundary.

(B) *Eccentricity Excitation Frequency*: The numerical predictions of periodic chip seizure ($\dot{y} \equiv V$) and cutting motions are presented over the range of excitation frequencies $\Omega \in [191.6, 1000]$ rad/s, which is presented in Fig. 5.7. The eccentricity amplitude e is directly related to the eccentricity frequency Ω and $A\Omega = e m_e \Omega^2$.

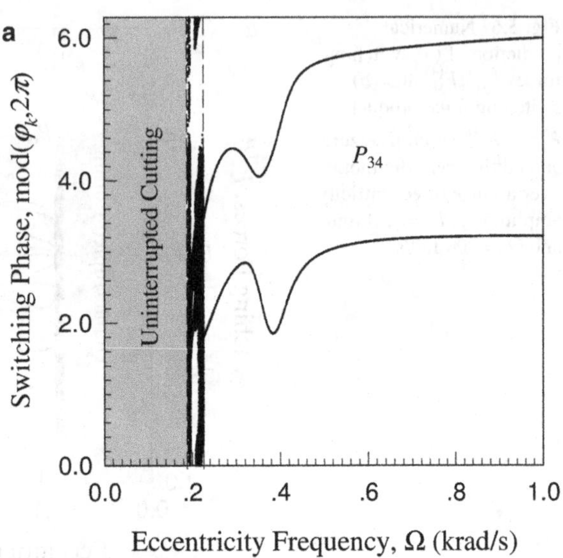

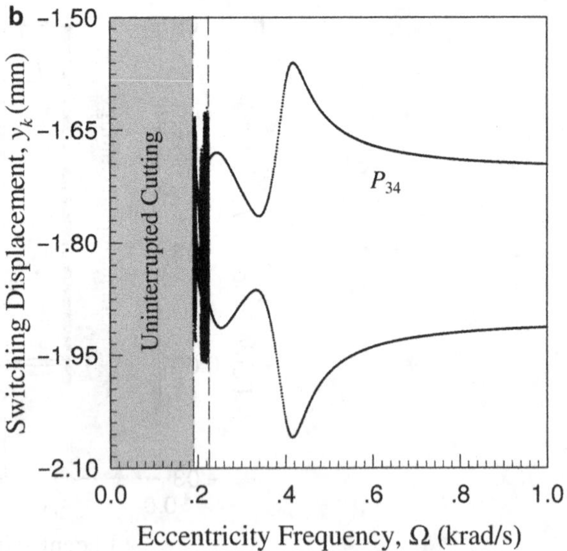

Fig. 5.7 Numerical prediction of (**a**) switching phase mod $(\varphi_k, 2\pi)$, (**b**) switching displacement $(y_k = \tilde{y})$, (**c**) switching forces $F_{y_k}^{(3)}, F_{y_k}^{(4)}$, and (**d**) switching force product $F_{y_k}^{(3)} \times F_{y_k}^{(4)}$ over a range of eccentricity frequency Ω; eccentricity amplitude e; $L_c = 1.0$ mm; and $e = 1$ mm

The switching phase mod $(\varphi_k, 2\pi)$, switching displacement y_k versus excitation frequency Ω are illustrated in Fig. 5.7a and b, respectively. In Fig. 5.7c, d, the switching forces $F_{\tilde{y}_k}^{(3)}$ and $F_{\tilde{y}_k}^{(4)}$ and switching force products $(F_{\tilde{y}_k}^{(3)} \times F_{\tilde{y}_k}^{(4)})$ versus excitation frequency Ω are presented. The range of excitation frequency for mapping structure $P_{34} = P_3 \circ P_4$ are $\Omega \in (0.2256, 1.0] \times 10^3$ rad/s. Outside of

Fig. 5.7 (continued)

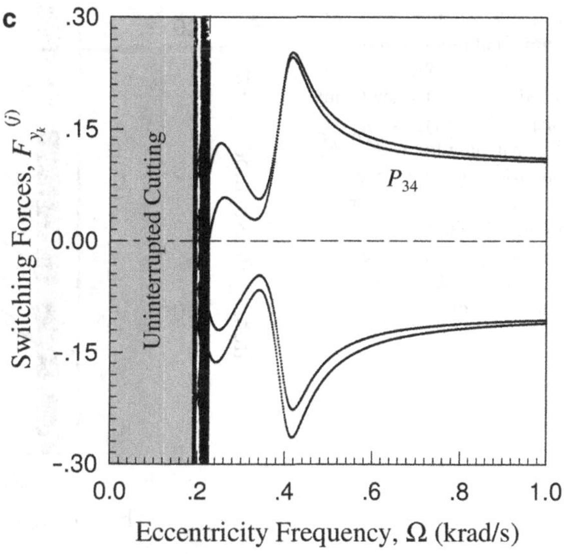

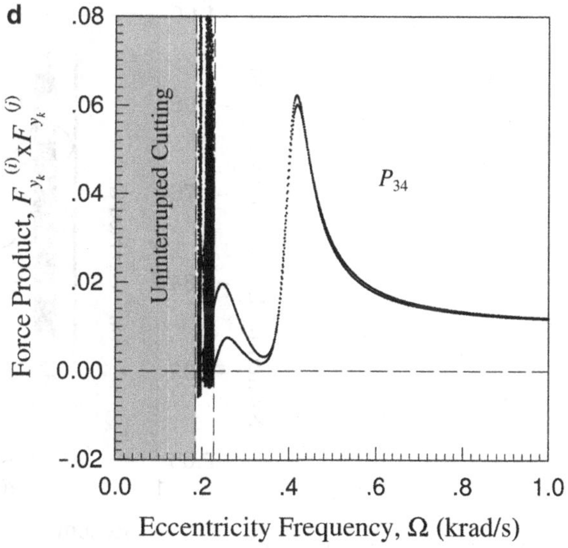

these intervals, the periodic motions are of period-k ($k = 1, 2, 3$), etc. and the interruption of cutting, chip seizure is satisfied on the interval as

$$
\begin{aligned}
&P_{(0(34)^m)^n} : (0.1912, 0.1916] \times 10^3 \text{ rad/s},\\
&P_{(034)^2} : (0.1916, 0.1940] \times 10^3 \text{ rad/s},\\
&P_{(0(34)^m)^n} : (0.1941, 0.1969] \times 10^3 \text{ rad/s},\\
&P_{0(34)^2} : (0.1970, 0.2028] \times 10^3 \text{ rad/s},\\
&P_{(0(34)^m)^n} : (0.2028, 0.2256] \times 10^3 \text{ rad/s},\\
&P_{34} : (0.2256, 1.0] \times 10^3 \text{ rad/s}.
\end{aligned}
\tag{5.87}
$$

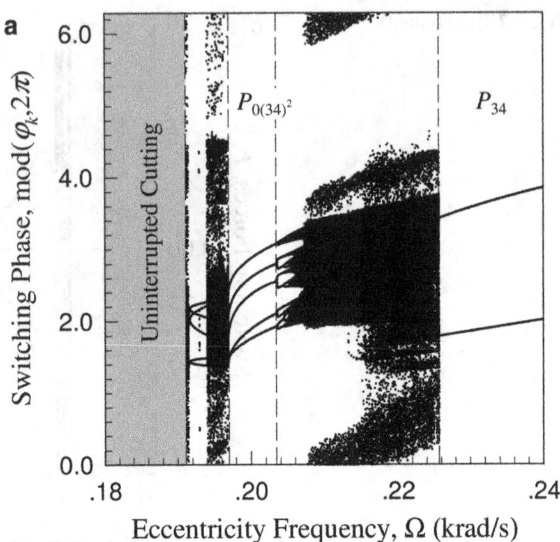

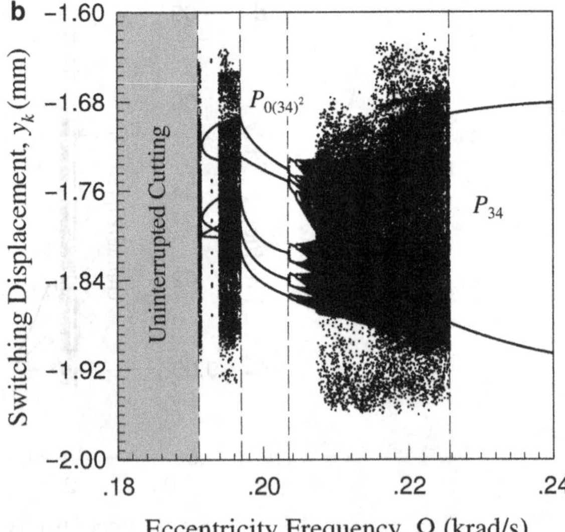

Fig. 5.8 Zoomed view for numerical prediction of (**a**) switching phase mod $(\varphi_k, 2\pi)$, (**b**) switching displacement $(y_k = \tilde{y})$, (**c**) switching forces $F_{y_k}^{(3)}, F_{y_k}^{(4)}$, and (**d**) switching force product $F_{y_k}^{(3)} \times F_{y_k}^{(4)}$ over a range of eccentricity frequency Ω; eccentricity amplitude e; $L_c = 1.0$ mm; and $e = 1$ mm

For $\Omega < 191.6$ rad/s, no motions intersect the discontinuity (or pure cutting occurs, no interruptions). The lower boundary of the eccentricity frequency Ω range exhibits complex motions which are presented in a detail view in Fig. 5.8. The switching phase is observed to have a quite dense area of switching points which originate from the more simplified motion.

Since the eccentricity frequency Ω continues to change, the interruptions of the motions become more frequent. Hence, the motions are complicated due to the interaction with *chip seizure*. This *qualitative* observation has also been observed in

Fig. 5.8 (continued)

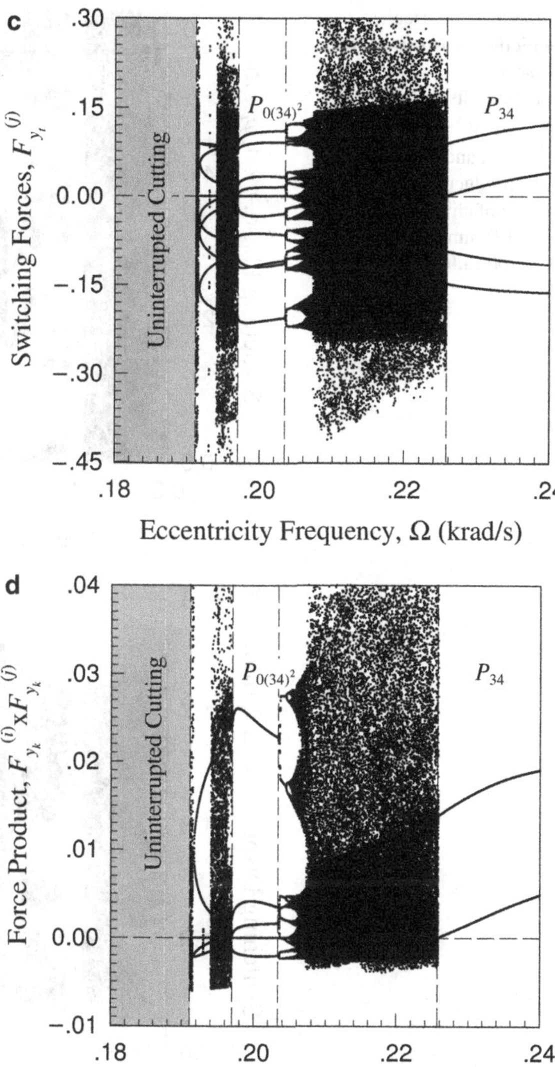

theory and experiment in Astakhov et al. (1997). Such chip seizure ($\dot{\tilde{y}} \equiv V$) is validated to occur at $\Omega \approx 225.6$ rad/s where complexity in the motion structure appears as shown in Fig. 5.8a, b. The forces and force product distributions show the onset of chip seizure ($\dot{\tilde{y}} \equiv V$) at this point in Fig. 5.8c, d. The addition of the chip seizure dynamics induces very complex motion which is inherently detrimental to the surface finish of the work-piece and wear of the tool-piece. The transients associated with the entry of the tool into the cutting process are critical in the amount of tool wear as in Chandrasekaran and Thoors (1994). The bifurcations observed in Fig. 5.7 produce transient effects which lead to such wear.

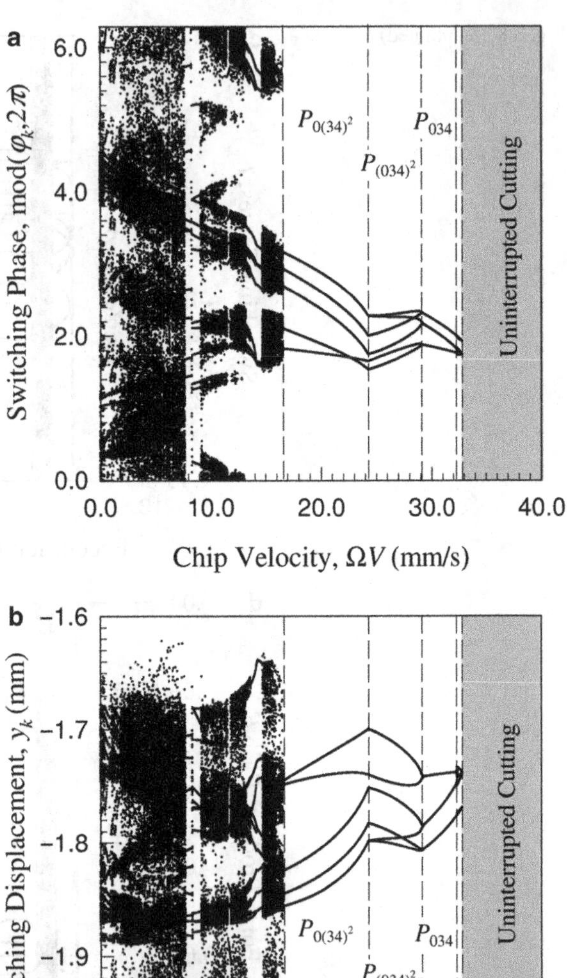

Fig. 5.9 Numerical prediction of (**a**) switching phase mod $(\varphi_k, 2\pi)$, (**b**) switching displacement $(y_k = \tilde{y})$, (**c**) switching forces $F_{y_k}^{(3)}, F_{y_k}^{(4)}$, and (**d**) switching force product $F_{y_k}^{(3)} \times F_{y_k}^{(4)}$ over a range of chip velocity (ΩV); $L_c = 1.0$ mm; and $\Omega = 200$ rad/s

(C) *Chip Velocity*. The numerical predictions of periodic chip seizure ($\dot{\tilde{y}} \equiv V$) and cutting motions are presented over the range of chip velocity $\bar{V} = \Omega V \in [0, 32.82]$ mm/s in Fig. 5.9. The eccentricity amplitude e is set to 0.1 mm for numerical predictions with $A\Omega = m_e\Omega^2/10$. The switching phase mod $(\varphi_k, 2\pi)$, and switching displacement y_k versus chip velocity ΩV are given in Fig. 5.9a and b, respectively. The switching forces $F_{\tilde{y}_k}^{(3)}$ and $F_{\tilde{y}_k}^{(4)}$ and switching

Fig. 5.9 (continued)

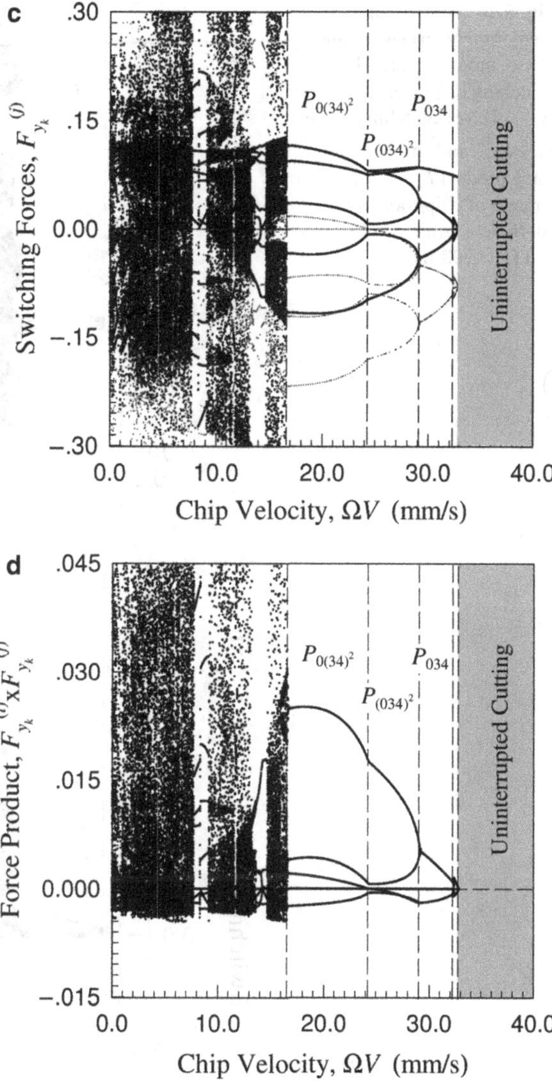

force products $(F_{\tilde{y}_k}^{(3)} \times F_{\tilde{y}_k}^{(4)})$ versus excitation frequency ΩV are presented in Fig. 5.9c, d. The distribution of motions over the range of chip velocity ΩV is

$$\begin{aligned}
&P_{(0(34)^2)^2} : (16.66, 17.08] \text{ mm/s}, \quad P_{0(34)^2} : (17.08, 24.26] \text{ mm/s}, \\
&P_{(034)^2} : (24.26, 29.14] \text{ mm/s}, \quad P_{034} : (29.14, 32.28] \text{ mm/s}, \\
&P_{(0(34)^m)^n} : (32.28, 32.82] \text{ mm/s}.
\end{aligned} \tag{5.88}$$

There are no interruptions of the motion for $\Omega V > 32.82$ mm/s, and such motions are not studied. Observe the chip seizure bifurcation at $\Omega V \approx 24.26$ mm/s, and the

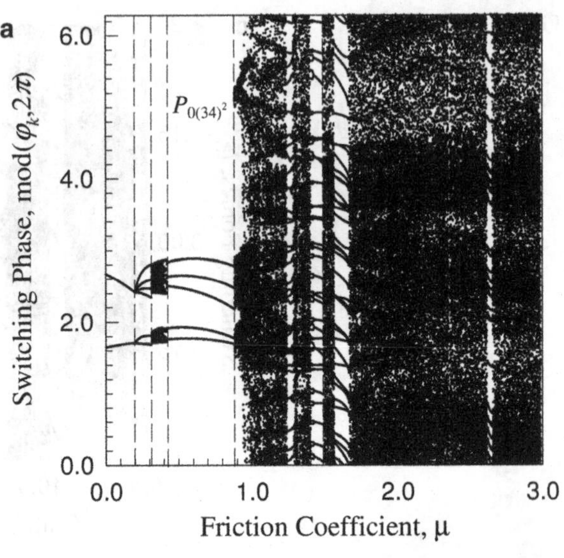

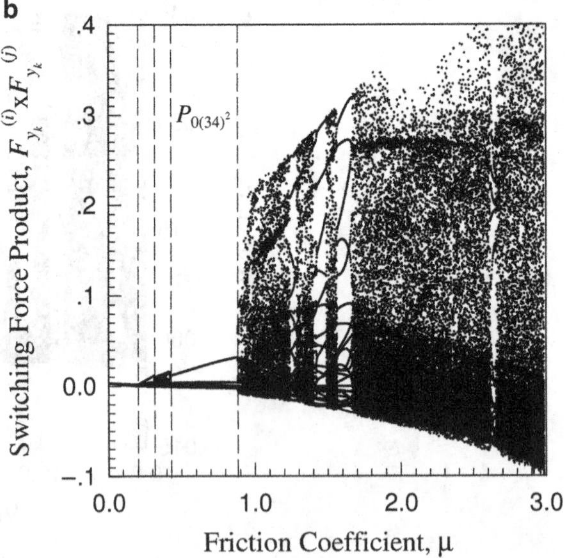

Fig. 5.10 Numerical prediction of (**a**) switching phase mod $(\varphi_k, 2\pi)$, (**b**) switching displacement $(y_k = \tilde{y})$, (**c**) switching forces $F_{y_k}^{(3)}, F_{y_k}^{(4)}$, and (**d**) switching force product $F_{y_k}^{(3)} \times F_{y_k}^{(4)}$ over a range of chip/tool friction coefficient (μ); $L_c = 1.0$ mm; and $\Omega = 200$ rad/s

grazing (tangential or saddle-node) bifurcations at $\Omega V \approx 17.08, 29.14$ mm/s^2. The frequency of loss of contact/cutting is directly related to the cutting velocity in a machine tool in Chandiramani and Pothala (2006). The interruption frequency of the periodic motion is directly related to the eccentricity frequency Ω and chip velocity ΩV in this study and is verified herein. The tool life of cutting motions varies inversely with both the interruption frequency and cutting velocity in Chou and Evans (1999). The velocity range considered shows the complexity of motions which not only increase the cutting velocity and interruption frequency, but also create material build up on the rake surface (further increasing wear during shearing of this material).

Fig. 5.10 (continued)

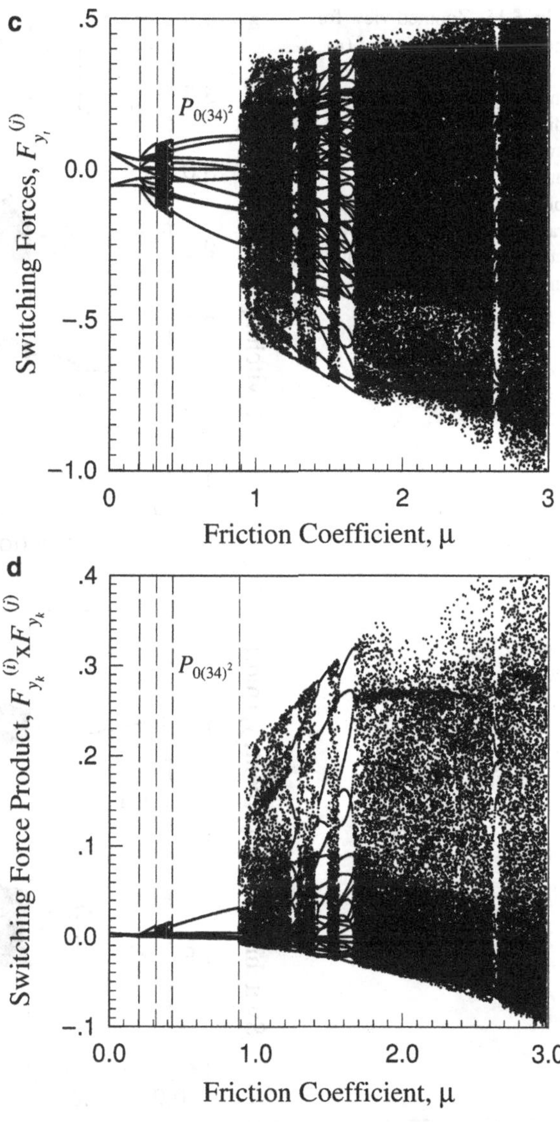

(D) *Friction Coefficient.* The numerical predictions of the periodic chip seizure and cutting motions are presented over the range of chip/tool friction coefficient $\mu \in [0, 3.0]$ in Fig. 5.10. The friction coefficient range associated with interrupted machining range from 0.33 to 2 in Chandrasekaran and Thoors(1994). The switching phase mod $(\varphi_k, 2\pi)$and switching displacement y_k versus chip stiffness k_2 are illustrated in Fig. 5.10a and b, respectively. The switching forces $F_{\tilde{y}_k}^{(3)}$ and $F_{\tilde{y}_k}^{(4)}$ and switching force products $(F_{\tilde{y}_k}^{(3)} \times F_{\tilde{y}_k}^{(4)})$ versus chip/tool friction coefficient μ are

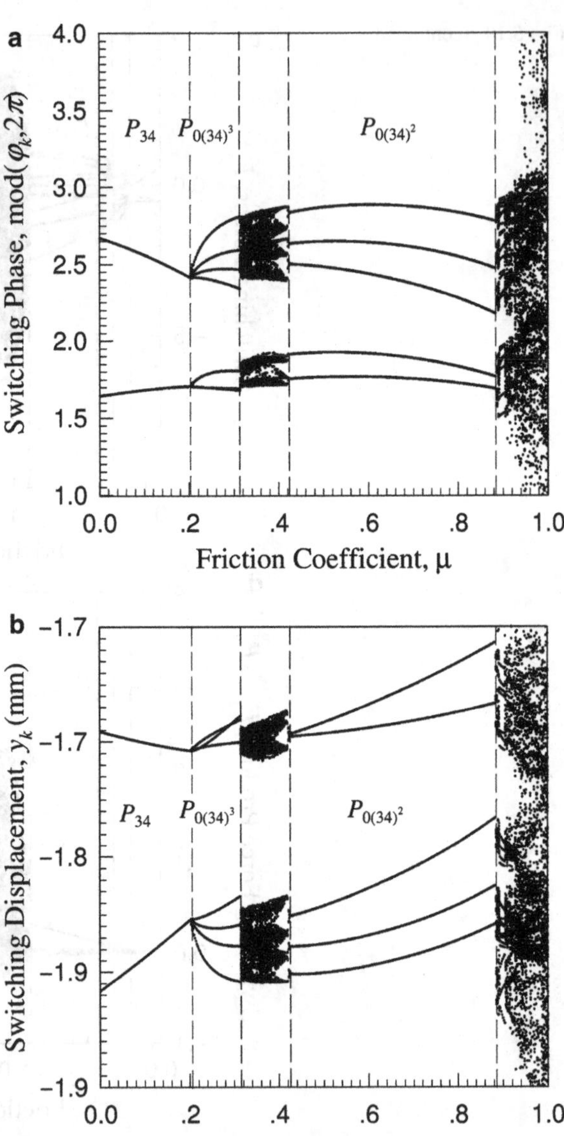

Fig. 5.11 Zoomed view for numerical prediction of (**a**) switching phase mod $(\varphi_k, 2\pi)$, (**b**) switching displacement $(y_k = \tilde{y})$, (**c**) switching forces $F_{y_k}^{(3)}, F_{y_k}^{(4)}$, and (**d**) switching force product $F_{y_k}^{(3)} \times F_{y_k}^{(4)}$ over a range of chip/tool friction coefficient μ; $L_c = 1.0$ mm; and $\Omega = 200$ rad/s

shown in Fig. 5.10c, b. The periodic motions of specific mapping structures are summarized with the certain ranges of friction coefficients as

$$\begin{aligned}
&P_{34} : (0.0000, 0.2100], \qquad P_{0(34)^3} : (0.2100, 0.3140],\\
&P_{(0(34)^m)^n} : (0.3140, 0.4380], \quad P_{0(34)^2} : (0.4380, 0.8840],\\
&P_{(0(34)^m)^n} : (0.8840, 0.300].
\end{aligned} \tag{5.89}$$

Fig. 5.11 (continued)

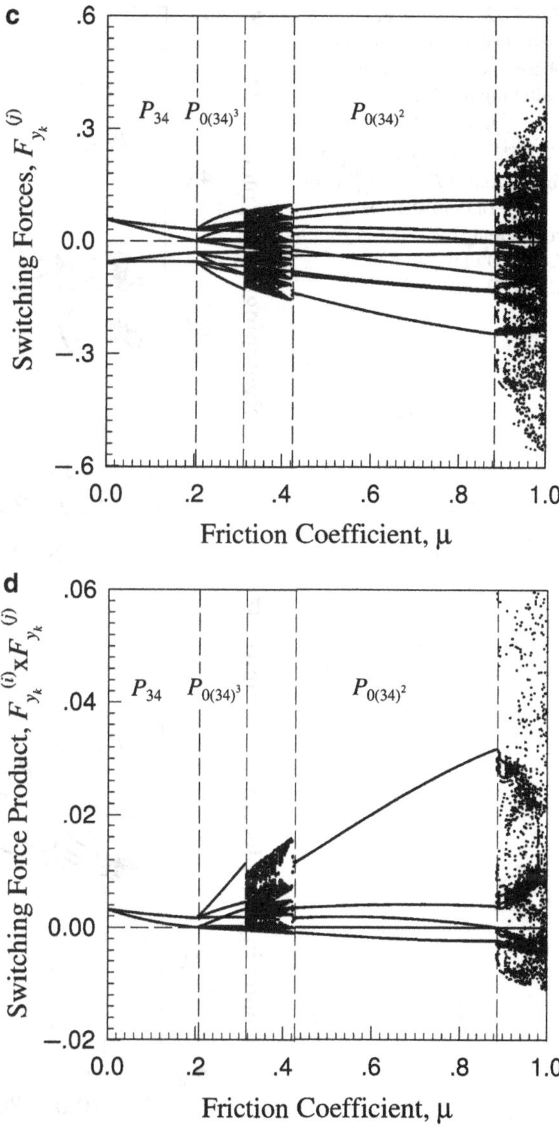

The lower range of frictional coefficient exhibits the transition from *semi-stable* interrupted cutting (no chip seizure) periodic motions to *chip seizure* periodic motions, as presented in Fig. 5.11. The switching phase is observed to have a dense area of switching point which originates from the more simplified motions. Hence, the motions are complicated due to the introduction of *chip seizure*. There are two chip seizure bifurcations well defined at $\mu = 0.210, 0.884$ and the grazing bifurcation at $\mu = 0.314$. Hence, the lower boundary for the semistable periodic

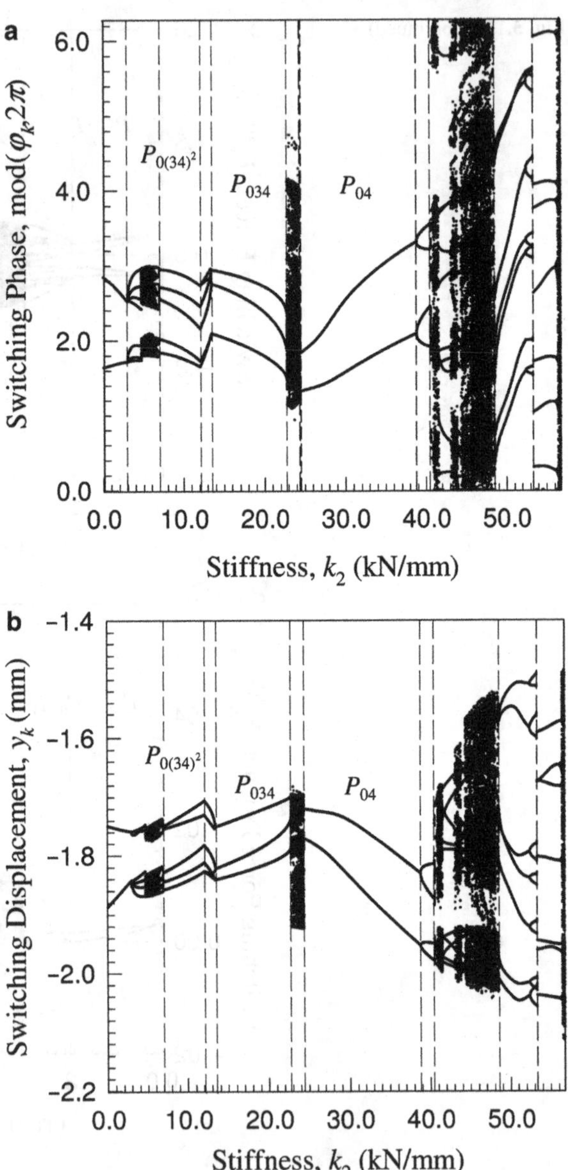

Fig. 5.12 Numerical prediction of (**a**) switching phase mod $(\varphi_k, 2\pi)$, (**b**) switching displacement $(y_k = \tilde{y})$, (**c**) switching forces $F_{y_k}^{(3)}, F_{y_k}^{(4)}$, and (**d**) switching force product $F_{y_k}^{(3)} \times F_{y_k}^{(4)}$ over a range of chip stiffness coefficient (k_2); $L_c = 1.0$ mm ; and $\Omega = 200$ rad/s

motion is at $\mu = 0.210$ for this machine-tool system. The friction coefficient on the chip/tool rake surface governs the stick–slip boundary for a machine tool in Maity and Das (2001). A phenomenon termed chatter is well known in manufacturing process and is in part a result of dry friction due to the velocity-dependent nature in Vela-Martinez et al. (2008).

Fig. 5.12 (continued)

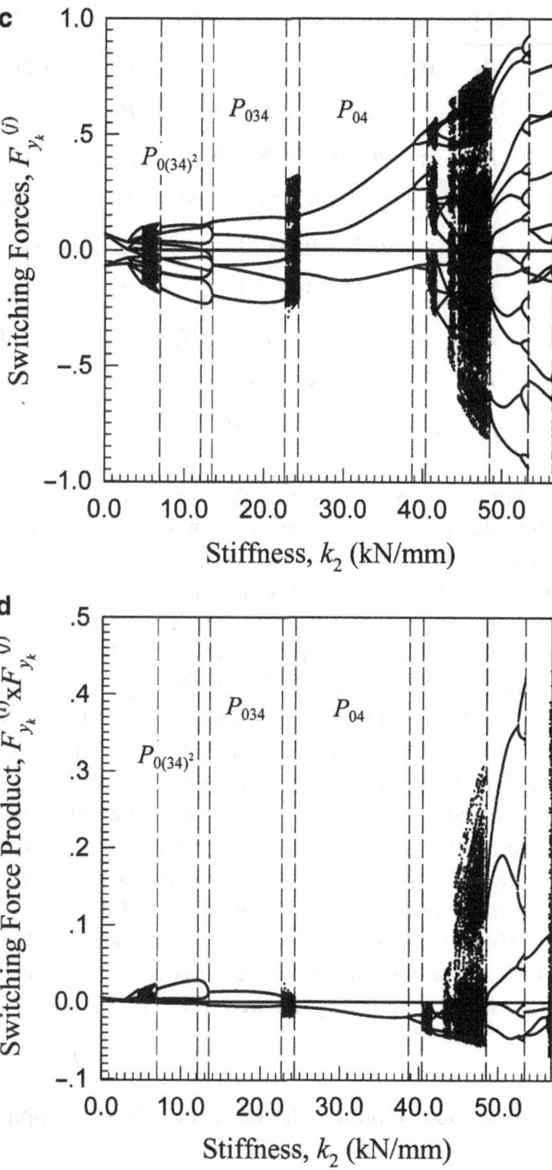

(E) *Chip Stiffness.* The numerical predictions of the periodic chip seizure and cutting motions are presented over the range of chip stiffness $k_2 \in [191.6, 1000]$ kN/mm, see Fig. 5.12. The switching phase $\mod(\varphi_k, 2\pi)$ and switching displacement y_k are illustrated in Fig. 5.12a and b, respectively. The switching forces ($F_{\tilde{y}_k}^{(3)}$ and $F_{\tilde{y}_k}^{(4)}$) and switching force products ($F_{\tilde{y}_k}^{(3)} \times F_{\tilde{y}_k}^{(4)}$) versus chip stiffness k_2 are shown in Fig. 5.12c, d.

The periodic motion is summarized by the specific ranges of mapping as

$$\begin{aligned}
&P_{34} : (0, 29.40] \text{ kN}, && P_{0(34)^3} : (29.40, 46.80] \text{ kN},\\
&P_{(0(34)^m)^n} : (46.80, 74.00] \text{ kN}, && P_{0(34)^2} : (74.00, 120.60] \text{ kN},\\
&P_{(034)^2} : (120.60, 133.20] \text{ kN}, && P_{034} : (133.20, 227.60] \text{ kN},\\
&P_{(0(34)^m)^n} : (227.60, 243.80] \text{ kN}, && P_{04} : (243.80, 386.20] \text{ kN},\\
&P_{(04)^2} : (386.20, 405.00] \text{ kN}, && P_{(04)^m} : (405.00, 415.40] \text{ kN},\\
&P_{(04)^5} : (415.40, 430.00] \text{ kN}, && P_{(04)^6} : (430.00, 431.20] \text{ kN},\\
&P_{(0(34)^m)^n} : (431.20, 486.60] \text{ kN}, && P_{(340)^2 4} : (486.60, 523.80] \text{ kN},\\
&P_{(4034043)^2} : (523.80, 533.40] \text{ kN}, && P_{0(40)^2 34} : (533.40, 526.00] \text{ kN},\\
&P_{(0(34)^m)^n} : (561.00, 567.60] \text{ kN}.
\end{aligned} \tag{5.90}$$

For $k_2 > 567.60$ kN/mm, the motion is not studied. The lower boundary of the chip stiffness range exhibits complex motions which are presented in a detail view in Fig. 5.13. The motion is quite simple until a chip seizure bifurcation occurs at $k_2 = 29.40, 120.60$ kN/mm and induces complicated motions. Hence, the motions are complicated due to the interaction with *chip seizure*. The forces and force product distributions which show the onset of chip seizure at this point are given in Fig. 5.13c and d, respectively. The grazing bifurcations occur at $k_2 = 46.80, 74.00,\ 133.20$ kN/mm which further attributes to the complexity of the motions. As a result of varying the chip stiffness, the natural frequencies of the machine-tool system vary and may move toward one or more of the exciting frequencies; hence, the system may experience more near interruption (grazing bifurcations) possibilities due to the added energy. The excitation amplitude and frequency in discontinuous systems widely affect the appearance and disappearance of grazing bifurcations in Gegg et al. (2008).

(F) *Parameter Map*. Consider a parameter (e, Ω) as an example to investigate cutting dynamics in machining process. Consider the following periodic motion to discuss the criteria as presented in Chap. 2.

$$P_{34} = P_3 \circ P_4. \tag{5.91}$$

The periodic motion relative to the above mapping structure implies that two switching points exist. For the two switching points, the corresponding switching force products are

$$\text{FP}^{(1)} = F_{\tilde{y}}^{(3,1)} \times F_{\tilde{y}}^{(4,1)} \text{ and } \text{FP}^{(2)} = F_{\tilde{y}}^{(3,2)} \times F_{\tilde{y}}^{(4,2)}. \tag{5.92}$$

The force components in the force product for this particular case are defined by domains Ω_α $(\alpha = 3, 4)$. The boundary of the two domains is the chip/tool friction boundary. Similar to Chap. 3, a zero force product indicates that the current motion will switch a new motion with a different mapping structure. Thus, the minimum of

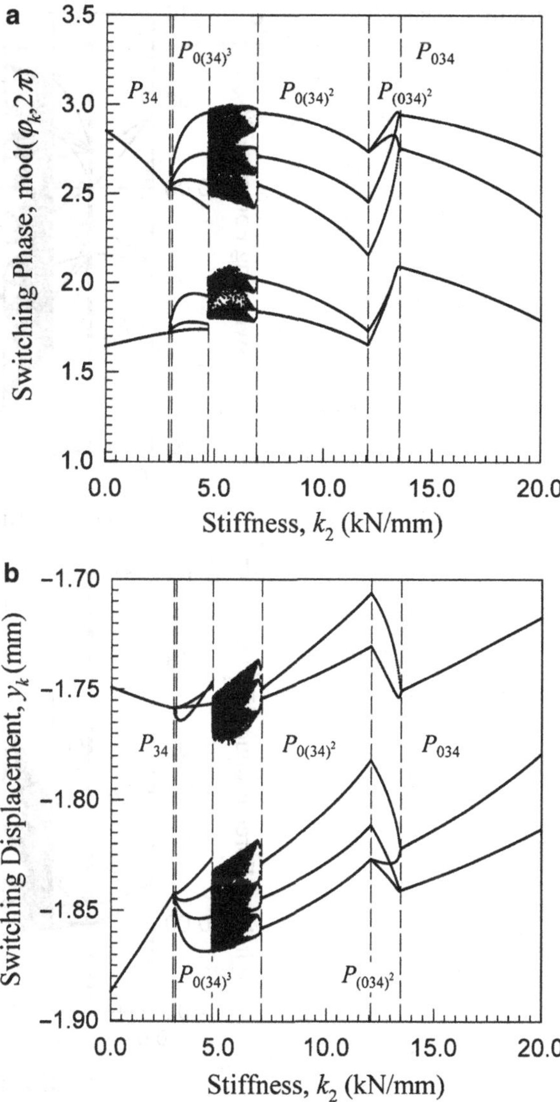

Fig. 5.13 Zoomed view for numerical prediction of (**a**) switching phase mod $(\varphi_k, 2\pi)$, (**b**) switching displacement $(y_k = \tilde{y})$, (**c**) switching forces $F_{y_k}^{(3)}, F_{y_k}^{(4)}$, and (**d**) switching force product $F_{y_k}^{(3)} \times F_{y_k}^{(4)}$ over a range of chip stiffness coefficient (k_2); $L_c = 1$ mm; and $\Omega = 200$ rad/s

the switching force products is presented as for the contour and three-dimensional plots. In general, the force components are

$$\mathrm{FP}_{\min} = \min_k \{\mathrm{FP}^{(k)}\} = \min_k \{F_{\tilde{y}}^{(i,k)} \times F_{\tilde{y}}^{(j,k)}\}, \tag{5.93}$$

where superscript index i and j are domains of the chip/tool friction boundary and the superscript index k is the kth switching force product for the steady-state motion of the machine-tool system. Consider the steady-state motion of a machine tool to

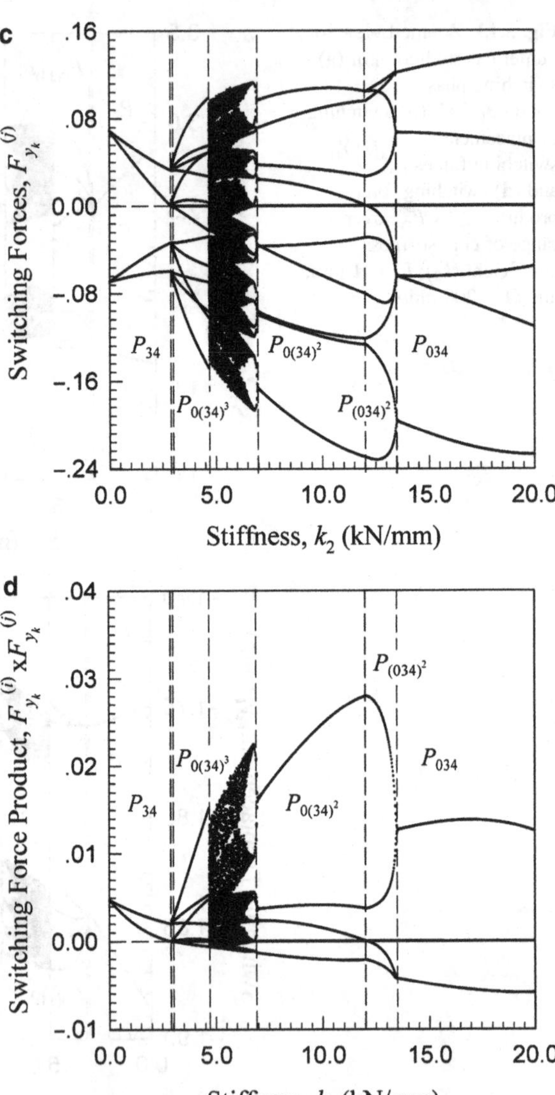

Fig. 5.13 (continued)

determine a parameter map (e, Ω). The parameters of most traditional reference in a parameter map are the frequency and amplitude. The related parameters in this study are the eccentricity frequency Ω and amplitude e. The dynamical system parameters for the following results are given in (5.83) and (5.84), and the corresponding initial contact conditions for the maximum frequency $\Omega = 600$ rad/s are

$$\begin{aligned} x_1^* &= 0.3941 \text{ mm}, \quad y_1^* = -4.4638 \text{ mm}, \\ x_2^* &= 0.0381 \text{ mm}, \quad y_2^* = -2.9276 \text{ mm}. \end{aligned} \tag{5.94}$$

The initial conditions for simulations are

$$\begin{aligned} x_k &= 0.5557 \text{ mm}, \quad y_k = -2.8190 \text{ mm}, \\ \dot{x}_k &= -24.2444 \text{ mm}, \quad \dot{y}_k = -4.0398 \text{ mm}. \end{aligned} \tag{5.95}$$

Once the motion becomes the steady state or a maximum number of periodic revolutions, the switching measures are recorded. The eccentricity amplitude is decreased and again excitation frequency starts at $\Omega = 600$ rad/s with the initial conditions of (5.95) are selected. The minimum force product in (5.93) is presented in the form of a contour plot in Fig. 5.14a. The color variation is determined on a logarithmic scale. The three-dimensional view of a mesh plot is shown in Fig. 5.14b where the MFP (minimum force product) is shown and it varies with eccentricity amplitude e and frequency Ω, respectively. Hence, in Fig. 5.14a, the largest MFP in

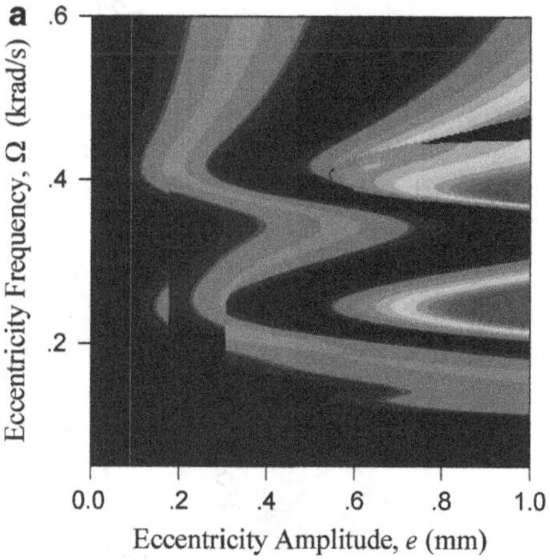

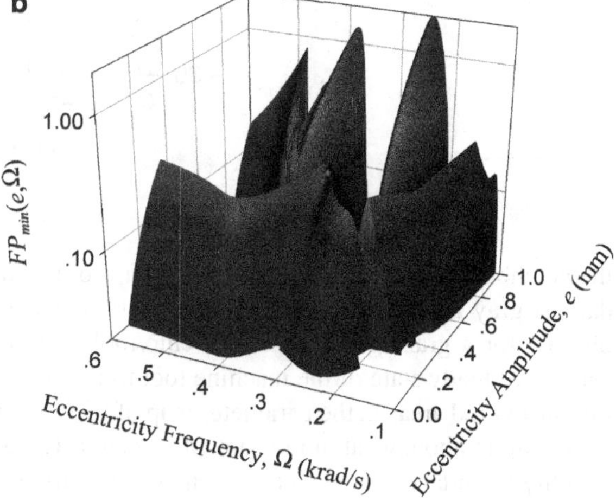

Fig. 5.14 Minimum force product study for a machine tool undergoing steady-state motion with eccentricity amplitude e versus eccentricity frequency Ω, (**a**) Contour$[e, \Omega] = \min(FP^{(k)})$, and (**b**) $\min(FP^{(k)})$ vs.e vs.Ω for $L_c = 1$(mm)

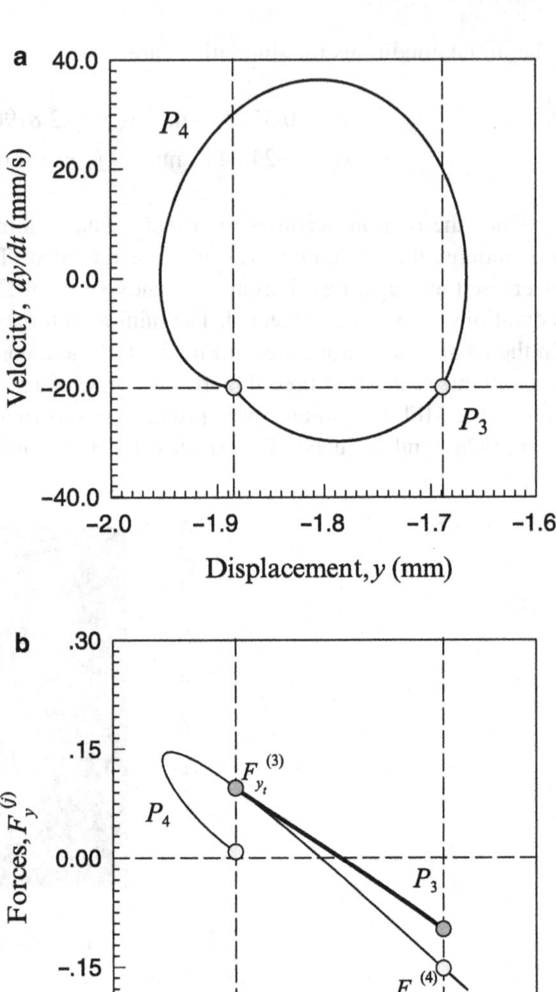

Fig. 5.15 Nonstick periodic motion (P_{34}): (**a**) phase trajectory in phase plane $(y, \dot{y})$, (**b**) forces ($F_y^{(3)}$ and $F_y^{(4)}$) versus displacement y, (**c**) forces ($F_{y_t}^{(3)}$ and $F_{y_t}^{(4)}$) versus velocity $\dot{y}$, (**d**) phase trajectory in phase plane $(x, \dot{x})$, (**e**) time history of velocity $\dot{y}$, and (**f**) time history of $F_y^{(3)}$ and $F_y^{(4)}$. $L_c = 1.0$ mm and $\Omega = 228$ rad/s

the neighborhood of natural frequency groups of this machine tool is observed. The darkest gray areas are for chip seizure motion. The discontinuities in the contour plot are for a grazing motion to the chip/tool friction boundary. This phenomena cause the steady state of the machine tool to jump to a new orbit with a zero MFP. For uncovered areas in the parameter map of (e, Ω), chip seizure may occur, but this says nothing about what motion actually occurs. The blue boundary is similar to the stability boundary for regenerative chatter in cutting dynamics in existing results.

Fig. 5.15 (continued)

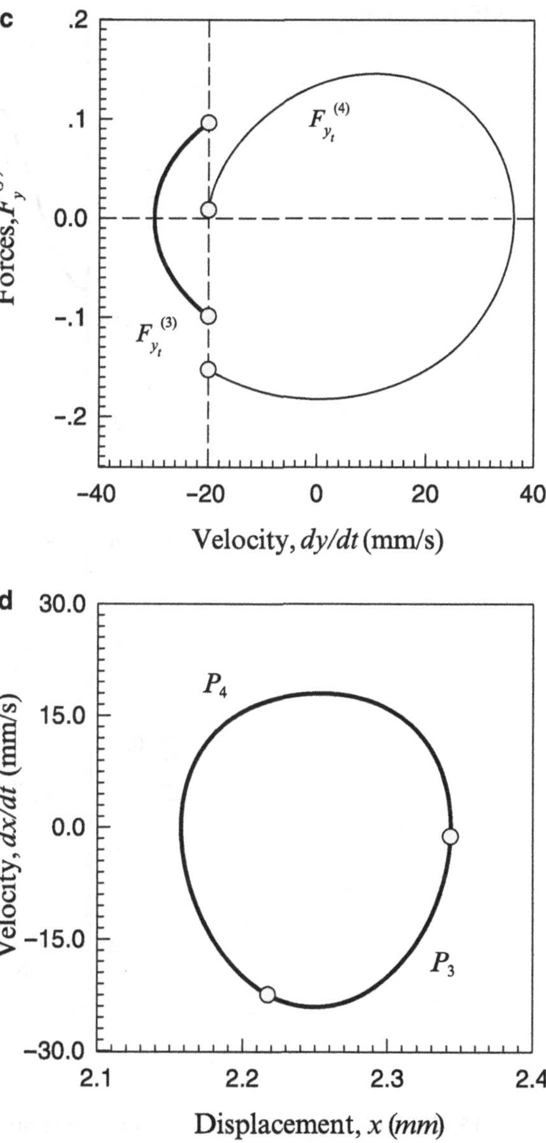

(G) *Numerical Illustrations*. The numerical simulation of the periodic chip seizure and cutting motion for this machine-tool system subject to an eccentricity force is presented in Figs. 5.15 and 5.16. The dynamical system parameters are given in (5.83) and the external force and geometry parameters are given in (5.84).

The initial conditions and input data for numerical illustrations are tabulated in Table 5.3. A regular cutting periodic motion for $\Omega = 228$ rad/s is presented in

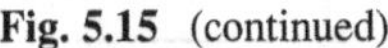

Fig. 5.15 (continued)

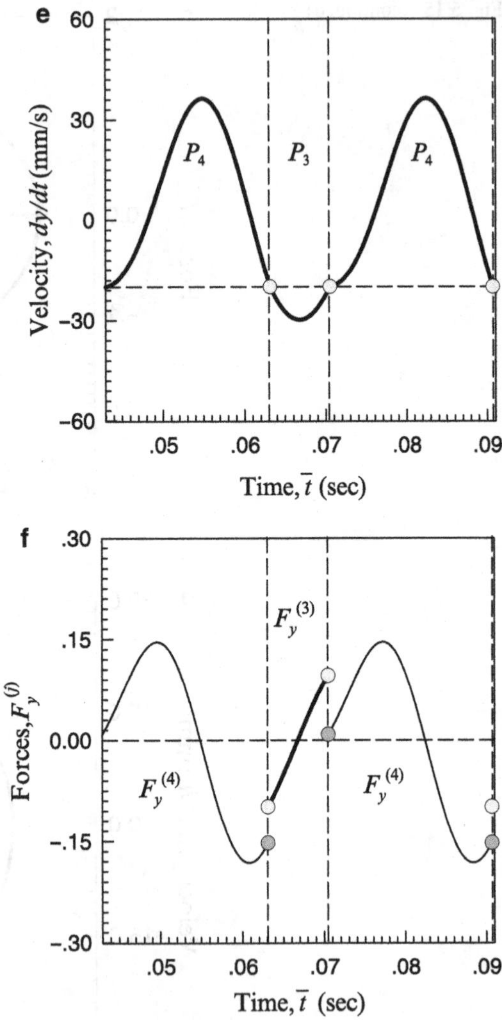

Fig. 5.15. The trajectory of the periodic motion relative to the mapping structure P_{34} is plotted in Fig. 5.15a. The switching points are denoted by circular symbols. The motion in domain Ω_3 is labeled by mapping P_3. Following intersection of the frictional boundary, the motion then moves into domain Ω_4 (labeled by mapping P_4). To verify the switchability of motion on the boundary, the forces $F_y^{(3)}$ and $F_y^{(4)}$ versus displacement y and velocity $\dot{y}$ are presented in Fig. 5.15b and c, respectively. The phase trajectory in phase plane $(x, \dot{x})$ is given in Fig. 5.15d. The time history of velocity $\dot{y}$ is presented in Fig. 5.15e because the switching occurs at $y = V$. For the observation of force conditions on the boundary, the time histories of $F_y^{(3)}$ and $F_y^{(4)}$ are presented in Fig. 5.15f.

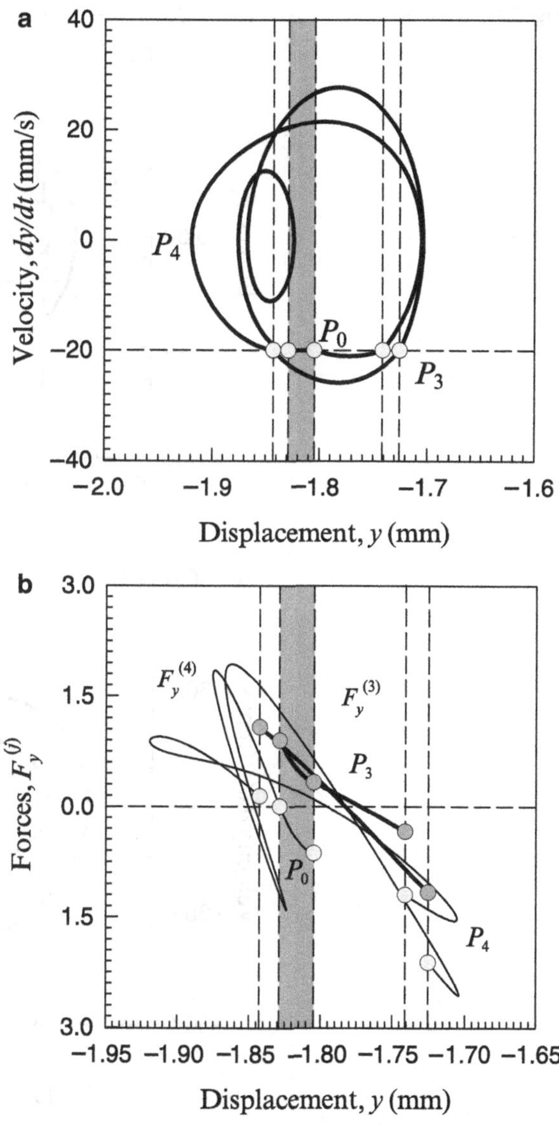

Fig. 5.16 Nonstick periodic motion ($P_{0(34)^2}$): (**a**) phase trajectory in phase plane $(y,\dot{y})$, (**b**) forces ($F_y^{(3)}$ and $F_y^{(4)}$) versus displacement y, (**c**) forces ($F_{y_t}^{(3)}$ and $F_{y_t}^{(4)}$) versus velocity $\dot{y}$, (**d**) phase trajectory in phase plane $(x,\dot{x})$, (**e**) time history of velocity $\dot{y}$ and (**f**) time history of $F_y^{(3)}$ and $F_y^{(4)}$. $L_c = 1.0$ mm and $\Omega = 200$ rad/s

An interrupted cutting periodic motion is illustrated in Fig. 5.16 for $\Omega = 200$ rad/s . In Fig. 5.16a, the trajectory of the periodic motion relative to the mapping structure $P_{0(34)^2}$ is illustrated. The motion in domain Ω_4 and Ω_3 is one portion of the total response which is followed by another motion in domain Ω_4 and Ω_3, then the chip seizure is labeled by mapping P_0 to complete the period-1 motion for three excitation periods. The motion switchability on the boundary can be given by force conditions. Thus, the distributions of force $F_{\tilde{y}}^{(3)}$ and $F_{\tilde{y}}^{(4)}$ along

Fig. 5.16 (continued)

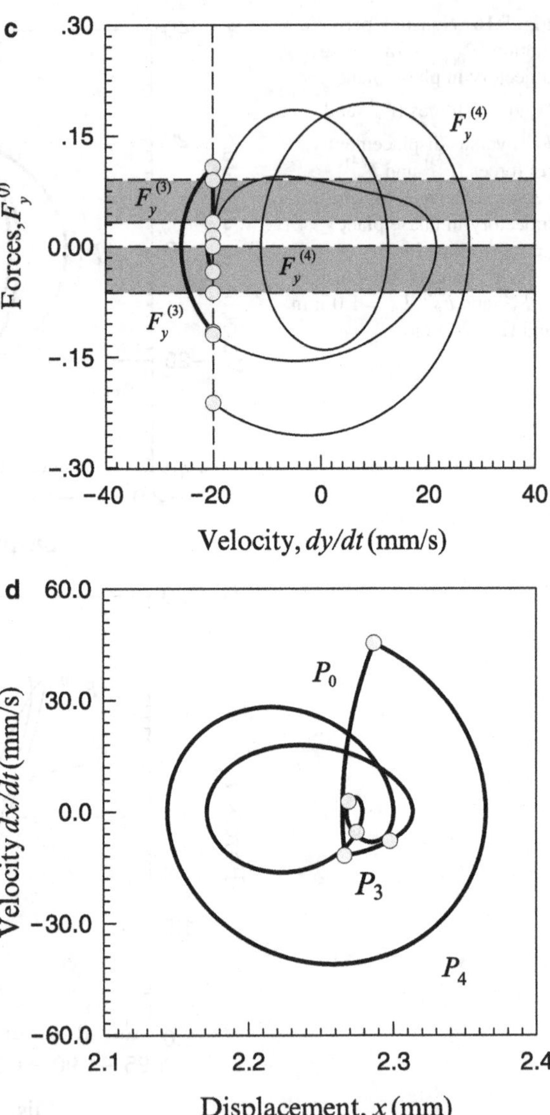

displacement y and velocity $\dot{y}$ are presented in Fig. 5.16b, c. The phase trajectory in phase plane $(x, \dot{x})$ is given in Fig. 5.16d. The time history of velocity $\dot{y}$ is presented in Fig. 5.16e. The time history of the forces is plotted in Fig. 5.16f. In Fig. 5.16b, f, the P_0 vanishes with $F_{\dot{y}}^{(4)} = F_y^{(4)} = 0$. The differences between the orbits of Figs. 5.15a and 5.16a depend on the period of the orbit and the existence of the chip seizure. As in Fig. 5.15a, the additional loop in the center of the phase will move toward the frictional boundary with a variation of the excitation frequency Ω.

Fig. 5.16 (continued)

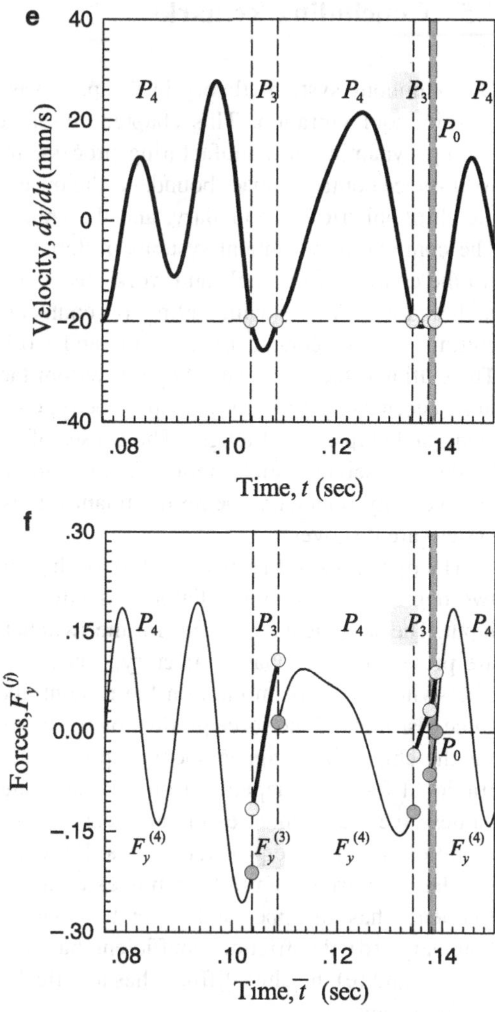

Table 5.3 Initial conditions and input data for numerical illustrations

Ω	L_c(mm)	x_k	y_k	$\dot{x}_k$	$\dot{y}_k$	Ωt_k	Mapping	Figure
228	1	2.2174	−1.8845	−2.2468	−20	3.5086	P_{34}	5.15
200	1	2.2751	−1.8427	−5.5565	−20	2.8721	$P_{0(34)^2}$	5.16

With decreasing excitation frequency Ω, the orbit associated with chip seizure induces higher-order periodic motions. From mapping structures, the analytical prediction can be completed as in Chap. 3. If readers are interested in analytical prediction, the reader can refer to Gegg et al. (2010a, b, c).

5.5 Concluding Remarks

Discontinuous systems theory in Chap. 2 was applied for cutting dynamics in a machine-tool vibration. This chapter presented an alternative way to look into cutting dynamics in manufacturing process. In machine-tool vibration, the tool-/work-piece contact/impact boundary, the onset/disappearance of cutting boundary, the chip/tool friction boundary, and the chip vanishing boundary are introduced. The continuous dynamical systems in domains include: (a) the tool-free running, (b) the contact of the tool- and work-pieces without cutting, (c) the contact of the tool- and work-pieces with cutting, (d) contact of the tool- and work-pieces without cutting, and (e) contact of the tool- and work-pieces with chip seizure motion. The cutting effects are caused by six system factors: eccentricity/excitation amplitude, eccentricity frequency, chip velocity, chip stiffness, chip/tool friction coefficient, and chip contact length. The causes of interruptions of the periodic motions by the chip seizure (stick motion) and grazing bifurcations of the frictional boundary (velocity boundary) being dominant routes to unstable motions in machining system are discovered.

The dynamics of the machine tool in the cutting process is described through a two-degree-of-freedom oscillator with a discontinuity, subject to a periodical force input. The analytical conditions for the switchability of motion on the discontinuity are presented. The phase trajectory, velocity, and force responses are presented. The switchability of motion on the discontinuous boundary is illustrated through force distribution and force product on the boundary. The parameter boundaries are for the chip seizure appearance/disappearance. Interruptions due to the chip/tool frictional force are dependent on critical values of the parameters. These critical values give the boundaries in parameter space for the interrupted cutting periodic motion through force product criteria. For chip seizure appearance/disappearance, (a) the eccentricity amplitude possesses a lower boundary; (b) the eccentricity frequency has an upper and lower boundary; (c) the chip velocity has an upper boundary; (d) the friction coefficient has apparent boundary noting semistable motion; and (e) the chip stiffness has a critical value dependent upon the eccentricity frequency.

The stick–slip combination is forced to purely slip (nonstick cutting) motion by appropriate eccentricity amplitudes. For small eccentricity amplitude, there are complex motions (e.g., pseudo-periodic/chaotic motion). Such motion complexity is attributed to the susceptibility of reduced operating contact forces. The chip/tool friction boundary has significant effects. The chip seizure motion has a negative force product along the boundary. In switching phase, a dense area of switching points originated from the onset of *chip seizure*.

For the eccentricity frequency, the appearance of chip seizure motion causes the chaotic/unstable motion which is not the route to chaotic/unstable motion. The onset of the complex motion is caused by chip seizure appearance. For reduction of chip velocity, the contact switching forces oppose each other, which implies the increased complexity at lower values. If the relative velocity of the chip and tool rake motion is high, the appearance of chip seizure can be avoided. In addition, the frictional coefficient must remain low. The significant difference between the opposing vector fields at the chip/tool friction boundary is the force due to friction rather than stiffness and damping parameters. Hence, if the chip/tool friction coefficient becomes larger, the chip seizure motion will be more apparent. With varying chip stiffness, the natural frequencies of the machine-tool system may change and may move toward one or more of the exciting frequencies. Thus, the system experienced more near interruption (grazing bifurcations) possibilities due to the pumped energy.

The parameter boundary for chip seizure appearance changes with frequency and amplitude of excitation. Reduction of the energy input from the natural characteristics of the system affects the contact forces and tool velocities. Hence, the chip seizure can be avoided if these boundaries are noted and effective manipulation or control is completed. The chip velocity directly controls interruption such a boundary as the chip/tool friction boundary will have on the motion. The critical chip velocity can be determined for a pure cutting parameter boundary with the chip/tool friction interaction, and such interaction is dependent on the natural frequency characteristics of the machine-tool system. With increasing stiffness coefficient, the potential stick–slip interruptions are sensitive to chip resistance and excitation frequency. For high excitation frequencies at or above the highest natural frequency, the chip seizure phenomenon is more likely to occur due to the grazing bifurcations of the chip/tool friction boundary. This can be attributed to the increase of chip resistance which prevents motion in the direction of the chip shearing. Hence, during interaction, the friction force decreases with the chip/tool friction boundary.

The interruptions due to the chip/tool friction and chip vanishing boundaries had no adverse effects on the machine-tool system, which are similar to the stability boundary in Gurney and Tobias (1962). The grazing bifurcation of the chip/tool friction boundary will appear since the amplitude of the motion decreases with eccentricity frequency. The duration of the noncutting phase of interrupted cutting periodic motion decreases with decreasing chip velocity or increasing relative velocity between the chip and tool-piece. When the contact length becomes longer than the width of the displacement orbit; the variation of the chip contact length will not have a profound effect on the motion.

References

Astakhov, V.P., Shvets, S.V. and Osman, M.O.M., 1997, Chip structure classification based on mechanics of its formation, *Journal of Materials Processing Technology*, **71**, pp. 247–257.

Chandiramani, N.K. and Pothala, T., 2006, Dynamics of 2-dof regenerative chatter during turning, *Journal of Sound and Vibration*, **290**, pp. 488–464.

Chandrasekaran, H. and Thoors, H., 1994, Tribology in interrupted machining: role of interruption cycle and work material, *Wear*, **179**, pp. 83–88.

Chou, Y.K. and Evans, C.J., 1999, Cubic boron nitride tool wear in interrupted hard cutting, *Wear*, 225–229, pp. 234–245.

Devaney, R.L., 2003, *An Introduction to Chaotic Dynamical Systems*, 2nd Ed., Westview Press: Advanced Book Program, Boulder.

Gegg, B.C., Suh, C.S. and Luo, A.C.J., 2007, "Periodic motions of the machine tools in cutting process," *Proceedings of the ASME International Design Engineering and Technical Conference*, DETC2007/VIB-35166, Las Vegas, Nevada, September 4th–7th.

Gegg, B.C., Suh, C.S. and Luo, A.C.J., 2008, "Chip stick and slip periodic motions of a machine tool in the cutting process," *Proceedings of the ASME International Manufacturing Science and Engineering Conference*, MSEC ICMP2008/DYN-72052, October 7th–10th.

Gegg, B.C., Suh, S.C.S. and Luo, A.C.J., 2010a, Analytical prediction of interrupted cutting periodic motion in a machine toll with a friction boundary, Nonlinear Science and Complexity (Eds, J.A. Tenreiro Machado, A.C.J. Luo, R.S. Barbosa, M.F. Silva, and L.B. Figueiredo, Springer: Dordrecht), 26–36.

Gegg, B.C., Suh, S.C.S. and Luo, A.C.J., 2010b, Modeling and theory of intermittent motions in a machine toll with a friction boundary, ASME *Journal of Manufacturing Science and Engineering*, **132**, 041001(1–9)

Gegg, B.C., Suh, S.C.S. and Luo, A.C.J., 2010c, A parameter study of the eccentricity frequency and amplitude and chip length effects on a machine tool with multiple boundaries, *Communications in Nonlinear Science and Numerical Simulation*, **15**, pp. 2575–2602.

Gurney, J.P. and Tobias, S.A., 1962, A graphical analysis of regenerative machine tool instability, Transaction ASME Journal of Engineering Industry, **84**, pp. 103–112.

Luo, A.C.J., 2006, *Singularity and Dynamics on Discontinuous Vector Fields*, Elsevier: Amsterdam.

Luo, A.C.J., 2011, *Discontinuous Dynamical Systems*, HEP-Springer: Heidelberg.

Luo, A.C.J. and Mao, T.T., 2010, Analytical conditions for motion switchability in a 2-DOF friction-induced oscillator moving on two constant speed belts, *Canadian Applied Mathematics Quarterly*, **17**, pp. 201–242.

Maity, K.P. and Das, N.S., 2001, A class of slipline field solutions for metal machining with slipping and sticking contact at the chip-tool interface, *International Journal of Mechanical Science*, **43**, pp. 2435–2452.

Vela-Martinez, L., Jauregui-Correa, J.C., Rubio-Cerda, E., Herrera-Ruiz, G. and Lozano-Guzman, A., 2008, Analysis of compliance between the cutting tool and the work piece on the stability of a turning process, *International Journal of Machine Tools & Manufacture*, 48, pp. 1054–1062.

ERRATUM TO:

Chapter 2
Discontinuous System Theory

Brandon C. Gegg, C. Steve Suh, and Albert C. J. Luo

B.C. Gegg et al., *Machine Tool Vibrations and Cutting Dynamics*,
DOI 10.1007/978-1-4419-9801-9, © Springer Science+Business Media, LLC 2011

DOI 10.1007/978-1-4419-9801-9_6

In Chapter 2, page 42, equation 2.71 is incorrectly displayed as:

$$\mathbf{n}^{\mathrm{T}}_{\partial\Omega_{ij}} \cdot \mathbf{F}^{(i)}(t_{m-}) \neq 0 \text{ and } \alpha = i, j. \tag{2.71}$$

The correct equation is:

$$\mathbf{n}^{\mathrm{T}}_{\partial\Omega_{ij}} \cdot \mathbf{F}^{(i)}(t_{m-}) \neq 0 \text{ and } L_{jj}(\mathbf{x}_{m\pm\varepsilon}, t_{m\pm\varepsilon}, \mathbf{p}_j) < 0. \tag{2.71}$$

The original online version for this chapter can be found at
http://dx.doi.org/10.1007/978-1-4419-9801-9_2

Appendix

In this Appendix, the closed-form solutions for dynamical systems with one degree of freedom and two degrees of freedom are listed.

A.1 Basic Solution

Consider a general solution for equation of motion for $(j = 1, 2, \ldots)$

$$\ddot{x} + 2d_j\dot{x} + c_jx = a_j \cos \Omega t + b_j. \tag{A.1}$$

With an initial condition $(t, x^{(j)}, \dot{x}^{(j)}) = (t_i, x_i^{(j)}, \dot{x}_i^{(j)})$, the general solution of (A.1) are given as follows:

Case I. For $d_j^2 - c_j > 0$

$$\left.\begin{aligned} x &= \mathrm{e}^{-d_j(t-t_i)}(C_1^{(j)}\mathrm{e}^{\lambda_d^{(j)}(t-t_i)} + C_2^{(j)}\mathrm{e}^{-\lambda_d^{(j)}(t-t_i)}) \\ &\quad + D_1^{(j)} \cos \Omega t + D_2^{(j)} \sin \Omega t + D_0^{(j)}, \\ \dot{x} &= \mathrm{e}^{-d_j(t-t_i)}[(\lambda_d^{(j)} - d_j)C_1^{(j)}\mathrm{e}^{\lambda_d(t-t_i)} - (\lambda_d^{(j)} + d_j)C_2^{(j)}\mathrm{e}^{-\lambda_d^{(j)}(t-t_i)}] \\ &\quad - D_1^{(j)}\Omega \sin \Omega t + D_2^{(j)}\Omega \cos \Omega t, \end{aligned}\right\} \tag{A.2}$$

where

$$\begin{aligned} C_1^{(j)} &= \frac{1}{2\omega_d^{(j)}}\{\dot{x}_i - [D_1^{(j)}(d_j + \omega_d^{(j)}) + D_2^{(j)}\Omega] \cos \Omega t_i \\ &\quad + [D_1^{(j)}\Omega - D_2^{(j)}(d_j + \omega_d)] \sin \Omega t_i + (x_i - D_0^{(j)})(d_j + \omega_d^{(j)})\}, \end{aligned}$$

DOI 10.1007/978-1-4419-9801-9,

$$C_2^{(j)} = \frac{1}{2\omega_d^{(j)}}\{-\dot{x}_i - [D_1^{(j)}\Omega + D_2^{(j)}(-d_j + \omega_d^{(j)})]\sin\Omega t_i \\ - [D_1^{(j)}(-d_j + \omega_d^{(j)}) - D_2^{(j)}\Omega]\cos\Omega t_i + (x_i - D_0^{(j)})(-d_j + \omega_d^{(j)})\} \tag{A.3}$$

$$\left.\begin{aligned} D_0^{(j)} &= -\frac{b_j}{c_j}, \quad D_1^{(j)} = \frac{a_j(c_j - \Omega^2)}{(c_j - \Omega^2)^2 + (2d_j\Omega)^2}, \\ D_2^{(j)} &= \frac{a_j(2d_j\Omega)}{(c_j - \Omega^2)^2 + (2d_j\Omega)^2}, \quad \lambda_d^{(j)} = \sqrt{d_j^2 - c_j}. \end{aligned}\right\} \tag{A.4}$$

Case II. For $d_j^2 - c_j < 0$

$$\left.\begin{aligned} x &= \mathrm{e}^{-d_j(t-t_i)}[C_1^{(j)}\cos\omega_d^{(j)}(t-t_i) + C_2^{(j)}\sin\omega_d^{(j)}(t-t_i)] \\ &\quad + D_1^{(j)}\cos\Omega t + D_2^{(j)}\sin\Omega t + D_0^{(j)}, \\ \dot{x} &= \mathrm{e}^{-d_j(t-t_i)}[(C_2^{(j)}\omega_d^{(j)} - d_jC_1^{(j)})\cos\omega_d^{(j)}(t-t_i) \\ &\quad - (C_1^{(j)}\omega_d^{(j)} + d_jC_2^{(j)})\sin\omega_d^{(j)}(t-t_i)] \\ &\quad - D_1^{(j)}\Omega\sin\Omega t + D_2^{(j)}\Omega\cos\Omega t, \end{aligned}\right\} \tag{A.5}$$

where

$$\left.\begin{aligned} C_1^{(j)} &= x_i - D_1^{(j)}\cos\Omega t_i - D_2^{(j)}\sin\Omega t_i - D_0^{(j)}, \\ C_2^{(j)} &= \frac{1}{\omega_d^{(j)}}[d_j(x_i - D_1^{(j)}\cos\Omega t_i - D_2^{(j)}\sin\Omega t_i - D_0^{(j)}) \\ &\quad + \dot{x}_i + D_1^{(j)}\Omega\sin\Omega t_i - D_2^{(j)}\Omega\cos\Omega t_i)], \\ \omega_d^{(j)} &= \sqrt{c_j - d_j^2}. \end{aligned}\right\} \tag{A.6}$$

Case III. For $d_j^2 - c_j = 0$

$$\left.\begin{aligned} x &= \mathrm{e}^{-d_j(t-t_i)}[C_1^{(j)}(t-t_i) + C_2^{(j)}] + D_1^{(j)}\cos\Omega t + D_2^{(j)}\sin\Omega t + D_0^{(j)}, \\ \dot{x} &= \mathrm{e}^{-d_j(t-t_i)}[C_1^{(j)} - C_1^{(j)}d_j(t-t_i) - d_jC_2^{(j)}] \\ &\quad - D_1^{(j)}\Omega\sin\Omega t + D_2^{(j)}\Omega\cos\Omega t, \end{aligned}\right\} \tag{A.7}$$

where

$$\begin{aligned} C_1^{(j)} &= x_i + \cos\Omega t_i(D_2^{(j)}\Omega - d_jD_1^{(j)}) \\ &\quad - \sin\Omega t_i(D_1^{(j)}\Omega + d_jD_2^{(j)}) - d_jD_1^{(j)}, \\ C_2^{(j)} &= x_i - D_1^{(j)}\cos\Omega t_i - D_2^{(j)}\sin\Omega t_i - D_0^{(j)}. \end{aligned} \tag{A.8}$$

Case IV. For $d_j \neq 0,\ c_j = 0$

$$\left.\begin{aligned} x &= C_1^{(j)} e^{-2d_j(t-t_i)} + D_1 \cos \Omega t + D_2^{(j)} \sin \Omega t + D_0^{(j)} t + C_2^{(j)}, \\ \dot{x} &= -2d_j C_1^{(j)} e^{-2d_j(t-t_i)} - D_1^{(j)} \Omega \sin \Omega t + D_2^{(j)} \Omega \cos \Omega t + D_0^{(j)}, \end{aligned}\right\} \tag{A.9}$$

where

$$\left.\begin{aligned} C_1^{(j)} &= -\frac{1}{2d_j}(\dot{x}_i + D_1^{(j)} \Omega \sin \Omega t_i - D_2^{(j)} \Omega \cos \Omega t_i - D_0^{(j)}), \\ C_2^{(j)} &= \frac{1}{2d_j}[2d_j x_i + \dot{x}_i + (D_1^{(j)} \Omega - 2d_j D_2^{(j)}) \sin \Omega t_i \\ &\quad - (2d_j D_1^{(j)} + D_2^{(j)} \Omega) \cos \Omega t_i - 2d_j D_0^{(j)} t_i - D_0^{(j)}]. \end{aligned}\right\} \tag{A.10}$$

Case V. $d_j = 0,\ c_j = 0$

$$\left.\begin{aligned} x &= -\frac{a_j}{\Omega^2} \cos \Omega t - \tfrac{1}{2} b_j t^2 + C_1^{(j)} t + C_2^{(j)}, \\ \dot{x} &= \frac{a_j}{\Omega^2} \sin \Omega t - b_j t + C_1^{(j)}, \end{aligned}\right\} \tag{A.11}$$

where

$$\left.\begin{aligned} C_1^{(j)} &= \dot{x}_i - \frac{a_j}{\Omega} \sin \Omega t_i + b_j t_i, \\ C_2^{(j)} &= x_i - \dot{x} t_i + \frac{a_j}{\Omega^2} \cos \Omega t_i + \frac{a_j}{\Omega} t_i \sin \Omega t_i - \frac{1}{2} b_j t_i{}^2. \end{aligned}\right\} \tag{A.12}$$

A.2 Stability and Bifurcation

Consider a periodic motion with a mapping of P as

$$\mathbf{x}_{i+1} = P\mathbf{x}_i, \mathbf{x}_i \in R^n. \tag{A.13}$$

With periodicity condition $\mathbf{x}_{i+1} = \mathbf{x}_i$, the periodic solution is obtained for the foregoing mapping. The stability of the periodic solutions can be determined by the eigenvalue analysis of the corresponding $n \times n$ Jacobian matrix DP, i.e.,

$$|DP - \lambda \mathrm{I}| = 0, \tag{A.14}$$

Therefore, there are n eigenvalues. If the n eigenvalues lie inside the unit circle, then the period-1 motion is stable. If one of them lies outside the unit circle, the periodic motion is unstable. Namely, the stable, periodic motion requires the eigenvalues to be

$$|\lambda_j|<1 \quad (j=1,2,\ldots,n). \tag{A.15}$$

If the magnitude of any eigenvalue is greater than one,

$$|\lambda_j|>1 \quad (j=1,2,\ldots,n). \tag{A.16}$$

the periodic motion is unstable.

For $|\lambda_j|<1 \quad (j=3,4,\ldots,n)$ and real $\lambda_j (j=1,2)$, if

$$\max\{\lambda_j,\, j=1,2\}=1,\ \min\{\lambda_j,\, j=1,2\}\in(-1,1), \tag{A.17}$$

then the saddle-node (SN) bifurcation occurs; if

$$\min\{\lambda_j,\, j=1,2\}=-1,\ \max\{\lambda_j,\, j=1,2\}\in(-1,1), \tag{A.18}$$

then the period-doubling bifurcation occurs.

For $|\lambda_j|<1 (j=3,4,\ldots,n)$ and complex $\lambda_j (j=1,2)$, if

$$|\lambda_j|=1 \quad (j=1,2), \tag{A.19}$$

then the Neimark bifurcation occurs.

A.3 Machine-Tool Systems

The dynamical system for the machine-tool system is

$$\begin{aligned}&\begin{bmatrix}1 & 0\\ 0 & 1\end{bmatrix}\begin{Bmatrix}\ddot{x}\\ \ddot{y}\end{Bmatrix}+\begin{bmatrix}D_{11}^{(i)} & D_{12}^{(i)}\\ D_{21}^{(i)} & D_{22}^{(i)}\end{bmatrix}\begin{Bmatrix}\dot{x}\\ \dot{y}\end{Bmatrix}+\begin{bmatrix}K_{11}^{(i)} & K_{12}^{(i)}\\ K_{21}^{(i)} & K_{22}^{(i)}\end{bmatrix}\begin{Bmatrix}x\\ y\end{Bmatrix}\\ &=\cos(t)\begin{Bmatrix}A_x^{(i)}\\ A_y^{(i)}\end{Bmatrix}+\begin{Bmatrix}C_x^{(i)}\\ C_y^{(i)}\end{Bmatrix}.\end{aligned} \tag{A.20}$$

The dynamical system parameters for this machine-tool system, in the case the tool does not contact the work-piece, in domain Ω_1 are

$$B_{11}^{(1)}=\frac{1}{m\Omega}d_x,\quad B_{12}^{(1)}=B_{21}^{(1)}=0,\quad B_{22}^{(1)}=\frac{1}{m\Omega}d_y, \tag{A.21}$$

$$K_{11}^{(1)} = \frac{1}{m\Omega^2} k_x, \quad K_{12}^{(1)} = K_{21}^{(1)} = 0, \quad K_{22}^{(1)} = \frac{1}{m\Omega^2} k_y, \tag{A.22}$$

and

$$C_x^{(1)} = C_y^{(1)} = A_x^{(1)} = A_y^{(1)} = 0. \tag{A.23}$$

The dynamical system parameters for this machine-tool system, in the case the tool contacts the work-piece where no cutting occurs, in domain Ω_2 are

$$\left.\begin{aligned} B_{11}^{(2)} &= \frac{1}{m\Omega}[d_x + d_1 \sin^2\beta], \quad B_{12}^{(2)} = -\frac{1}{m\Omega} d_1 \cos\beta \sin\beta, \\ B_{21}^{(2)} &= -\frac{1}{m\Omega} d_1 \cos\beta \sin\beta, \quad B_{22}^{(2)} = \frac{1}{m\Omega}[d_y + d_1 \cos^2\beta], \end{aligned}\right\} \tag{A.24}$$

$$\left.\begin{aligned} K_{11}^{(2)} &= \frac{1}{m\Omega^2}[k_x + k_1 \sin^2\beta], \quad K_{12}^{(2)} = -\frac{1}{m\Omega^2} k_1 \cos\beta \sin\beta, \\ K_{21}^{(2)} &= -\frac{1}{m\Omega^2} k_1 \cos\beta \sin\beta, \quad K_{22}^{(2)} = \frac{1}{m\Omega^2}[k_y + k_1 \cos^2\beta], \end{aligned}\right\} \tag{A.25}$$

and

$$\left.\begin{aligned} C_x^{(2)} &= \frac{1}{m\Omega^2}\{k_1[x_1^* \sin\beta - y_1^* \cos\beta] \sin\beta, \\ C_y^{(2)} &= \frac{1}{m\Omega^2}\{-k_1[x_1^* \sin\beta - y_1^* \cos\beta] \cos\beta, \\ A_x^{(2)} &= \frac{A}{m\Omega^2} \sin\eta, \quad A_y^{(2)} = \frac{A}{m\Omega^2} \cos\eta. \end{aligned}\right\} \tag{A.26}$$

The dynamical system parameters for this machine-tool system, in the case the tool contacts the work-piece where cutting occurs where $\dot{z}<0$ and $D_4>0$, in domains Ω_3, $\dot{z}>0$, and Ω_4 are

$$B_{11}^{(j)} = \frac{1}{m\Omega}[d_x + d_1 \sin^2\beta + d_2 \cos\alpha(\cos\alpha + (-1)^{j-1}\mu \sin\alpha)],$$

$$B_{12}^{(j)} = \frac{1}{m\Omega}[-d_1 \cos\beta \sin\beta - d_2 \sin\alpha(\cos\alpha + (-1)^{j-1}\mu \sin\alpha)],$$

$$B_{21}^{(j)} = \frac{1}{m\Omega}[-d_1 \sin\beta \cos\beta - d_2 \cos\alpha(\sin\alpha + (-1)^{j}\mu \cos\alpha)],$$

$$B_{22}^{(j)} = \frac{1}{m\Omega}[d_y + d_1 \cos^2\beta + d_2 \sin\alpha(\sin\alpha + (-1)^{j}\mu \cos\alpha)], \tag{A.27}$$

$$\left.\begin{aligned}
K_{11}^{(j)} &= \frac{1}{m\Omega^2}[k_x + k_1 \sin^2\beta + k_2 \cos\alpha(\cos\alpha + (-1)^{j-1}\mu\sin\alpha)],\\
K_{12}^{(j)} &= \frac{1}{m\Omega^2}[-k_1 \cos\beta\sin\beta + k_2 \sin\alpha(\cos\alpha + (-1)^{j-1}\mu\sin\alpha)],\\
K_{21}^{(j)} &= \frac{1}{m\Omega^2}[-k_1 \cos\beta\sin\beta - k_2 \cos\alpha(\sin\alpha + (-1)^{j}\mu\cos\alpha)],\\
K_{22}^{(j)} &= \frac{1}{m\Omega^2}[k_y + k_1 \cos^2\beta + k_2 \sin\alpha(\sin\alpha + (-1)^{j}\mu\cos\alpha)],
\end{aligned}\right\} \tag{A.28}$$

and

$$\begin{aligned}
C_x^{(j)} = \frac{1}{m\Omega^2}\{&k_1[x_1^* \sin\beta - y_1^* \cos\beta]\sin\beta\\
&+ k_2[x_2^* \cos\alpha - y_2^* \sin\alpha][\cos\alpha + (-1)^{j-1}\mu\sin\alpha]\},
\end{aligned} \tag{A.29}$$

$$\begin{aligned}
C_y^{(j)} = \frac{1}{m\Omega^2}\{&-k_1[x_1^* \sin\beta - y_1^* \cos\beta]\cos\beta\\
&+ k_2[-x_2^* \cos\alpha + y_2^* \sin\alpha][\sin\alpha + (-1)^{j}\mu\cos\alpha]\},
\end{aligned} \tag{A.30}$$

$$A_x^{(j)} = \frac{A}{m\Omega^2}\sin\eta, \quad A_y^{(j)} = \frac{A}{m\Omega^2}\cos\eta. \tag{A.31}$$

for $j = 3, 4$; respectively. The parameters for the machine-tool where the chip adheres to the tool-piece rake face ($\dot{z} \equiv 0$) are

$$d = \frac{1}{2m\Omega}\left[d_2 + d_1 \sin^2(\alpha+\beta) + d_x \cos^2\alpha + d_y \sin^2\alpha\right], \tag{A.32}$$

$$\omega^2 = \frac{1}{m\Omega^2}\left[k_1 \sin^2(\alpha+\beta) + k_2 + k_x \cos^2\alpha + k_y \sin^2\alpha\right], \tag{A.33}$$

$$A_0 = \frac{A}{m\Omega^2}\sin(\eta - \alpha), \tag{A.34}$$

$$B_0 = \frac{V}{m\Omega^2}\left[k_1 \cos(\alpha+\beta)\sin(\alpha+\beta) + (k_x - k_y)\cos\alpha\sin\alpha\right], \tag{A.35}$$

$$\begin{aligned}
C_0 = \frac{1}{m\Omega^2}(\{&[d_1 V - k_1(Vt_0 + \tilde{y}_0)]\cos(\alpha+\beta) + k_1[x_1^* \sin\beta - y_1^* \cos\beta]\}\\
&\times \sin(\alpha+\beta) + [V(d_x - d_y) + (Vt_0 + \tilde{y}_0)(k_y - k_x)]\cos\alpha\sin\alpha + k_2\tilde{x}_2^*).
\end{aligned} \tag{A.36}$$

A.4 Closed-Form Solutions

Consider a dynamical system

$$\begin{bmatrix} 1 & 0 \\ 0 & 1 \end{bmatrix}\begin{Bmatrix} \ddot{x} \\ \ddot{y} \end{Bmatrix} + \begin{bmatrix} D_{11} & D_{12} \\ D_{21} & D_{22} \end{bmatrix}\begin{Bmatrix} \dot{x} \\ \dot{y} \end{Bmatrix} + \begin{bmatrix} K_{11} & K_{12} \\ K_{21} & K_{22} \end{bmatrix}\begin{Bmatrix} x \\ y \end{Bmatrix}$$
$$= \cos\Omega t \begin{Bmatrix} A_x \\ A_y \end{Bmatrix} + \begin{Bmatrix} C_x \\ C_y \end{Bmatrix}. \tag{A.37}$$

with initial conditions

$$x = x_k, \quad y = y_k, \quad \dot{x} = \dot{x}_k, \quad \dot{y} = \dot{y}_k \text{ at } t = t_k. \tag{A.38}$$

The closed-form solution for this machine-tool system is presented as follows.

(a) For real λ_σ ($\sigma = 1, 2, \ldots, 4$),

$$\begin{Bmatrix} x \\ y \end{Bmatrix} = \sum_{\sigma=1}^{4} \begin{Bmatrix} 1 \\ r_\sigma \end{Bmatrix} C_\sigma e^{\lambda_\sigma (t-t_k)} + \begin{Bmatrix} A_{px} \\ A_{py} \end{Bmatrix} \cos t + \begin{Bmatrix} B_{px} \\ B_{py} \end{Bmatrix} \sin t + \begin{Bmatrix} C_{px} \\ C_{py} \end{Bmatrix}, \tag{A.39}$$

where

$$\begin{bmatrix} 1 & 1 & 1 & 1 \\ r_1 & r_2 & r_3 & r_4 \\ \lambda_1 & \lambda_2 & \lambda_3 & \lambda_4 \\ r_1\lambda_1 & r_2\lambda_2 & r_3\lambda_3 & r_4\lambda_4 \end{bmatrix} \begin{Bmatrix} C_1 \\ C_2 \\ C_3 \\ C_4 \end{Bmatrix} = \begin{Bmatrix} A \\ B \\ C \\ D \end{Bmatrix}, \tag{A.40}$$

$$\begin{Bmatrix} A \\ B \\ C \\ D \end{Bmatrix} = \begin{Bmatrix} x_k - A_{px}\cos\Omega t_k - B_{px}\sin\Omega t_k - C_{px} \\ y_k - A_{py}\cos\Omega t_k - B_{py}\sin\Omega t_k - C_{py} \\ \dot{x}_k + \Omega A_{px}\sin\Omega t_k - B_{px}\Omega\cos\Omega t_k \\ \dot{x}_y + \Omega A_{py}\sin\Omega t_k - B_{py}\Omega\cos\Omega t_k \end{Bmatrix}, \tag{A.41}$$

$$\begin{bmatrix} K_{11} - m\Omega^2 & K_{12} & \Omega D_{11} & \Omega D_{12} \\ K_{21} & K_{22} - m\Omega^2 & \Omega D_{21} & \Omega D_{22} \\ -\Omega D_{11} & -\Omega D_{12} & K_{11} - m\Omega^2 & K_{12} \\ -\Omega D_{21} & -\Omega D_{22} & K_{21} & K_{22} - m\Omega^2 \end{bmatrix} \begin{Bmatrix} A_{px} \\ A_{py} \\ B_{px} \\ B_{py} \end{Bmatrix} = \begin{Bmatrix} A_x \\ A_y \\ 0 \\ 0 \end{Bmatrix}, \tag{A.42}$$

$$\begin{Bmatrix} C_{px} \\ C_{py} \end{Bmatrix} = \frac{1}{K_{11}K_{22} - K_{12}K_{21}} \begin{bmatrix} K_{22} & -K_{12} \\ -K_{21} & K_{11} \end{bmatrix} \begin{Bmatrix} C_x \\ C_y \end{Bmatrix}. \tag{A.43}$$

(b) For complex $\lambda_{1,2} = \alpha_1 \pm \beta_1\mathbf{i}$ and $\lambda_{3,4} = \alpha_2 \pm \beta_2\mathbf{i}$,

$$\begin{aligned}\begin{Bmatrix} x \\ y \end{Bmatrix} &= \left[\begin{Bmatrix} C_1 \\ A_1C_1 + B_1C_2 \end{Bmatrix} \cos\beta_1(t-t_k) + \begin{Bmatrix} C_2 \\ A_1C_2 - B_1C_1 \end{Bmatrix} \sin\beta_1(t-t_k)\right] \mathrm{e}^{\alpha_1(t-t_k)} \\ &+ \left[\begin{Bmatrix} C_3 \\ A_2C_3 + B_2C_4 \end{Bmatrix} \cos\beta_2(t-t_k) + \begin{Bmatrix} C_4 \\ A_2C_4 - B_2C_3 \end{Bmatrix} \sin\beta_2(t-t_k)\right] \mathrm{e}^{\alpha_2(t-t_k)} \\ &+ \begin{Bmatrix} A_{px} \\ A_{py} \end{Bmatrix} \cos t + \begin{Bmatrix} B_{px} \\ B_{py} \end{Bmatrix} \sin t + \begin{Bmatrix} C_{px} \\ C_{py} \end{Bmatrix},\end{aligned} \tag{A.44}$$

where

$$\begin{bmatrix} 1 & 0 & 1 & 0 \\ A_1 & B_1 & A_2 & B_2 \\ \alpha_1 & \beta_1 & \alpha_2 & \beta_2 \\ A_1\alpha_1 - B_1\beta_1 & B_1\alpha_1 + A_1\beta_1 & A_2\alpha_2 - B_2\beta_2 & B_2\alpha_2 + A_2\beta_2 \end{bmatrix} \begin{Bmatrix} C_1 \\ C_2 \\ C_3 \\ C_4 \end{Bmatrix} = \begin{Bmatrix} A \\ B \\ C \\ D \end{Bmatrix}, \tag{A.45}$$

where A_i and B_i $(i = 1, 2)$ are the real and imaginary parts of the modal ratios from eigenvalues of the machine-tool system,

$$r_{1,2} = A_1 \pm B_1\mathbf{i} \text{ and } r_{3,4} = A_2 \pm B_2\mathbf{i}. \tag{A.46}$$

(c) For complex $\lambda_{1,2} = \alpha_1 \pm \beta_1\mathbf{i}$ and real $\lambda_{3,4}$

$$\begin{aligned}\begin{Bmatrix} x \\ y \end{Bmatrix} &= \left[\begin{Bmatrix} C_1 \\ A_1C_1 + B_1C_2 \end{Bmatrix} \cos\beta_1(t-t_k) + \begin{Bmatrix} C_2 \\ A_1C_2 - B_1C_1 \end{Bmatrix} \sin\beta_1(t-t_k)\right] \mathrm{e}^{\alpha_1(t-t_k)} \\ &+ \begin{Bmatrix} 1 \\ r_3 \end{Bmatrix} C_3\mathrm{e}^{\lambda_3(t-t_k)} + \begin{Bmatrix} 1 \\ r_4 \end{Bmatrix} C_4\mathrm{e}^{\lambda_4(t-t_k)} \\ &+ \begin{Bmatrix} A_{px} \\ A_{py} \end{Bmatrix} \cos t + \begin{Bmatrix} B_{px} \\ B_{py} \end{Bmatrix} \sin t + \begin{Bmatrix} C_{px} \\ C_{py} \end{Bmatrix},\end{aligned} \tag{A.47}$$

where

$$\begin{bmatrix} 1 & 0 & 1 & 1 \\ A_1 & B_1 & r_3 & r_4 \\ \alpha_1 & \beta_1 & \lambda_3 & \lambda_4 \\ A_1\alpha_1 - B_1\beta_1 & B_1\alpha_1 + A_1\beta_1 & r_3\lambda_3 & r_4\lambda_4 \end{bmatrix} \begin{Bmatrix} C_1 \\ C_2 \\ C_3 \\ C_4 \end{Bmatrix} = \begin{Bmatrix} A \\ B \\ C \\ D \end{Bmatrix}. \tag{A.48}$$

Consider the stick motion with

$$\ddot{\tilde{x}} + 2d\dot{\tilde{x}} + \omega^2\tilde{x} = A_0 \cos \Omega t + B_0 t + C_0. \tag{A.49}$$

The solution for the chip seizure motion is given as follows

(a) For $d > \omega^2$

$$\tilde{x} = \tilde{C}_1 \mathrm{e}^{-(d-\lambda_d)(t-t_k)} + \tilde{C}_2 \mathrm{e}^{-(d+\lambda_d)(t-t_k)} + A_5 \cos \Omega t + B_5 \sin \Omega t + C_5 t + D_5, \tag{A.50}$$

where

$$\lambda_d = \sqrt{d^2 - \omega^2}, \tag{A.51}$$

$$\begin{aligned} \tilde{C}_1 =& \frac{1}{2\lambda_d}\{(\tilde{x}_k - C_5 t_k - D_5)(d + \lambda_d) - [A_5(d + \lambda_d) + B_5] \cos \Omega t_k \\ &+ [A_5\Omega - B_5(d + \lambda_d)] \sin \Omega t_k - C_5 + \dot{\tilde{x}}_k, \\ \tilde{C}_2 =& \frac{1}{2\lambda_d}\{(C_5 t_k + D_5 - \tilde{x}_k)(d - \lambda_d) + [B_5\Omega + A_5(d - \lambda_d)] \cos \Omega t_k \\ &+ [B_5(d - \lambda_d) - A_5\Omega] \sin \Omega t_k + C_5 - \dot{\tilde{x}}_k \end{aligned} \tag{A.52}$$

and

$$\begin{aligned} A_5 &= \frac{A_F(\omega^2 - \Omega^2)\sin(\eta - \alpha)}{m[(\omega^2 - \Omega^2)^2 + (2d\Omega)^2]}, \\ B_5 &= \frac{2A_F d\Omega \sin(\eta - \alpha)}{m[(\omega^2 - \Omega^2)^2 + (2d\Omega)^2]}, \\ C_5 &= \frac{B_0}{\omega^2}, D_5 = \frac{1}{\omega^2}(C_0\omega^2 - 2B_0 d). \end{aligned} \tag{A.53}$$

(b) For $d > \omega^2$,

$$\tilde{x} = \mathrm{e}^{-d(t-t_k)}\{\tilde{C}_1 \cos\omega_d(t-t_k) + \tilde{C}_2 \sin\omega_d(t-t_k)\} \\ + A_5 \cos\Omega t + B_5 \sin\Omega t + C_5 t + D_5, \tag{A.54}$$

where

$$\omega_d = \sqrt{\omega^2 - d^2}, \tag{A.55}$$

$$\begin{aligned} \tilde{C}_1 &= \tilde{x}_k - (A_5 \cos\Omega t_k + B_5 \sin\Omega t_k + C_5 t_0 + D_5), \\ \tilde{C}_2 &= \frac{1}{\omega_d}[(A_5\Omega - B_5 d] \sin\Omega t_k + [B_5\Omega + A_5 d] \cos\Omega t_k \\ &\quad + \dot{\tilde{x}}_k + \tilde{x}_k d - C_5 t_k d + D_5 d - C_5. \end{aligned} \tag{A.56}$$

(c) For $d = \omega^2$,

$$\tilde{x} = \mathrm{e}^{-d(t-t_k)}\{\tilde{C}_1 + \tilde{C}_2(t-t_k)\} + A_5 \cos\Omega t + B_5 \sin\Omega t + C_5 t + D_5, \tag{A.57}$$

where

$$\begin{aligned} \tilde{C}_1 &= \tilde{x}_k - (A_5 \cos\Omega t_k + B_5 \sin\Omega t_k + C_5 t_k + D_5), \\ \tilde{C}_2 &= [\dot{\tilde{x}}_k + \tilde{x}_k d - (B_5\Omega + A_5 d) \cos\Omega t_k + (A_5\Omega - B_5 d) \sin\Omega t_k \\ &\quad - C_5 t_k d + D_5 d + C_5. \end{aligned} \tag{A.58}$$

Index